MATLAB® Programming for Biomedical Engineers and Scientists

MATLAB® Programming for Biomedical Engineers and Scientists

Second Edition

Andrew P. King

Paul Aljabar

ACADEMIC PRESS

An imprint of Elsevier

Academic Press is an imprint of Elsevier
125 London Wall, London EC2Y 5AS, United Kingdom
525 B Street, Suite 1650, San Diego, CA 92101, United States
50 Hampshire Street, 5th Floor, Cambridge, MA 02139, United States
The Boulevard, Langford Lane, Kidlington, Oxford OX5 1GB, United Kingdom

Notices

ISBN: 978-0-323-85773-4

For information on all Academic Press publications
visit our website at https://www.elsevier.com/books-and-journals

Publisher: Katey Birtcher
Acquisitions Editor: Steve Merken
Editorial Project Manager: Alice Grant
Publishing Services Manager: Shereen Jameel
Production Project Manager: Kamatchi Madhavan
Designer: Ryan Cook

Typeset by VTeX

Printed in United Kingdom

Last digit is the print number:
9 8 7 6 5 4 3 2 1

Working together
to grow libraries in
developing countries

www.elsevier.com • www.bookaid.org

Dedication

A.P.K.: For Elsa, Helina, and Nathan

P.A.: For Jo, Owen, and Leila

"When somebody has learned how to program a computer ... You're joining a group of people who can do incredible things. They can make the computer do anything they can imagine."

Sir Tim Berners-Lee

Contents

About the authors

Andrew P. King has over 20 years of experience of teaching computing courses at university level. He is currently a Reader in the Biomedical Engineering department at King's College London. With Paul Aljabar, he designed and developed the computer programming module for biomedical engineering students, upon which this book was based. The module has been running since 2014, and Andrew still co-organizes and teaches on it. Between 2001 and 2005, Andrew worked as an Assistant Professor in the Computer Science department at Mekelle University in Ethiopia, and was responsible for curriculum development, and design and delivery of a number of computing modules. Andrew's research interests focus mainly on the use of machine learning and artificial intelligence techniques to tackle problems in medical imaging, with a special focus on dynamic imaging data, i.e., moving organs (Google Scholar: https://goo.gl/ZZGrGr, group web site: http://kclmmag.org).

Paul Aljabar is a mathematician who enjoys using computer programming to address health and biomedical problems. He taught high school mathematics in London for 12 years before taking up a research career. Since then, his work has focused on the analysis of large collections of medical images for a range of applications, for example, to build anatomical atlases or distinguish normal from pathological physiology. As described above, Paul and Andrew developed this book and learning materials together while teaching biomedical engineering undergraduates. Paul has taught on a range of undergraduate and graduate programs focusing on the analysis and interpretation of medical and biomedical data, carried out through modeling, programming, and the application of methods that are also used in his research (Google Scholar: https://goo.gl/jAgPru).

Preface

Aims and motivation

This book aims to teach the fundamental concepts of computer programming to students and researchers who are studying or working in the biomedical sciences. The book and associated materials grew out of the authors' experience of teaching a first-year undergraduate module on computer programming to biomedical engineering students at King's College London for a number of years. Such students have a strong interest in the biomedical applications of computing, and we felt that it was important to provide not only a concise introduction to the key concepts of programming, but also to make these concepts relevant through the use of biomedical examples and case studies.

The book is primarily aimed at undergraduate students of biomedical sciences, but we hope that it will also act as a concise introduction for students in other disciplines wishing to familiarize themselves with the important field of computer programming.

Our emphasis throughout is on practical skill acquisition. We introduce all concepts accompanied by real code examples and step-by-step explanations. These examples are reinforced by learning activities in the chapter text, as well as further exercises for self-assessment at the end of each chapter. Code for all examples, activities, and exercises is available for download from the book's web site.

Practical skills will be learned using the MATLAB software package, focusing on the procedural programming paradigm. All code is compatible with MATLAB 2020b. Note that we do not intend this text to be an exhaustive coverage of MATLAB as a software package. Rather, we view it as a useful tool with which to teach computer programming. We are always interested in hearing feedback and suggestions from students or teachers (particularly those with an interest in biomedical applications) about material that could be added or omitted.

Learning objectives

On completion of the book the reader should be able to:

- Analyze problems and apply structured design methods to produce elegant, efficient, and well-structured program designs.
- Implement a structured program design in MATLAB, making use of incremental development approaches.
- Write code that makes good use of MATLAB programming features, including control structures, functions, and advanced data types.
- Write MATLAB code to read in medical data from files and write data to files.
- Write MATLAB code that is efficient and robust to errors in input data.
- Write MATLAB code to analyze and visualize medical data, including images.

What's new in the second edition?

To readers familiar with the first edition of this book, we would like to highlight some of the significant changes and additions that were made for the second edition. We would also like to thank the instructors and reviewers who provided the valuable feedback that helped to shape these changes.

First, there are two entirely new chapters in the second edition: Chapter 14 on Machine Learning and Chapter 15 on Engineering Mathematics. The Machine Learning chapter was introduced in response to a surge in the popularity of such techniques in recent years. We believe that more and more people will be motivated to learn computer programming, because it affords an opportunity to explore the latest machine learning and artificial intelligence techniques. The Engineering Mathematics chapter was introduced to broaden the appeal of the book and to more completely cover areas of interest, and the typical syllabi, in undergraduate modules at different institutions.

We have also restructured the book to improve the ordering of key concepts. The chapter in the first edition on Data Types has been divided into two chapters in this edition: Chapter 3 on Basic Data Types and Chapter 6 on Advanced Data Types. We felt that it was important to introduce the basics of data types earlier in the book to enable subsequent material to be tackled with the appropriate prerequisite knowledge. More advanced data types (coverage of which is expanded in the second edition) can then be introduced later.

Other, less substantial changes have been made, which we do not attempt to describe exhaustively here. A number were made to render the content up-to-date with the latest version of MATLAB (v2020b), for example, in Chapter 12 on graphical user interface (GUI) development. Changes were also made in response to feedback received, as well as our own personal experience of using the book as a teaching aid. We hope that all these revisions will be beneficial to teachers and students alike.

How to use this book

The book is primarily intended as a companion text for a taught course on computer programming. Each chapter starts with clear learning objectives, and the students are helped to meet these objectives through the text, examples, and activities that follow. The objectives associated with each example or activity are clearly indicated in the text. Exercises at the end of each chapter enable students to self-assess whether they have met the learning objectives.

The book comes with an on-line teaching pack containing teaching materials for a taught undergraduate module to accompany the book (sample lesson plans, lecture slides, and suggestions for coursework projects), as well as MAT-LAB code and data files for all examples, activities, and exercises in the book. All materials are available from the book's web site at: https://educate.elsevier.com/book/details/9780323857734.

The book consists of 15 chapters:

- *Chapter 1: Introduction to computer programming and MATLAB®*: This chapter introduces the reader to the fundamental concepts of computer programming and provides a hands-on introduction to the MATLAB software package. At this stage, we go into just enough detail to enable the reader to start using MATLAB to perform simple operations and produce simple visualizations. We also start to show the reader how to create and debug simple computer programs using MATLAB.
- *Chapter 2: Control structures*: Here, we introduce the concept of control structures, which can be seen as the basic building blocks of computer programs. We show how they can be used to create more complex code, resulting in more powerful and flexible programs. Conditional and iteration control structures are introduced, and we describe the MATLAB commands that are provided to implement them.
- *Chapter 3: Basic data types*: In this chapter, we discuss the way in which MATLAB stores data. We describe what is meant by a data type and introduce the basic data types that MATLAB provides. We also show how to find out the type of a variable and convert between data types.
- *Chapter 4: Functions*: In this chapter, we introduce functions, an important means for programmers to split large programs into smaller modules. As we start to tackle more complex problems, functions are an essential part of the programmer's toolbox. We show how data can be passed into and returned from functions, and also introduce special topics, such as variable scope and recursion.
- *Chapter 5: Program development and testing*: Next, we cover the important (but often overlooked) topic of how to go about developing a program

and checking that it does what it is supposed to do. We introduce the concept of incremental development, in which we always maintain a working version of the program, even if it is trivial, and then gradually add more functionality to it. We also give further detail on debugging and describe how we can make our programs robust to errors.

- *Chapter 6: Advanced data types*: Up until this point, we will have been dealing with programs that use fairly simple data, such as numbers or letters. In this chapter, we introduce some of the more complex (and powerful) data types, such as cell arrays, structures, and tables.
- *Chapter 7: File input/output*: Many biomedical applications involve reading data from external files and/or writing data to them. In this chapter, we describe the commands that MATLAB provides for these operations. We describe the different types of file that might be encountered, and cover the commands intended for use on each type.
- *Chapter 8: Program design*: In addition to the necessity of knowing how to write working code, when tackling larger and more complex problems, it becomes important to be able to break a program down into a number of simpler modules that can interact with each other by passing data. Such an approach to coding makes the process quicker, and the resulting code is likely to contain fewer errors. Program design refers to the process of deciding which modules to define and how they interact. We illustrate this process using the commonly used approach of top-down step-wise refinement, and we show how structure charts and pseudocode can be used to document a design.
- *Chapter 9: Visualization*: In this chapter, we go into more detail about how to produce visualizations of data. We extend the basic coverage introduced in Chapter 1 and describe how to create plots containing multiple datasets, and plots that visualize multivariate data and images.
- *Chapter 10: Code efficiency*: There are typically several different ways that a computer program can solve a given problem, and some may be more (or less) efficient. Here we consider how the way in which a program solves a problem affects its efficiency, in terms of time and the computer's memory. We show how efficiency can be measured and discuss ways in which program efficiency can be improved.
- *Chapter 11: Signal and image processing*: Many biomedical applications involve processing of large amounts of data. This chapter describes techniques for processing signals (i.e., one-dimensional data) or images (two-dimensional, or even three- or four-dimensional data). We introduce the powerful techniques of filtering and convolution and demonstrate how they can be applied to one- or higher-dimensional data. We return to the topic of images (first introduced in Chapter 9) and further show how pipelines may be constructed from simple operations to build more complex and powerful image processing algorithms.
- *Chapter 12: Graphical user interfaces*: As well as the functionality of a program (i.e., what it does), it is important to consider how it will in-

teract with its users. This can be done using a simple command window interface, but more powerful and flexible interfaces can be developed using a GUI. This chapter describes how GUIs can be created and designed in MATLAB, and how we can write code to control the way in which they interact with the user.

- *Chapter 13: Statistics*: This chapter introduces the basics of statistical data analysis with MATLAB. The chapter does not cover statistical theory in depth, but the intention is rather to describe how the most common statistical operations can be carried out using MATLAB. Some knowledge of statistics is assumed, although fundamental concepts are briefly reviewed. We cover the fields of descriptive statistics (summarizing and visualizing data) and inferential statistics (answering questions about data).

- *Chapter 14: Machine learning*: The field of machine learning has grown greatly in popularity in recent years. In this chapter, we review some of the fundamental concepts of machine learning, and then describe some of the features that MATLAB provides to enable users to develop and evaluate their own machine learning models. We look at functions provided for unsupervised learning, as well as supervised learning using the *Classification Learner* and *Regression Learner* apps. Finally, we provide a brief introduction to developing deep learning models in MATLAB.

- *Chapter 15: Engineering mathematics*: The final chapter gives an overview of some of the mathematical operations available in MATLAB that can be useful in engineering applications. This is a very broad topic, so only a subset of topics are covered in this chapter: scalars and vectors, complex numbers, matrices, including systems of equations, numerical differentiation, and integration. A list of possible further topics and sources is given at the end of the chapter.

We believe that the first 10 or 11 chapters of the book are suitable content for an undergraduate module in computer programming for biomedical scientists. The remaining chapters are included as more advanced topics, which may be useful for students' future work in using MATLAB as a tool for biomedical problem-solving.

We end each chapter of the book with a brief pen portrait of a famous computer programmer from the past or present. As well as providing a bit of color and interest, we hope that this will help the reader to see themselves as following in the footsteps of a long line of pioneers who have developed the field of computer programming and made it what it is today: an intellectually stimulating and incredibly powerful tool to assist people working in a wide range of fields to further the limits of human achievement.

Finally, a word about data. We have made every effort to make our biomedical examples and exercises as realistic as possible. Where possible, we have used real biomedical data. On the occasions where this was not possible, we have tried to generate realistic synthetic data. We apologize if this synthetic data lacks realism or plausibility in any respect, or gives a misleading impression about the application that the exercise deals with. Our intention has been to provide examples that are relevant to those with an interest in biomedical applications, and we accept responsibility for any flaws or errors in the data we provide.

Acknowledgments

The authors would like to thank the following people for their contributions to the development of this book:

- All biomedical engineering students at King's College London who have taken our Computer Programming module over the past few years, for their constructive feedback, which has improved the quality of our teaching materials that formed the basis for this book. Also, for the same reasons, all teaching assistants and lecturers who have contributed to the delivery of the module.
- Dr. Adam Shortland of the One Small Step Gait Laboratory at Guy's Hospital, London, for insightful discussions and providing the data for a number of activities and exercises throughout the book: Activity 1.6, Exercise 2.9, Exercise 4.7, Activity 6.2, Activity 9.3, and Exercise 9.5.
- The people in charge of the ACDC challenge[1] (Associate Professor O. Bernard, Professor P. M. Jodoin, and Associate Professor A. Lalande) and the University Hospital of Dijon (France) for their kind permission to use the ACDC MRI dataset [1] in Example 14.5 and Exercise 14.5.
- From the publishers, Elsevier: Tim Pitts for prompting us to write the first edition of the book, Stephen Merken for prompting us to expand it to form this second edition, and Ruby Gammell and Andrae Akeh for encouraging us to produce it on time.

---

[1] h:tps://www.creatis.insa-lyon.fr/Challenge/acdc/.

CHAPTER 1

Introduction to computer programming and MATLAB®

LEARNING OBJECTIVES

At the end of this chapter you should be able to:

O1.A Describe the broad categories of programming languages and the difference between compiled and interpreted languages

O1.B Use and manipulate the MATLAB® environment

O1.C Form simple expressions using scalars, arrays, variables, built-in functions, and assignment in MATLAB

O1.D Describe the different basic data types in MATLAB and be able to determine the type of a MATLAB variable

O1.E Perform simple input/output operations in MATLAB

O1.F Perform basic data visualization in MATLAB by plotting 2-D graphs and fitting curves to the plotted data

O1.G Form and manipulate matrices in MATLAB

O1.H Write MATLAB scripts, add meaningful and concise comments to them, and use the debugger and code analyzer to identify and fix coding errors

1.1 Introduction

The processing power of digital computers has increased massively since their invention in the 1950s. Today's computers are about a trillion times more powerful than those available 70 years ago. Likewise, the use of computers (and technology in general) in biology and medicine (commonly termed *biomedical engineering*) has expanded hugely over the same time period. Today, diagnosis and treatment of many diseases is reliant on the use of technology such as imaging scanners, robotic surgeons, and cardiac pacemakers. Computers and computer programming are at the heart of this technology.

The simultaneous rise in computing power and the biomedical use of technology is not a coincidence: advances in technology often drive advances in biology and medicine, enabling more sensitive diagnostic procedures and more accurate and effective treatments to be developed. In recent years, the ever-closer relationship between technology and medicine has resulted in the

1

MATLAB® Programming for Biomedical Engineers and Scientists. https://doi.org/10.1016/B978-0-32-385773-4.00010-1

emergence of a new breed of medical professional: someone who both understands the biological and medical challenges, but also has the skills needed to apply technology to address them and to exploit any opportunities that arise. If this sounds like you (or perhaps the person you would like to be), then this book is for you. Over the next fifteen chapters, we hope to cover the important topics in computer programming, but with a strong biomedical focus, which we hope will give you the skills you need to tackle the increasingly complex problems faced in modern medicine.

In this first chapter of the book, we will introduce some fundamental concepts about computers and computer programming, and then focus on familiarizing ourselves with the *MATLAB* software application. This is the software that will be used throughout the book to learn how to develop computer programs.

1.2 Computers and computer programming

The development of the modern digital computer marked a step change in the way that people use machines. Until then, mechanical and electrical machines had been proposed and built to perform *specific operations*, such as solving differential equations. In contrast, the modern digital computer is a *general-purpose machine* that can perform any calculation that is computable. The way in which we instruct it what to do is by using computer programming.

The development of modern digital computers has a rich history, and we discuss some of its important contributions and contributors in the "Famous Computer Programmers" sections at the end of each chapter of this book. These days, almost all modern computers are based upon the same basic architecture, known as the *Von Neumann architecture* (named after John Von Neumann, see Chapter 2's Famous Computer Programmer). In simple terms, the Von Neumann architecture refers to a computer in which both the computer program and the data it uses and produces are stored in the same computer memory, i.e., the computer program is *software*. In the early days of computer science, the introduction of the Von Neumann architecture was an important change from the previous approach to programming by making *hardware* changes, such as setting switches and inserting patch cables on the computer. This increased flexibility in the coding process made developing computer programs much easier and faster.

All computer programs are written in a programming language, and there are many different programming languages. We can make a number of distinctions between different programming languages, depending on how we write and use the programs. First, a useful distinction is between an *interpreted* language and a *compiled* language. A program written in an interpreted language can be run (or *executed*) immediately after it has been written. For compiled languages, we need to perform an intermediate step, called *compilation*. Compiling a program simply means translating it into a language that the computer

Table 1.1 Examples of different types of computer programming language.

	Interpreted	Compiled
Procedural	Perl, MATLAB	C
Object-Oriented	Python	C++, Java, C#
Declarative	PROLOG	LISP

can understand (known as *machine code*). Once this is done, the computer can execute our program. With an interpreted language this "translation" is done in real-time, as the program is executed, and a compilation step is not necessary.

In this book, we will be learning how to program computers using a software application called *MATLAB*. MATLAB is (usually) an *interpreted* language. Although it is also possible to use it as a compiled language, we will be using it solely as an interpreted language.

Another distinction we can make between different programming languages relates to how the programs are written and what they consist of. Historically, most programming languages have been *procedural* languages. This means that programs consist of a sequence of instructions provided by the programmer. Until the 1990s almost all programming languages were procedural. More recently, an alternative called *object-oriented programming* has gained in popularity. Object-oriented languages allow the programmer to break down the problem into *objects*: self-contained entities that can contain both data and operations to act upon data. Finally, some languages are *declarative*. In principle, when we write a declarative program it does not matter what order we write the program statements in; we are just "declaring" something about the problem or how it can be solved. The compiler or interpreter will do the rest. Currently, declarative languages are mostly used only for computer science research.

MATLAB is (mostly) a procedural language. It does have some object-oriented features, but we will not cover them in this book. Although the distinction between interpreted/compiled and procedural/object-oriented/declarative is not always clear-cut, we present in Table 1.1 a simple summary of some examples of the different types of programming language.

1.3 MATLAB

MATLAB is a commercial software package developed by MathWorks[1] for performing a range of mathematical operations. It was first developed in 1984, and included mostly linear algebra operations (hence the name, MATrix LABoratory). It has since been expanded to include a wide range of functionality,

---

[1] https://mathworks.com/.

including visualization, statistics, and algorithm development using computer programming. It has more than 4 million users in academia, research, and industry, and has found particular application in the fields of engineering, science, and economics. As well as its core functionality, there are a number of specific toolboxes that are available to extend MATLAB's capabilities.

Although MATLAB is a very powerful software package, it is a commercial product, and a license is required to be able to use it. You may already have access to MATLAB through an institutional license. Alternatively, student licenses can be obtained relatively cheaply. However, full licenses are more expensive. Details of how to download and install a trial version of MATLAB, and to buy a license, can be found online from the MathWorks web site:

- Download trial version: https://mathworks.com/campaigns/products/trials.html
- Installation and licensing information: https://mathworks.com/help/install/

If you do not have an institutional license, and you are not eligible for a student license, you may want to consider one of the free, open-source alternatives to MATLAB. Some of the most common are:

- *Octave*: Good compatibility with MATLAB, but includes a subset of its functionality: http://www.gnu.org/software/octave
- *Freemat*: Good compatibility with MATLAB, but includes a subset of its functionality: http://freemat.sourceforge.net
- *Scilab*: Some differences from MATLAB: http://scilab.org

1.4 The MATLAB environment

After we start MATLAB, the first thing we see is the MATLAB *environment window*. To begin with, it is important to spend some time familiarizing ourselves with the different components of this window and how to use and manipulate them. Fig. 1.1 shows a screenshot of the MATLAB environment window (if yours does not look exactly like this, don't worry; as we will see shortly, it is possible to customize its appearance).

The *command window* is where we enter our MATLAB commands and view the responses MATLAB gives (if any). As we enter commands, they will be added to our *command history*. We can always access and repeat old commands through the *command history* window. Another way of cycling through old commands in the *command window* is to use the up and down arrow keys. Although the *command window* is the main way in which we instruct MATLAB to perform the operations we want, it is often also possible to execute commands via the *menu bar* at the top of the environment window.

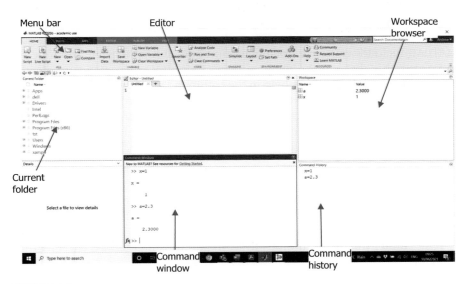

FIGURE 1.1
The MATLAB environment.

The MATLAB environment window can also be used to navigate through our file system. The *current folder* window shows our current folder and its contents. It can also be used for simple navigation of the file system.

As we use the *command window* to create and manipulate data, these data are added to our *workspace*. The workspace is basically a collection of all of the data that have been created or loaded into MATLAB. The current contents of the workspace are always displayed in the *workspace browser*.

When we start creating scripts to perform more complex MATLAB operations (see Section 1.12), we will use the *editor* window to create and edit our script files. This can be useful if we want to perform the same sequence of operations several times or save a list of operations for future use.

We can customize the MATLAB environment by clicking on the *Layout* button in the menu bar (under the *Home* tab).

■ **Activity 1.1**

Try customizing the MATLAB environment window. For example, switch between *Default* and *Two Column* layouts; hide the *Current Folder* and/or *Workspace* panels, then make them reappear again; try switching the *Command History* between the *Popup* and *Docked* modes, and see the effect this has. If you want to get rid of any changes you've made and return to the original layout, then simply select *Layout → Default*.   ■

01.B

1.5 Help

MATLAB includes extensive documentation and online help. At any time we can access this through the menu system (*Help → Documentation* under the *Home* tab). We can also access help in the command window (type doc or help, followed by the command we want information about). All documentation is also available online, so we can simply use our favorite search engine and enter the keyword "MATLAB" together with some keywords related to the task we wish to perform.

■ Activity 1.2

01.B   Try working through some of the simple MATLAB tutorials that are freely available as part of its documentation. For example, select *Help → Documentation*, then click on *MATLAB*, followed by *Get Started*. Click on *Desktop Basics* under the list of tutorials to learn about using the MATLAB environment.   ■

1.6 Variables, arrays, and simple operations

1.6.1 MATLAB variables

Now we will start to type some simple commands into the command window. The first thing we will do is to create some *variables*. Variables can be thought of as named containers for holding information, often numbers. They can hold information about anything, for example, clinical data, such as heart rate, blood pressure, height, and weight. They are called variables, because the values (but not the names) can vary. For example, the following commands create two new variables named a and y that have values 1 and 3, respectively.

```
>> a = 1
a =
    1

>> y = 3
y =
    3
```

Note that >> is the MATLAB *command prompt*, i.e., it will be displayed in the command window to "prompt" you to enter some commands. We do not need to type this in ourselves, just the text after it, e.g., a = 1, and then hit <RETURN>. The lines after each command show MATLAB's response. Again, this is not something we type in; it is an "answer" produced by MATLAB to confirm what it has done in response to our command. For example, after the first command, it is confirming that it has created a variable called a and given it the value 1.

When we enter these commands into the MATLAB command window, we see the new variables appear in the *workspace browser*. These variables can then be used in forming subsequent expressions for evaluation by MATLAB, e.g.,

```
>> c = a + y
c =
     4

>> d = a ^ y
d =
     1

>> e = y / 2 - a * 4
e =
    -2.5000
```

Again, as we type these commands MATLAB gives responses in the command window, and we create new variables, which are added to the workspace. This time, they are called c, d, and e (the ^ symbol raises a number to the power of another number, i.e., a^y means a^y).

We can also ask MATLAB to evaluate expressions without specifying a variable to assign the answer to. Try typing the following command and look at MATLAB's response and the workspace.

```
>> (c + d) / e
ans =
    -2
```

We can see that MATLAB has created a new variable called ans and used it to store the result of our expression. MATLAB will always use this "default" variable name when we do not specify a variable ourselves.

1.6.2   Variable names

In general, we have a lot of flexibility in deciding what names to give our variables. In all of the examples we have seen above, the variable names were single letters, but this does not have to be the case. Variable names in MATLAB can have multiple characters, but there are some rules and guidelines that we should be aware of:

- Names must start with a letter, and this can be followed by multiple letters, digits or underscores.
- We cannot use MATLAB *keywords* as variable names (keywords are names that have special meaning in MATLAB, as we will see later).
- It is also not a good idea to give a variable the same name as a built-in MATLAB function (see next section); if we do this, the variable will "hide" the function, meaning that it can no longer be called.
- Finally, note that MATLAB is *case-sensitive*. For example, the variable names d and D would represent different variables in the MATLAB workspace.

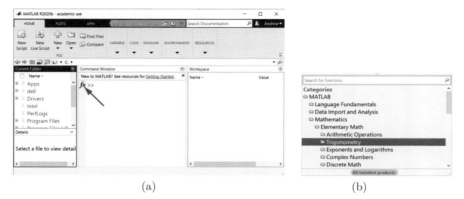

(a)   (b)

FIGURE 1.2
Accessing MATLAB's built-in functions using the graphical menu: (a) click on the *fx* symbol (indicated with an arrow); (b) browse through available functions in the pop up menu.

1.6.3   Built-in MATLAB functions

MATLAB contains a large number of built-in functions for operating on numbers or variables. If we know which function we want to use, we can just type it into the command window, e.g.,

```
>> sin(pi/4)
ans =
    0.7071

>> z = tan(a)
z =
    1.5574,
```

and the function will be evaluated using *argument(s)* provided in brackets. Note that the name `pi` is a built-in MATLAB variable that contains the value 3.14159...

Alternatively, if we want to browse through all available functions, we can click on the *fx* symbol immediately to the left of the command prompt (see Fig. 1.2). A menu will pop up allowing us to navigate through a range of categories of functions. For example, to find trigonometric functions, choose *Mathematics, Elementary Math,* and then *Trigonometry*. A list of commonly used mathematical functions that operate on numerical values is given in Table 1.2.

1.6.4   Creating and manipulating arrays

As well as containing single values, variables can also contain lists of values. In this case, they are known as *arrays*. It is in the creation and manipulation of arrays that MATLAB can be particularly powerful. To illustrate this, consider the following commands. First, we create a sample array variable,

Table 1.2 A list of common MATLAB numerical functions.

`sin(x)`	Sine (input in radians)	`log10(x)`	Base 10 logarithm
`cos(x)`	Cosine (input in radians)	`sqrt(x)`	Square root
`tan(x)`	Tangent (input in radians)	`abs(x)`	Absolute value
`asin(x)`	Arc sine (result in radians)	`floor(x)`	Round to next smallest integer
`acos(x)`	Arc cosine (result in radians)	`ceil(x)`	Round to next largest integer
`atan(x)`	Arc tangent (result in radians)	`round(x)`	Round to nearest integer
`exp(x)`	Exponential of x, i.e. e^x	`fix(x)`	Round to integer towards zero
`log(x)`	Natural logarithm	`mod(x,y)`	Modulus (remainder)

```
>> d = [1 3 4 2 5];
```

We define arrays by listing all values of the array, separated by spaces, enclosed by square brackets. Alternatively, we can separate the values by commas, so the following two commands are equivalent:

```
>> d = [1 3 4 2 5];
>> d = [1,2,3,4,5];
```

Note that adding a semicolon at the end of the command causes MATLAB to suppress its response to the command. The variable will be created and added to the workspace, but MATLAB will not display anything in the command window. Note also that by entering the above command, we will overwrite our previous definition of d = a^y.

Now, suppose that we wish to compute the value of each array element raised to the power of 3. We simply type the following command:

```
>> d.^3
ans =
      1      8     27     64    125
```

If we perform an arithmetic operation like this on an array variable, MATLAB will automatically apply the operation to each element of the array. This can make processing of large amounts of data very easy. However, note the ".^" symbol before the caret (^). This tells MATLAB to apply the ^ operator to each element *individually*, or *element-wise*. Without the ".", common arithmetic operators will be interpreted differently, and can have special meanings when applied to arrays. We will see this later in Section 1.7.

Note how spaces are treated in MATLAB. In particular, spaces are used to separate elements of arrays, as in the command d = [1 3 4 2 5] we saw above. Otherwise, when evaluating expressions, spaces are largely ignored by MATLAB, e.g., as we can see, the following two commands are identical:

```
>> d.^3
ans =
      1      8     27     64    125
```

```
>> d .^ 3
ans =
     1     8    27    64   125
```

1.6.5 Regularly-spaced arrays

It is quite common to want to create a large array of numbers with values that are regularly spaced. In this case, there is a shorthand way to enter the values, as an alternative to typing in each value individually. The ":" operator allows us to create an array by specifying its first value, the spacing between adjacent values, and the last value. For example, try typing the following:

```
>> x = 0:0.25:3;
```

and look at the resulting array. It consists of values separated by 0.25, starting with 0 and ending with 3. The colon operator is a much easier way to create arrays with a large number of elements.

An alternative way to create such regularly spaced arrays is to use the MATLAB linspace command. Consider the following code:

```
>> x = linspace(0, 3, 13);
```

which has the same effect as the colon operator example shown above. The three arguments passed to the linspace function (inside the brackets) are the first value, the last value, and the number of values in the array.

1.6.6 Accessing array elements

So far, we have seen how to create arrays and to perform operations on the array elements. Often we will then want to access individual array elements. In MATLAB, this can be done using round brackets, e.g.,

```
>> a = x(3);
```

This command accesses the third element of the array x and assigns it to the variable a. We also refer to this operation as *array indexing*, and the value in brackets (3 in this case) is known as the array *index*. In MATLAB, the integers used for array indices start at 1, meaning that the first element of any standard array is indexed by the integer 1.[2]

In a similar way to accessing array elements, we can use array indexing to assign values to individual array elements without affecting the rest of the array:

```
>> x(3) = 10;
```

---

[2]This property of starting array indices at 1 is called *1-indexing*. It is fairly uncommon for programming languages to use 1-indexing. MATLAB is a little unusual in this regard, and many languages, such as Java, Python and C++ use 0-indexing instead, meaning that the first element of an array is indexed by the integer 0.

Table 1.3 A list of common MATLAB array functions.

`max(x)`	Maximum value	`cross(x,y)`	Vector cross product
`min(x)`	Minimum value	`dot(x,y)`	Vector dot product
`range(x)`	Range of values	`disp(x)`	Display an array
`sum(x)`	Sum of all values	`mean(x)`	Mean value of array
`prod(x)`	Product of all values	`std(x)`	Standard deviation of array
`length(x)`	Length of array		

1.6.7 Array functions

We have already seen that MATLAB has a number of built-in functions for operating on numbers and numerical variables. There are also a number of built-in functions specifically for operating on arrays. A list of common MATLAB array functions is given in Table 1.3. For example, try typing the following:

```
>> a = [1 5 3 9 11 3 23 6 13 14 3 9];
>> max(a)
ans =
    23

>> min(a)
ans =
    1

>> sum(a)
ans =
    100
```

In addition to the array functions listed in Table 1.3, all of the numerical functions summarized in Table 1.2 can be used with array arguments. In this case, the functions will be applied to each array element separately and the answers combined into an array result. For example, try typing the following commands:

```
>> x = [10 3 2 7 9 12];
>> sqrt(x)
ans =
    3.1623    1.7321    1.4142    2.6458    3.0000    3.4641

>> y = [−23 12 0 1 −1 43 −100];
>> abs(y)
ans =
    23    12    0    1    1    43    100
```

■ Activity 1.3

Perform the following tasks using MATLAB arrays: *01.C*

1. Create an array called `celsius_temps` containing regularly spaced numbers, starting with 36.1 and ending with 37.2, in steps of 0.1.

These represent the normal range of body temperatures in degrees Celsius.

2. Create a new array called `fahrenheit_temps`, which contains the Fahrenheit equivalents of the `celsius_temps` values, using the following conversion equation:

$$\text{Fahrenheit} = 32 + \text{Celsius} \times 9/5$$

3. Access the sixth value from the `fahrenheit_temps` array (i.e., the Fahrenheit equivalent of a body temperature of 36.6 °C).
4. Using a built-in MATLAB function (see Table 1.3), find the range of normal body temperatures, in both Celsius and Fahrenheit.
5. Compute the mean value of normal body temperatures in Celsius. To begin with, try storing this in a variable called `mean`. Next, try computing the mean normal body temperature in Fahrenheit. What problem arises and why? How can you fix this problem? ■

■ Activity 1.4

O1.C

The probability of getting cancer has been shown to have an approximately linear relationship with the number of cigarettes smoked. Enter the array [5 2 14 7 9 11] into MATLAB, which represents the daily number of cigarettes smoked each by a group of six smokers. Assuming that the relationship between smoking and cancer risk is

$$\text{cancer probability} = 0.023 \times \text{number of cigarettes} + 0.08,$$

perform the following operations:

1. Compute the probability of each of the six smokers getting cancer.
2. Replace the second element of the array with the number 12, then recompute the probabilities for the entire array.
3. Add a new number 20 to the end of the array, and recompute all probabilities again.
   (Hint: In MATLAB, the word `end`*, when used as an array index, refers to the last element of the array.)* ■

1.7   Matrices

At its simplest, a MATLAB array is a *one-dimensional* (1-D) list of data elements. *Matrices* can also be defined, which are *two-dimensional* (2-D) arrays. In this case, we use semicolons to separate the rows in the matrix, for example,

```
>> a = [1 2; 3 4];
>> b = [2, 4;
        1, 3];
```

Table 1.4 Some MATLAB built-in matrix functions.

`eye(n)`	Create n by n identity matrix
`zeros(m,n)`	Create m by n matrix of zeros
`ones(m,n)`	Create m by n matrix of ones
`rand(m,n)`	Create m by n matrix of random numbers in range [0-1]
`size(a)`	Return the number of rows/columns in the matrix a
`inv(a)`	Return the inverse of a square matrix a
`transpose(a)`	Return the transpose of the matrix a

As with 1-D arrays, the row elements of matrices can be delimited by either spaces or commas. Note also that we can press <RETURN> to move to a new line in the middle of defining an array (either 1-D or 2-D). MATLAB will not process the array definition until we have closed the array with the] character.

MATLAB matrix elements can be accessed in the same way as 1-D array elements, except that two indices need to be specified, one index for the row and one index for the column:

```
>> a(1,2)
ans =
    2

>> b(2,2)
ans =
    3
```

If we want to access an entire row or column of a matrix, we can use the colon operator.[3]

```
>> a(:,2)
ans =
    2
    4

>> b(1,:)
ans =
    2    4
```

MATLAB also provides a number of built-in functions specifically intended for use with matrices, and these are summarized in Table 1.4.

■ Activity 1.5

Write MATLAB code to define the following matrices: 01.6

$$X = \begin{pmatrix} 4 & 2 & 4 \\ 7 & 6 & 3 \end{pmatrix} \quad Y = \begin{pmatrix} 1 & 2 & 3 \\ 1 & 0 & 0 \end{pmatrix} \quad Z = \begin{pmatrix} 2 & 0 & 9 \\ 1 & 5 & 7 \end{pmatrix}$$

---

[3]This is a different use of the colon operator from previous uses we have seen (i.e., for generating arrays of numbers).

Now perform the following operations:

1. Replace the element in the second row and second column of Z with 1.
2. Replace the first row of X with [2 2 2].
3. Replace the first row of Y with the second row of Z. ■

■ Activity 1.6

01.C, 01.E, 01.G

Gait analysis is the study of human motion with the aim of assessing and treating individuals with conditions that affect their ability to walk normally. One way of measuring human motion is to attach markers to a person's joints (hip, knee, ankle) and optically track them using cameras while the person walks. From these tracking data, important information, such as knee *flexion* (i.e., angle) can be derived. The peak knee flexion and the range of knee flexion values are indicators that are of particular interest in gait analysis.

The file *knee_flexion.mat* contains flexion data acquired from the left knee of a patient with a neurological disease while walking. The file contains one variable called `left_knee_flexion`. This is a matrix in which the first column represents time (in seconds), and the second column represents knee flexion values (in degrees).

Write MATLAB code to load in the *knee_flexion.mat* file, and compute and display to the command window the following values:

- The range of knee flexion values (i.e., the difference between the maximum and minimum values).
- The peak knee flexion value.
- The time at which peak knee flexion occurs.

(Hints: Files with a ".mat" extension are MATLAB data files, and they can be read in using the `load` *command. If you run the command* `load('knee_flexion.mat')` *in the command window, a variable called* `left_knee_flexion` *will be available in the workspace. We will discuss reading and writing data in more detail in Section 1.9.*

To find the peak knee flexion and its time, look at the MATLAB documentation for the `max` *function.)* ■

As well as storing and accessing 2-D data, matrices allow us to perform a range of different *linear algebra* operations. The following code illustrates the use of some of common MATLAB matrix operations:

```
>> c = a * b;
>> d = a + b;
>> e = inv(a);
>> f = transpose(b);
```

Here, the * and + operators automatically perform *matrix* multiplication and addition, because their arguments are both matrices. To explicitly request that an operation is carried out *element-wise*, we use a " . " before the operator. For example, note the difference between the following two commands:

```
>> c = a * b
c =
       4      10
      10      24

>> d = a .* b
d =
       2       8
       3      12
```

Here, d is the result of element-wise multiplication of a and b, whereas c is the result of carrying out matrix multiplication.

Note that element-wise addition/subtraction are the same as matrix addition/subtraction, so there is no need to use a dot operator with + and −.

■ Activity 1.7

Write MATLAB commands to define the matrices A, B, and C as follows: 01.E, 01.G

$$A = \begin{pmatrix} 1 & 2 & 1 \\ 3 & 2 & 1 \\ 3 & 2 & 4 \end{pmatrix} \quad B = \begin{pmatrix} 4 & 4 & 4 \\ 2 & 2 & 2 \\ 1 & 1 & 1 \end{pmatrix} \quad C = \begin{pmatrix} 1 & 0 & 1 \\ 2 & 1 & 3 \\ 1 & 0 & 1 \end{pmatrix}$$

Now compute the following:

1. $D = (AB)C$
2. $E = A(BC)$
3. $F = (A + B)C$
4. $G = A + BC$ ■

Finally, matrices can also be used to solve systems of linear equations such as the following:

$$5x_1 + x_2 = 5$$
$$6x_1 + 3x_2 = 9$$

We approach this problem by first expressing the equations in matrix form, i.e.,

$$\begin{pmatrix} 5 & 1 \\ 6 & 3 \end{pmatrix} \begin{pmatrix} x_1 \\ x_2 \end{pmatrix} = \begin{pmatrix} 5 \\ 9 \end{pmatrix}.$$

Then, the system of equations can be solved by premultiplying by the inverse of the first matrix to solve for x_1 and x_2:

$$\begin{pmatrix} x_1 \\ x_2 \end{pmatrix} = \begin{pmatrix} 5 & 1 \\ 6 & 3 \end{pmatrix}^{-1} \begin{pmatrix} 5 \\ 9 \end{pmatrix}.$$

■ Example 1.1

01.G

The following MATLAB code implements the solution to the system of linear equations given above. Enter the commands in MATLAB to find the values of x_1 and x_2.

```
>> M = [5 1; 6 3];
>> y = [5; 9];
>> x = inv(M) * y
```

■

The following activity is one that may be solved using simultaneous equations and matrices.

■ Activity 1.8

01.G

A doctor wishes to administer 2 ml of a drug in a solution, which has a 10% concentration. However, only concentrations of 5% and 20% of the drug are available. Use MATLAB to determine how much of the 5% and 20% solutions the doctor should mix to produce 2 ml of 10% concentration solution.

■

More advanced uses of matrices can be found in Section 15.4.

1.8 Data types

So far, all variables and values we have used have been numeric. Although MAT-LAB is mostly used for numerical calculations, it is possible, and sometimes very useful, to manipulate data of other types. MATLAB allows the following basic data types (the names in brackets are the terms MATLAB uses for the type):

- Floating point, e.g., 1.234 (`single`, `double`, depending on the precision required)
- Integer, e.g., 3 (`int8`, `int16`, `int32`, `int64`, depending on the number of bits to be used to store the number)
- Character, e.g., "a" (`char`)
- Boolean, e.g., true (`logical`)

Other, more advanced types are also possible, but we won't discuss them just yet. (We'll return to the concept of data types in Chapter 6.) Note that numeric values can be either floating point or integer types. In particular, unless we explicitly specify otherwise, all numeric values in MATLAB will be `double` precision floating point values. We can always find out the type of a variable(s) using the `whos` command. If we just type `whos` in the command window, we obtain a list of all variables in the workspace and their types. If we give specific variable names, it will only display the types of those given.

■ Example 1.2

For example, try typing the following in the command window: *01.D*

```
>> a = 1
>> b = 'x'
>> c = false
>> whos
>> d = [1 2 3 4]
>> e = ['a' 'b' 'c' 'd']
>> f = ['abcd']
>> whos d e f
```

Note how the *character* type is indicated by enclosing characters in single quotes (double quotes have a different meaning in MATLAB, as we will see in Section 3.5.1). Note also that the definitions of the variables e and f are equivalent: the form used in assigning to f is short-hand for that used in assigning to e. We will see a practical use for arrays of characters in the next section.   ■

Sometimes we want to convert between different data types. We will discuss this topic in detail in Section 3.7.3, but for now we introduce an example usage of type conversion. When using the `disp` function to display an array (see Table 1.3), we may want to mix different types, e.g., a character array and a number. Normally, in MATLAB this is not possible, because arrays can only contain data of a single type. However, we can get around this by converting the types of some of our data, as shown below.

```
>> sys = 100;
>> str = ['Systolic B.P. = ' num2str(sys)];
>> disp(str);
```

Here, we want to mix character array data with a numeric value, representing systolic blood pressure. We use the built-in MATLAB function `num2str` to convert the numeric value into its character array equivalent, and then combine the character array that forms the start of the message with the converted numeric data. This results in a single character array, which is assigned to `str`, which is then displayed using the `disp` command.

■ Activity 1.9

Type the following in the MATLAB command window: *01.D*

```
>> heart_rate = 65
>> blood_sugar = 6.2
>> name = 'Joe Bloggs'
>> rate = num2str(heart_rate)
```

Predict what the data types of the four variables will be. Verify your predictions by using the `whos` command. ∎

1.9   Loading and saving data

When using MATLAB for biomedical problems, we may be working with large amounts of data that are saved in an external file. When using the `disp` function to display an array (see Table 1.3), we may want to mix different types, e.g., a character array and a number. When writing data items to a text file, we normally need a special character called a *delimiter*, which is used to separate one data item from the next. We can save data delimited in this way with the `writematrix` command, as illustrated in the following example:

```
>> f = [1 2 3 4 9 8 7 6 5];
>> writematrix(f, 'test.txt');
```

Here, the two arguments of `writematrix` specify the variable containing the data to be saved and the file name (a character array). By default, `writematrix` uses commas as the delimiters (i.e., the file will be a comma-delimited text file). We can tell MATLAB to use a different delimiter as shown in the code below.

```
>> writematrix(f, 'testtab.txt', 'Delimiter', 'tab');
>> writematrix(f, 'testspace.txt', 'Delimiter', ' ');
```

In these commands, the delimiters are a *tab* and a space character. Note that the delimiters specified in this way are used to separate the different *columns* of the array. MATLAB will always use a "new line" character to separate *rows* when writing matrices (i.e., 2-D arrays).

The `load` and `save` commands are probably the most commonly used commands for reading and writing data (you already used the `load` command in Activity 1.6). The following commands illustrate the use of `load` and `save`:

```
>> d = load('test.txt');
>> e = d / 2;
>> save('newdata.mat', 'd', 'e');
>> save('alldata.mat');
>> clear;
>> load('newdata.mat');
```

When using the `save` command the first argument specifies the name of the file to be saved and any subsequent arguments specify which variable(s) to save inside the file. Omitting the variable name(s) will cause the entire workspace to be saved. `save` will create a MATLAB-specific binary file containing the variables and their values that can be loaded in again at any time in the future.

These binary files are called *MAT* files, and it is common to give them the file extension "*.mat*." The `load` command can be used to read either text files created by `writematrix` or binary *MAT* files created by `save`. Note that when using the `load` command with a text file, we should assign the result of the `load` into a variable for future use (e.g., see the difference between the first and the last commands above). The `clear` command removes all current variables from the workspace.

An alternative, and easier, way of loading *MAT* files into the workspace is to simply use the MATLAB environment to drag-and-drop the appropriate file icon from the current folder window to the workspace browser (or just double-click on it in the current folder window).

Finally, the *import data* wizard can often be the quickest and easiest way to read data from text files (i.e., not *MAT* files). Selecting the *Import Data* button on the *Home* tab will bring up a wizard in which we can choose the file and interactively specify delimiters and the number of header lines in the file.

We will look again in more detail at the subject of MATLAB file input/output functions in Chapter 7.

■ Activity 1.10

Continuing with the application from Activity 1.4, create a regularly spaced   *01.C, 01.E*
array of numbers of cigarettes smoked between zero and twenty in steps of
two. Use the linear equation provided earlier to compute the probabilities
of getting cancer for each of these numbers. Save the probabilities only to a
text file, called *cancer.txt*, using the tab character as the delimiter.   ■

1.10 Visualizing data

As well as accessing and manipulating data, MATLAB can also be used for visualization of data. The `plot` command is used for simple data visualization. Consider the following code:

```
>> x = 0:0.1:2*pi;
>> y = sin(x);
>> plot(x,y,'-b');
```

This should produce the output shown in Fig. 1.3a. The first line in the code uses the colon operator to create an array, called x, containing numbers that start with 0, and go up to 2π in steps of 0.1. This array acts as the x coordinates for the plot. The second line applies the built-in MATLAB `sin` function to every element of x. The resulting array acts as the y coordinates for the plot. Finally, the third line of code displays the graph using the `plot` command. The first two arguments are the x and y coordinate arrays (which must be of the same

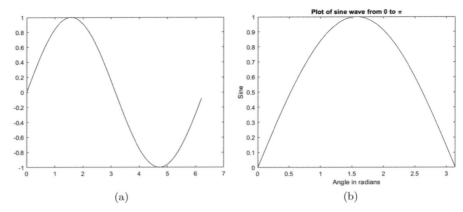

FIGURE 1.3
Plots of a sine wave between (a) 0 and 2π; (b) 0 and π. Plot (b) also has annotations on the axes and a title.

Table 1.5 Codes that can be used to specify the appearance of markers and lines in MATLAB plots. They are used in combination and passed to the `plot` command, e.g., `'o:r'` gives a plot with circular markers and a dotted line colored red.

Marker		Line		Color	
.	Dot	—	Solid line	k	Black
x	Cross	— —	Dashed line	b	Blue
o	Circle	:	Dotted line	r	Red
d	Diamond	—.	Dash-dot line	g	Green
s	Square			m	Magenta
+	Plus			c	Cyan
*	Asterisk			y	Yellow

length). The third argument specifies the appearance of the line. Table 1.5 summarizes the different symbols that can be used to control the appearance of the line and the markers. All of these symbols can be included in the third character array argument to the `plot` command. Try experimenting with some of these to see the effect they have.

Next, we will add some text to our plot. Try typing the following commands to add a title and labels for the x and y axes, respectively:

```
>> title('Plot of sine wave from 0 to \pi');
>> xlabel('Angle in radians');
>> ylabel('Sine');
```

If we only wish to visualize the plot for values of x between 0 and π, the `axis` command can be used to specify minimum and maximum values for the x and y axes:

```
>> axis([0 pi 0 1]);
```

This should produce the output shown in Fig. 1.3b, i.e., the same plot, but zoomed to the x range from 0 to π, and the y range from 0 to 1.

Sometimes we may want to display two or more curves on the same plot. This is possible with the `plot` command as the following code illustrates:

```
>> y2 = cos(x);
>> plot(x,y,'-b', x,y2,'—r');
>> legend('Sine','Cosine');
```

We can display multiple curves by just concatenating groups of the three arguments in the list given to the `plot` command, i.e., x data, y data, line/marker style, etc. The `legend` command adds a legend to identify the different curves. The `legend` command should have one character array argument for each curve plotted.

Finally, note that we can close down all figure windows that are currently open by typing

```
>> close all
```

We will return to the subject of data visualization in Chapter 9, and discuss more ways of visualizing multiple datasets in Section 9.2.2.

1.11 Curve fitting

As well as plotting data, MATLAB allows us to perform curve fitting to help identify relationships between data. As we will see, this can be done in two ways: either by calling the built-in functions `polyfit` and `polyval` from the command window, or by using the menu options on a figure window.

■ Example 1.3

To illustrate the use of `polyfit` and `polyval`, consider the following example: When diagnosing heart disease, cardiologists observe and measure the motion of the heart as its beats. Radial myocardial displacements are measured for the left ventricle of a patient's heart using a dynamic magnetic resonance (MR) scan. The data for one segment of the ventricle are contained in a file called *radial.mat*, which can be downloaded from the book's web site. This *MAT* file contains two variables: `radial` represents the radial displacement measurements, and `t` represents the time (in milliseconds)

01.E, 01.F

of each measurement. The following code will load the data, plot `radial` against `t`, and then fit a cubic polynomial curve to the data:

```
>> load('radial.mat');

>> plot(t, radial, '-r');
>> title('Radial myocardial displacement versus time');
>> xlabel('Time (ms)');
>> ylabel('Radial displacement (mm)');

>> p = polyfit(t, radial, 3);

>> radial_100ms = polyval(p, 100);
```

Note that the `polyfit` function takes three arguments: the x and y data, which we are fitting the curve to, and the order of the polynomial (3, in this case, for a cubic polynomial). The value returned (`p`) is an array containing the cubic polynomial coefficients. This is then used as one of the inputs to the `polyval` function. `polyval` computes the value of a polynomial curve at a given value. In the example code, we compute the value of the fitted cubic polynomial (i.e., the predicted radial displacement) at $t = 100$ milliseconds. ■

Alternatively, the same curve fitting can be carried out using the menu options on a figure plot. Try displaying a plot of `t` against `radial` using the code example given above (see Fig. 1.4a), and then selecting the *Tools* menu on the figure window, followed by the *Basic Fitting* option. This opens a new window, where we can choose which curve(s) to fit, as well as to predict values using the fitted curve(s) (see Fig. 1.4b). Try experimenting with this functionality.

■ Activity 1.11

01.C, 01.F
The injury severity score (ISS) is a medical score to assess trauma severity. Data on ISS and hospital stay (in days) have been collected from a number of patients who were admitted to hospital after accidents. The ISS data are [64 35 50 46 59 41 27 39 66], and the length of stay data are [8 2 5 5 4 3 1 4 6]. Use MATLAB to plot the relationship between ISS and hospital stay. Then, fit a linear regression line to these data, and estimate what length of hospital stay would be expected for a patient with an ISS of 55. ■

1.12   MATLAB scripts

As we start to write slightly longer pieces of MATLAB code, we will find it useful to be able to save sequences of commands for later reuse. MATLAB *scripts* (or script *m*-files) are the way to do this. To create a new MATLAB script file, we can use the *editor* window. First, we click on the *New* button (under the *Home* tab)

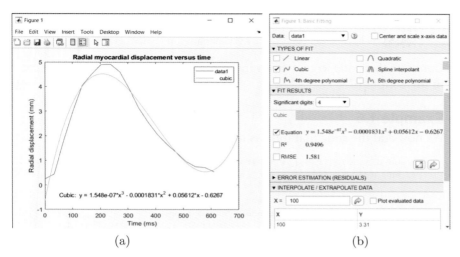

(a)   (b)

FIGURE 1.4

Curve fitting using the figure window: (a) plot of time against radial myocardial displacement; (b) fitting a cubic polynomial to the data using the graphical user interface.

and choose *Script*. This will create a blank file. We can now type our MATLAB commands into this file. To save the file, we click on the *Save* button, 🖫 (under the *Editor* tab). The first time that we save the file we will be prompted for a file name. It is important to save MATLAB script files with the file extension ".*m*" (hence the name *m*-file). To run the commands in our script, we click on the *run* icon ▷.

■ **Example 1.4**

The code listing shown below is an extended version of Example 1.3 (plotting and fitting a curve to radial displacements of the left ventricle of the heart). We can enter this code into a MATLAB script *m*-file, save it with the file name *example_1_4.m*, and then run the script, as described above. (Note that we do *not* include the MATLAB command prompt in this listing, because we are entering the code into the editor window, not the command window.)

01.C, 01.F, 01.H

```
load('radial.mat');

p = polyfit(t, radial, 3);

t_curve=linspace(min(t), max(t), 20);
radial_curve=polyval(p, t_curve);

plot(t, radial, 'or', t_curve, radial_curve, '-b');
title('Radial myocardial displacement versus time');
```

```
xlabel('Time (ms)');
ylabel('Radial displacement (mm)');
legend('Displacement data', 'Fitted cubic polynomial');
```

Compare this code to that given in Example 1.3. What does the extra code in this script do? ∎

■ Example 1.5

O1.C, O1.F, O1.H

We now add the code that follows to Example 1.4.

```
[v,end_diastole_frame] = max(radial);
[v, end_systole_frame] = min(radial);

est_end_dia = polyval(p, t(end_diastole_frame));
est_end_sys = polyval(p, t(end_systole_frame));

radial_difference = est_end_dia − est_end_sys;
```

The code has been extended to compute the difference between the maximum and minimum displacements using the fitted curve. Note how we use the optional second return value of the max and min functions to get the array *indices* of the maximum and minimum displacements. These represent the *end diastole* and *end systole* frames, respectively (i.e., at maximum relaxation and maximum contraction, respectively). ∎

Sometimes we may write a script in which a line of code becomes quite long. For MATLAB, there is no problem with having long lines of code. The MATLAB interpreter will execute them just like any other line. But when it comes to reading and understanding a script, very long lines can make things difficult. In such cases, we can use the special MATLAB symbol "..." (i.e., three dots). This tells MATLAB that we wish to continue the current command on the next line. For instance, we can add the code that follows to the end of the listing from Example 1.5.

```
disp(['Radial difference between systole/diastole = ' ...
     num2str(radial_difference) 'mm']);
```

■ Activity 1.12

O1.C, O1.E, O1.H

Write a MATLAB script *m*-file to read in a height (in meters) and a weight (in kilograms) from the keyboard, then compute and display the body mass index (BMI), according to the formula

$$BMI = mass/height^2.$$

(Hint: Look at the MATLAB documentation for the input *command.)* ∎

■ **Activity 1.13**

Important information about the function of the heart can be learned by   *01.C, 01.E, 01.H*
measuring the volume of its chambers over the cardiac cycle. The left ventricle is of particular interest, as it acts to pump blood around the body. As
a very simple approximation, the left ventricle can be considered to be a
half-ellipsoid.

The equation for the volume of an ellipsoid is

$$\frac{4}{3}\pi abc,$$

where a, b, and c are the radii of the ellipsoid along its three axes (see illustration below).

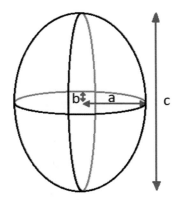

Estimates of these radii can be made using an imaging device, such as an
echocardiography (i.e., cardiac ultrasound) machine.

Write a MATLAB script *m*-file to read in values from the keyboard for the
three radii of a left ventricle (i.e., the radii a and b and the height of the
half-ellipsoid, $\frac{c}{2}$, all in mm), and estimate its volume by approximating it
as a half-ellipsoid. The volume should be reported in mm^3 and also in
ml (1000 mm^3 = 1 ml). You can test your script using typical values for
the left ventricle radii at diastole (i.e., maximum relaxation), which are approximately 20–30 mm for the two short axes and 60–90 mm for the long
axis. The range of volumes that correspond to these typical dimensions is
50–170 ml.   ■

1.13   Comments

When writing MATLAB scripts, it is possible to add *comments* to our code. A
comment is a piece of text that will be ignored by MATLAB when running our

script. Comments in MATLAB are specified by the "%" symbol: any text after a "%" symbol on any line of a script will not be interpreted by MATLAB. It is good practice to add comments to our code to explain what the code is doing. This is useful if we or other people need to look at our code in the future with a view to modifying it or fixing errors.

■ Example 1.6

01.H   Let's add some comments to the script we looked at in Example 1.4 to see how it improves code readability.

```
% load data
load('radial.mat');

% fit curve
p = polyfit(t, radial, 3);

% determine data used to visualise curve
t_curve=linspace(min(t), max(t), 20);
radial_curve=polyval(p, t_curve);

% plot original data and curve
plot(t, radial, 'or', t_curve, radial_curve, '-b');
title('Radial myocardial displacement versus time');
xlabel('Time (ms)');
ylabel('Radial displacement (mm)');
legend('Displacement data', 'Fitted cubic polynomial');
```

■

■ Activity 1.14

01.H   Go through some of the example and activity solution files and add meaningful but concise comments to help the readability of the code.   ■

1.14   Debugging

Unfortunately, often the programs we write will not work the first time we run them. If they have errors, or *bugs*, in them, we must first *debug* them. MATLAB has two useful features to help us in the debugging process.

1.14.1   MATLAB debugger

MATLAB has a built-in debugger to help us identify and fix bugs in our code. One of the most useful features of the MATLAB debugger is the ability to set *breakpoints*. A breakpoint at a particular line in a program will cause execution of the program to automatically stop at that line. We can then inspect the values of variables to check that they are as we expect, and step through the code, line by line. To set a breakpoint on a particular line, in the *editor* window left-click once in the gray vertical bar, just to the right of the line number. To confirm

```
PUBLISH        VIEW
                                                Step In
      Go To ▼                                   Step Out      Function Call Stack:              Quit
      Find ▼     Breakpoints  Continue  Step   Run to Cursor  test                    ▼        Debugging
      NAVIGATE     BREAKPOINTS                            DEBUG
```

```
 Editor - /home/ak06/test.m                                              ⊙ ×
  test.m  ×  +
 1
 2●⇨    theta = pi / 6; % angle of rotation
 3 -    pts = [1 1; 1 2; 2 1; 2 2]; % four points defining a square
 4
 5       % need to transpose to make them suitable for a
 6       % matrix-based coordinate transformation
 7 -    pts = transpose(pts);
 8
 9       % define a rotation matrix by theta radians
10 -    R=[cos(theta) sin(theta);
11        -sin(theta) cos(theta)];
12
13       % rotate the points about the origin
14 -    pts_t = R * pts;
15
16       % plot the original and transformed points
17 -    plot(pts(1,:),pts(2,:),'ok', pts_t(1,:),pts_t(2,:),'ob');
18 -    legend('Initial points', 'Transformed points');
19 -    axis([-1 3 -1 3]);
20 -    axis square;
```

FIGURE 1.5
The MATLAB debugger.

that the breakpoint has been created MATLAB will display a small red circle ●
in the gray bar (see Fig. 1.5, line 2). Now, when we click on the *run* icon ▷
to start the script, execution will stop at the line with the breakpoint. This fact
will be indicated by a green arrow, just to the right of the breakpoint: ⇨ (see
Fig. 1.5).

Once execution has been stopped, we are free to inspect variable values, as we
normally would in MATLAB, i.e., by typing in the command window or by
looking at the workspace browser. To continue executing the program, there
are a number of options, each available from an icon in the *Debug* section of
the menu under the *Editor* tab:

- *Continue*: Execute the program from the breakpoint until program termi-
  nation.
- *Step*: Execute the current line of code, then pause again.
- *Step in*: If the current line is a function, step into that function.
- *Step out*: If we are currently inside a function, step out of the function.
- *Run to cursor*: Execute the program until the line where the cursor is posi-
  tioned is reached, then pause.
- *Quit debugging*: Exit the debugger.

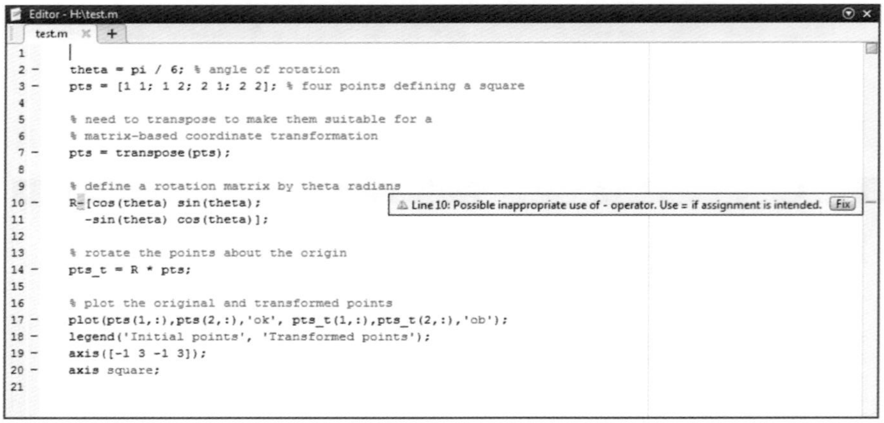

```
Editor - H:\test.m                                                      ⊙ x
 test.m  ×   +
 1    │
 2 -  theta = pi / 6; % angle of rotation
 3 -  pts = [1 1; 1 2; 2 1; 2 2]; % four points defining a square
 4
 5    % need to transpose to make them suitable for a
 6    % matrix-based coordinate transformation
 7 -  pts = transpose(pts);
 8
 9    % define a rotation matrix by theta radians
10 -  R=[cos(theta) sin(theta);        ⚠ Line 10: Possible inappropriate use of - operator. Use = if assignment is intended.  Fix
11      -sin(theta) cos(theta)];
12
13    % rotate the points about the origin
14 -  pts_t = R * pts;
15
16    % plot the original and transformed points
17 -  plot(pts(1,:),pts(2,:),'ok', pts_t(1,:),pts_t(2,:),'ob');
18 -  legend('Initial points', 'Transformed points');
19 -  axis([-1 3 -1 3]);
20 -  axis square;
21
```

FIGURE 1.6
The MATLAB code analyzer.

1.14.2   MATLAB code analyzer

The second MATLAB feature that is useful for finding and fixing errors is the *code analyzer*. This can be used to automatically check our code for possible problems. It can be run in two ways: either on demand by requesting a code analyzer report, or in real time in the editor window.

Your MATLAB installation is probably set up to analyze your code in real time (i.e., whenever you save a script file in the editor window). To view or change this preference, click on *Preferences* under the *Home* tab, and select *MATLAB* → *Code Analyzer*. Then, either check or uncheck the option labeled *Enable integrated warning and error messages*.

We can see an example of the real time code analyzer in Fig. 1.6. A potential error has been spotted on line 10 and highlighted. It shows that a minus sign has been used where an assignment was expected. A small orange horizontal bar to the right of the text window indicates a warning (a red bar would indicate an error). When the user hovers their mouse pointer over this bar, a message pops up, in this case, warning of a "Possible inappropriate use of - operator." Clicking on the *Fix* button will perform the fix recommended by the code analyzer, in this case, to replace the "−" with a "=."

Alternatively, if we don't want such real time warning and error messages, we can change the preference described above, and just request a code analyzer report whenever we want one. To do this, click on the ⊙ symbol at the top right of the editor window, and select *Show Code Analyzer Report*.

We can also request a code analyzer report from the command window by using either the command `checkcode` or the command `mlintrpt`. In each case, we need to specify the file that we want analyzed:

```
>> checkcode('test.m')
L 10 (C 2): Possible inappropriate use of — operator.
Use = if assignment is intended.
```

The command `mlintrpt('test.m')` can also be invoked from the command window and will produce the same report, but it will be given in a separate window.

We will return to the important topic of code debugging in Chapter 5.

■ Activity 1.15

The MATLAB *m*-file shown below is intended to read a matrix from a text file and display how many coefficients the matrix has. However, the code does not currently work correctly. Use the MATLAB code analyzer and debugger to identify the bug(s) and fix the code.

01.C, 01.E, 01.G, 01.H

```
% clear workspace
clear

% load matrix from text file
m = load('m.txt');

% compute number of coefficients
s = size(m);
n = s^2;

% display answer
disp(['The matrix has ' num2str(n) ' elements']);
```

(This file *activity_1_15.m*, and the test file *m.txt* are available for download from the book's web site.)   ■

1.15   Summary

Learning how to program a computer opens up a huge range of possibilities: in principle, we can write a program to perform any computable operation. Programming languages are the means with which we instruct the computer what to do. There are several ways of distinguishing between different types of programming language, e.g., interpreted/compiled, procedural/object-oriented/declarative. In this book, we will learn procedural programming using MATLAB, which is normally used as an interpreted programming language.

As well as being a programming language, MATLAB is a powerful software package that implements a wide range of mathematical operations. It can be used to create and assign values to variables, which can have different types, such as numbers, characters, Booleans, or 1-D or 2-D arrays (matrices). Basic arithmetic operations and a large number of built-in functions can be applied to values, as well as to arrays and matrices. Variables can be saved to external files and loaded from such files. MATLAB can also be used to visualize data by plotting one array of values against another, and also to fit curves to the data.

MATLAB scripts (or script *m*-files) allow us to save a sequence of MATLAB commands for repeated use. It is a good idea to add comments to scripts to make them easier to understand. The MATLAB debugger and code analyzer can be used to track down and fix bugs in our code.

1.16 Further resources

- The MATLAB documentation contains extensive information and examples about all commands and built-in functions. Just type doc, followed by the command/function name in the command window. This documentation is also available on-line at the MathWorks web-site.
- The MathWorks web-site contains a number of useful video tutorials: http://mathworks.com/products/matlab/videos.html.
- The File Exchange section of the same web-site can be useful for downloading functions and MATLAB codes that have been shared by others: http://mathworks.com/matlabcentral/fileexchange/.

1.17 Exercises

■ Exercise 1.1

O1.B, O1.C, O1.D

First, write a command to clear the MATLAB workspace. Then, create the following variables:

1. A variable called a, which has the character value 'q'
2. A variable called b, which has the Boolean value true
3. A variable called c, which is an array of integers between 1 and 10
4. A variable called d, which is an array of characters with the values "h," "e," "l," "l," "o"

Now find out the data types of all variables you just created.　■

■ Exercise 1.2

O1.C

A group of patients have had their heights in centimeters measured as: 159, 185, 170, 169, 163, 180, 177, 172, 168, and 175. The same patients' weights in kilograms are the following: 59, 88, 75, 77, 68, 90, 93, 76, 70, and 82. Use MATLAB to compute and display the mean and standard deviation of the patients' heights and weights.　■

■ Exercise 1.3

O1.C

Let a = [1 3 1 2] and b = [7 10 3 11].

1. Sum all elements in a and add the result to each element of b.

2. Raise each element of b to the power of the corresponding element of a.
3. Divide each element of b by 4.   ■

■ Exercise 1.4

Write MATLAB commands to compute the *sine*, *cosine*, and *tangent* of an array of numbers between 0 and 2π (in steps of 0.1). Save the original input array and all three output arrays to a single *MAT* file, and then clear the workspace.   ■

01.C, 01.E

■ Exercise 1.5

Write a MATLAB script *m*-file to read in three floating point values from the user (a, b, and c), which represent the lengths of three sides of a triangle. Then compute the area of the triangle according to the equation:

$$\text{Area} = \sqrt{s(s-a)(s-b)(s-c)},$$

where s is half of the sum of the three sides.   ■

01.C, 01.E, 01.H

■ Exercise 1.6

A quick way of determining if any integer up to 90 is divisible by 9 is to sum the digits of the number: if the sum is equal to 9, then the original number is also divisible by 9. Write a MATLAB script *m*-file to verify the rule, i.e., create an array of all multiples of 9 between 9 and 90, and compute the sums of their two digits.

01.C, 01.H

(Hint: Look at the MATLAB documentation for the idivide *and* mod *commands. These can be used to find the result and remainder after integer division, so if you use them with a denominator of 10, they will return the two digits of a decimal number. Because these two commands only work with integer values, you will also need to know how to convert from floating point numbers to integers: type* doc int16.*)*   ■

■ Exercise 1.7

Write a MATLAB script *m*-file to plot curves of the *sine*, *cosine*, and *tangent* functions between 0 and 2π. Your script should display all three curves on the same graph and annotate the plot. Limit the range of the *y*-axis to between -2 to 2.

01.C, 01.F, 01.H

(Hint: Look at the MATLAB documentation for the axis *or* ylim *commands.)*   ■

■ Exercise 1.8

01.G

Create the following matrices *using a single line of MATLAB code* each:

1. A 4×4 matrix containing all zeros apart from 3s on the diagonal.
2. A 3×4 matrix, in which all elements have the value -3.   ■

■ Exercise 1.9

01.G

A magnetic resonance (MR) scanner acquires three image slices with their plane equations in scanner coordinates (in mm) given by the following:

$$-5x + 5y + 11z = 21$$
$$-2x - 4y - 2z = -24$$
$$32x + 15y - 29z = 115$$

It is known that the slices intersect at a point. Find the coordinates of the point.   ■

■ Exercise 1.10

01.E, 01.F, 01.H

Data have been collected for the ages and diastolic/systolic blood pressures (both in units of mmHg) of a group of patients, who are taking part in a clinical trial. All data are available to you in the files *age.txt*, *dbp.txt*, and *sbp.txt*. Write a MATLAB script *m*-file that reads in the data and creates a single figure containing two plots: one of age against systolic blood pressure and one of age against diastolic blood pressure. Your script should annotate the plots.   ■

■ Exercise 1.11

01.C, 01.E, 01.F

Positron emission tomography (PET) imaging involves the patient being injected with a radioactive substance (a *radiotracer*), the concentration of which is then imaged using a scanner. Staff involved in performing PET scans receive some radiation dose as a result of being in the vicinity of the patient while the radiotracer is inside their body.

A study is investigating the effective radioactive dose to which staff are exposed as a result of performing PET scanning (e.g., see [2]). Data indicating the effective radiation dose (in microsieverts, μSv) as a function of distance (in meters) from the patient are available in the file *dose_data.mat*. Write a MATLAB script *m*-file that loads in this file and fits a polynomial of an appropriate order to the data. The script should then produce a plot of distance against effective dose and display the fitted curve on the same figure. Finally, the fitted curve should be used to compute the expected effective radiation dose at a distance of 3.25 m.   ■

FAMOUS COMPUTER PROGRAMMERS: ADA LOVELACE

Augusta Ada King, the Countess of Lovelace (commonly known these days as Ada Lovelace), was an English mathematician and writer. She was born in 1815 and was the daughter of the famous English poet, Lord Byron. Her father left her mother a month after Ada was born, eventually dying of disease in the Greek War of Independence when Ada was eight. Ada's mother encouraged her daughter to become interested in mathematics, in an effort to prevent her succumbing to what she saw as Lord Byron's insanity.

As a young adult, Ada Lovelace developed a friendship with the British mathematician, Charles Babbage. Babbage is well known for designing the analytical engine, which is commonly recognized as the world's first mechanical computer. In 1842–1843 Ada translated an article on the analytical engine, supplementing it with some notes of her own. It is these notes that are the source of Ada's fame today. They contained what is generally considered to be the first computer algorithm, i.e., a sequence of steps that can be executed by a computing machine. It is this vision of a computer, as being more than just a numerical calculator, that has had a major influence on the historical development of computers, and has led to her being considered as the world's first computer programmer. However, her program has never been run, because the analytical engine was never actually built in its designer's lifetime. (See http://plan28.org for details of a modern-day project to actually build the analytical engine and run Ada's program.)

As well as her early work on computer programming, in 1851 Ada was involved in a disastrous attempt to build a mathematical model for making successful bets. This left her thousands of pounds in debt and resulted in her being blackmailed by one of her colleagues, who threatened to tell Ada's husband of her involvement.

Ada Lovelace died of cancer in 1852 at the age of 36. As a result of her importance in the history of computing, one of the early computer programming languages, developed in 1980, was named Ada.

"The Analytical Engine weaves algebraic patterns, just as the Jacquard loom weaves flowers and leaves."

Ada Lovelace

Control structures

2.1 Introduction

In Chapter 1, we introduced the fundamental concepts of computer programming, as well as the basics of the MATLAB software package. In particular, we saw how to create, save, and run MATLAB script *m*-files. Scripts allow us to save sequences of commands for future use, allowing time savings to be made for code that may be needed again. MATLAB scripts become more powerful as we learn how to do some simple computer programming. Computer programming involves not only executing a number of commands, but controlling the sequence of execution (or *control flow*) using special programming *constructs*. In this chapter, we introduce some of the programming constructs provided by MATLAB.

2.2 Conditional `if` statements

The normal flow of control in a MATLAB script, and procedural programming in general, is *sequential*, i.e., each program statement is executed in sequence, one after the other. This is illustrated in Fig. 2.1.

35

MATLAB® Programming for Biomedical Engineers and Scientists. https://doi.org/10.1016/B978-0-32-385773-4.00011-3

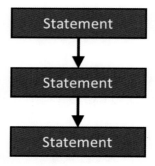

FIGURE 2.1
The normal sequential control flow in procedural programming.

For example, to use a medical analogy, a treatment plan might consist of per-forming a blood test, followed by analyzing the results, and finally administer-ing a particular dose of a drug. Each action is performed one after the other, in sequence. However, with purely sequential control flow, we are quite limited in what we can achieve. For instance, we might want to only administer the drug if particular conditions are met in the blood test results. Or, we might want to repeat the blood test after drug delivery to check the patient's response to the drug, and, potentially, alter the dose accordingly. We might even want to do this many times until the blood test shows a satisfactory response to the drug.

Similarly, in computer programming, if all of our programs featured only se-quential control flow, they would be limited in their power. To write more complex and powerful programs, we need to make use of programming con-structs that alter this normal control flow. One type of programming construct is the *conditional* statement, and the simplest type of conditional statement in MATLAB is the `if` statement. The control flow of an `if` statement is illustrated in Fig. 2.2a, in which we use a diamond-shaped box to indicate a decision point. We can see that there are now two possible paths through the program, involving execution of different statements. Which path is taken depends upon the result of applying a test *condition*. Returning to our medical analogy, the condition might be based on the results of the blood test, and the two paths of execution might be administering the drug or not.

■ Example 2.1

O2.A   The use of an `if` statement is illustrated in the example below.

```
a = input('Enter a number:');
if (a >= 0)
    root = sqrt(a);
    disp(['Square root = ' num2str(root)]);
else
    disp(['Number is negative, there is no square root']);
end
```
■

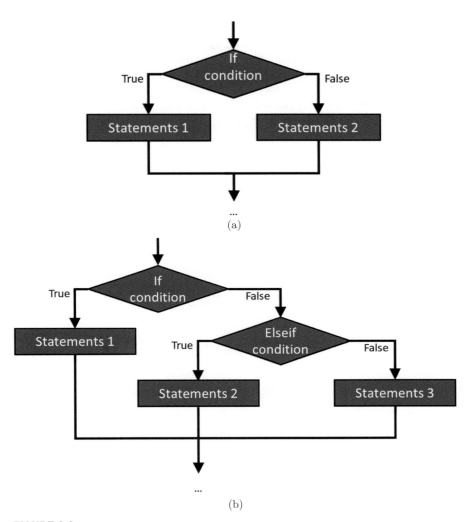

FIGURE 2.2
The control flow of conditional statements: (a) if—else; (b) if—elseif—else.

Try entering these statements and saving them as a MATLAB script *m*-file. Examine the code to make sure that you understand what it does (recall that the disp statement displays an array in the command window; in this case, an array of characters). You can even step through the code using the MATLAB debugger. Pay particular attention to the alternative *paths of execution* that the program can take, depending on the result of the comparison (a>=0). This comparison corresponds to the *condition* box in Fig. 2.2a, and it should always follow the `if` *keyword*.[1] The two statements after the `if` and its condition will be executed only if the condition is true. The statement after the `else` will be executed if the condition is false. After one of these two paths has been taken, control will resume immediately after the `end` statement. Note that the `if` and `end` keywords are both compulsory and should always come in pairs. The `else` keyword is optional and if omitted, the program execution path will jump to after the `end` if the condition is false.

■ Activity 2.1

02.A

In Activity 1.12 you wrote MATLAB code to compute a patient's body mass index (BMI), according to the formula

$$BMI = mass/height^2,$$

based on a height (in meters) and weight (in kilograms) entered using the keyboard. Modify this code (or write a new version) to subsequently display a text message based on the computed BMI, as indicated in the table below.

Range	Message
BMI ≤ 18.5	"Underweight"
18.5 < BMI ≤ 25	"Normal"
25 < BMI ≤ 30	"Overweight"
30 < BMI	"Obese"

■

In Activity 2.1 you will have seen that it is sometimes necessary to test multiple conditions and execute different pieces of code, depending on the results. MATLAB provides an (optional) additional clause to the `if` statement to help with such situations: the `elseif` clause. A flow diagram illustrating the use of the `elseif` clause is shown in Fig. 2.2b. We can see that it combines an `else` statement with a subsequent `if` statement, as illustrated in the following piece of code:

```
x = input('Enter number: ');
if (x < 0)
    disp('Strictly negative number');
elseif (x > 0)
    disp('Strictly positive number');
else
    disp('Zero');
end
```

Here, we can see that the `elseif` clause allows us to test the condition `(x>0)`, but *only* if the first condition of `(x<0)` was false. Note that the above code is

---

[1] Recall the discussion from Section 1.6 about variable naming in MATLAB, in particular that we cannot use MATLAB *keywords* as variable names. The `if` statement is an example of such a keyword.

equivalent to the version shown below, but results in a less deep "nesting" of statements (see Section 2.9).

```
x = input('Enter number: ');
if (x < 0)
    disp('Strictly negative number');
else
    if (x > 0)
        disp('Strictly positive number');
    else
        disp('Zero');
    end
end
```

We can actually have as many `elseif` clauses as we like inside an `if` statement, but only one `else` clause, since this acts to "catch" the control flow when no other condition(s) have been matched.

■ Activity 2.2

Rewrite your code for Activity 2.1 to use `elseif` clauses, instead of nested `if` statements. Test your new code using different input values to make sure that each BMI range leads to the correct output.   ■

02.A

SYNTAX: IF STATEMENT

Formally, we can define the syntax rule for a MATLAB `if` statement as follows:

```
if condition1
    statements1
elseif condition2
    statements2
else
    statements3
end
```

Note the following points about this syntax rule:

- The `if`/`end` keywords are compulsory.
- `statements1`, `statements2`, and `statements3` each refer to one or several statements.
- We can have multiple (or no) `elseif` clauses.
- The `else` clause is optional.

2.3   Comparison/logical operators

Example 2.1 introduced the concept of a *condition*. Conditions commonly involve comparisons between expressions involving variables and values. For example, we saw the `>=` comparison operator in the example. A list of common comparison (or *relational*) operators, such as `>=`, is shown in Table 2.1,

Table 2.1 Common relational and logical operators.

Relational		Logical	
==	Equal to	~	NOT
~=	Not equal to	&&	AND (scalar *short-circuit* operation)
>	Greater than	&	AND (scalar/array operation)
<	Less than	||	OR (scalar *short-circuit* operation)
>=	Greater than or equal to	|	OR (scalar/array operation)
<=	Less than or equal to		

along with common *logical* operators for combining the results of different comparisons.

Most of these operators are fairly intuitive. However, note the distinction between & and && (and likewise between | and ||). The & and | operators perform AND and OR operations, respectively. They will evaluate the expressions on both sides of the operator and return a `true` or `false` value, depending on whether both (&) or either (|) of them evaluated to `true`. These operators will work with either scalar (i.e., single value) logical expressions or array expressions (i.e., that evaluate to an array of `true`/`false` values).

The && and || operators perform the same AND and OR operations, but using what is known as a *short-circuiting behavior*. This means that, if the result of the overall AND/OR operation can be determined from the left-hand expression alone, then the right-hand expression will not be evaluated. For example, if the left-hand expression of an AND operation is `false`, then the result of the AND will also be `false`, regardless of the value of the right-hand expression. Therefore the advantage of short-circuiting is that unnecessary operations are not performed. However, *note that && and || can only be used with scalar values, not arrays.*

■ Example 2.2

02.A, 02.B The use of these relational and logical operators is illustrated in the following code excerpt:

```
...
at_risk = false;
if (gender == 'm') && (calories > 2500)
    at_risk = true;
end
if (gender == 'f') && (calories > 2000)
    at_risk = true;
end
...
```

Here, in both `if` statements, the Boolean `at_risk` variable is set to true only if both comparisons evaluate to true, e.g., if `gender` is equal to "m"

<u>and</u> `calories` is greater than 2500. If either comparison evaluates to false, the flow of execution will pass to after the corresponding `end` statement. We can use the short-circuiting version of the AND operator, as both sides of the operator are scalar (logical) values (i.e., not arrays). ■

Note that there were round brackets around the comparisons in Example 2.2. These indicate to MATLAB that the expressions inside the brackets should be evaluated first. These are not essential, but they may change the meaning of the condition. For example, suppose the brackets were not included, i.e.,

```
...
if gender == 'm' && calories > 2500
...
```

There are three different operators in this condition: `==`, `&&` and `>`. In what order would MATLAB evaluate them? The answer to this question lies in the rules for *operator precedence*. Whenever MATLAB needs to evaluate an expression containing multiple operators, it will use the order of precedence shown in Table 2.2. We have come across all of these operators before with a few exceptions. The `'` operator, when placed <u>after</u> a matrix, is just a short way of calling the `transpose` function, which we introduced in Section 1.7. The *unary* plus and minus operators refer to a + or − sign being placed before a value or variable to indicate its sign.

Now let's return to our condition without brackets. Of the three operators present (`==`, `&&`, and `>`), we can see from Table 2.2 that the ones with the highest precedence are `==` and `>`. Therefore these will be evaluated first. As they are equal in precedence, they will be evaluated from left to right, i.e., the `==` first followed by the `>`. After these evaluations, the condition is simplified to just the `&&` operator, with a Boolean value either side of it (whether they are true or false will depend on the values of the `gender` and `calories` variables). Finally, this `&&` operator is evaluated.

In this example, removing the brackets still resulted in the desired behavior. However, this will not always be the case. Furthermore, even if the brackets are not essential, it is still a good idea to include them to make our code easier to understand and less prone to errors.

■ Activity 2.3

If `a=1`, `b=2`, and `c=3`, use your knowledge of the rules of operator precedence to predict the values of the following expressions (i.e., work them out in your head and write down the answers). Verify your predictions by evaluating them using MATLAB.   *02.B*

1. `a+b * c`
2. `a ^ b+c`

Table 2.2 MATLAB operator precedence.

Precedence	Operator	Meaning
1	( )	Brackets
2	'	Matrix transpose
	.^	Power
	^	Matrix power
3	+	Unary plus
	−	Unary minus
	~	Logical not
4	.*	Element-wise multiplication
	./	Element-wise division
	*	Matrix multiplication
	/	Matrix division
5	+	Addition
	−	Subtraction
6	:	Colon operator
7	<	Less than
	<=	Less than or equal to
	>	Greater than
	>=	Greater than or equal to
	==	Equal to
	~=	Not equal to
8	&&	Logical and
9	||	Logical or

```
3. 2 * a == 2 && c − 1 == b
4. b + 1:6
```

∎

2.4 Conditional `switch` statements

Sometimes we may have many `if` statements, which all use conditions based on the same variable. It is not incorrect to use `if` statements in such cases, but it can lead to a large number of consecutive `if` statements in our code, making it harder to read and more prone to errors. In this case, it is preferable to use a `switch` statement. The `switch` statement offers an easy way of writing code, where the same variable needs to be checked against a number of different values.

∎ Example 2.3

02.A   The following example illustrates the use of a `switch` statement:

```
switch day
    case 1
```

```
            day_name = 'Monday';
        case 2
            day_name = 'Tuesday';
        case 3
            day_name = 'Wednesday';
        case 4
            day_name = 'Thursday';
        case 5
            day_name = 'Friday';
        case 6
            day_name = 'Saturday';
        case 7
            day_name = 'Sunday';
        otherwise
            day_name = 'Unknown';
    end
```

MATLAB will compare the *switch expression* (in this case, `day`) with each *case expression* in turn (the numbers 1–7). When a comparison evaluates to true, MATLAB executes the corresponding statements, and then exits the `switch` statement, i.e., control flow passes to after the `end` statement. The `otherwise` block is optional and executes only when no comparison evaluates to true. ∎

SYNTAX: SWITCH STATEMENT

The control flow of a `switch` statement is illustrated in Fig. 2.3 and the formal syntax is shown below.

```
    switch switch_expression
        case case_expression1
            statements1
        case case_expression2
            statements2
        ...
        case case_expressionN
            statementsN
        otherwise
            default statements
    end
```

Remember that the `switch` statement is used *only* for equality tests; we cannot use it for other types of comparison (e.g., >, <, etc.).

In Example 2.3, the switch expression was compared to a single value in each case. It is possible to compare the expression to multiple values by enclosing them within curly brackets and separating them by commas. The corresponding statements are executed if *any* of the values are matched. This is equivalent to an `if` statement with multiple equality tests combined using a || operator.

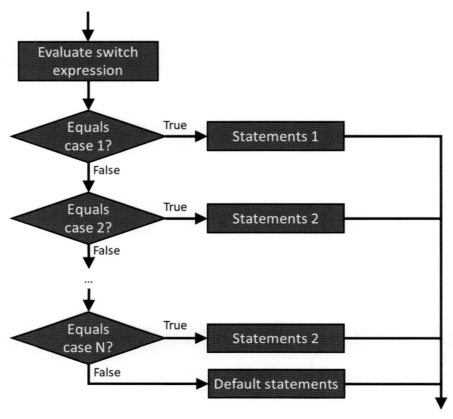

FIGURE 2.3
The control flow of a `switch` statement.

■ Example 2.4

O2.A

A `switch` statement with multiple values in each case is illustrated below.

```
switch day
    case {1,2,3,4,5}
        day_type = 'Weekday';
    case {6,7}
        day_type = 'Weekend';
    otherwise
        day_type = 'Unknown';
end
```

In this code, `day_type` will be assigned the character array `'Weekday'` if `day` has *any of* the values 1, 2, 3, 4, or 5. Likewise the value `'Weekend'` will be assigned if `day` is either 6 or 7. If none of these cases match, the value `'Unknown'` will be assigned to `day_type`. ■

■ **Activity 2.4**

Write a MATLAB script *m*-file to read a character from the keyboard, then read a number. Based on the value of the character, the program should compute either the *sine* of the number (if the character was a "s"), the *cosine* (if it was a "c"), or the *tangent* (if it was a "t"). If none of these three characters was entered, a suitable error message should be displayed. Use a `switch` statement in your program.   02.A

(Hint: If you use the `input` *command to read a character, by default it will try to "evaluate" it as a variable. To prevent this, add a second argument "s." See the MATLAB documentation for* `input` *for details.)*

Once you have completed your solution, you could try to modify your code to make it more robust. For example, try to allow it to accept either upper case or lower case letters (e.g., "s" or "S") for each option.   ■

■ **Activity 2.5**

The normal human heart rate ranges from 60 to 100 beats per minute (BPM) at rest. *Bradycardia* refers to a slow heart rate, defined as below 60 BPM. *Tachycardia* refers to a fast heart rate, defined as above 100 BPM. Write a MATLAB script *m*-file to read in a heart rate from the keyboard, and then display the text "Warning, abnormal heart rate" if it is outside of the normal range.   ■   02.A, 02.B

■ **Activity 2.6**

Now modify your solution to Activity 2.5 to display one of two warning messages, depending on whether the heart rate indicates bradycardia or tachycardia. If the heart rate is within the normal range, the text "Normal heart rate" should be displayed.   ■   02.A

2.5   Iteration: `for` loops

In addition to conditional statements, the second fundamental type of programming construct that we will consider in this chapter is the *iteration* statement. Iteration statements are intended for use in situations in which one or more operations need to be repeatedly performed a number of times. In computer programming, iteration statements are often known as *loop* statements. Returning to the medical analogy that we introduced in Section 2.2, iteration could involve performing blood tests and modifying the drug dosage repeatedly until certain conditions in the blood test result are met (i.e., the patient has responded satisfactorily).

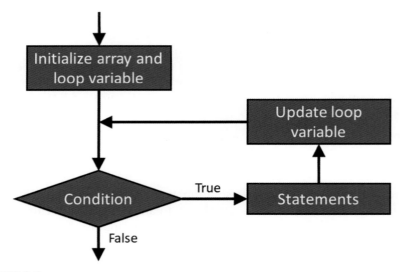

FIGURE 2.4
The control flow of a `for` loop.

There are two types of iteration statement provided in MATLAB. In this section, we consider the `for` loop. The control flow of a MATLAB `for` loop is shown in Fig. 2.4. Iteratively, the same *statements* are executed a number of times until a certain *condition* becomes false. A *loop variable* will be updated in each iteration to take on a different value from an *array*, which has been initialized at the beginning of the `for` loop. When the statement has been executed once for each element of the array, the condition becomes false.

■ Example 2.5

02.C   `for` loops can be useful when we want to execute some statements a fixed number of times. Consider the following example first, then we will return to the flow chart in Fig. 2.4.

```
% ask user for input
n = input('Enter number:');
f = 1;
if (n >= 0)
    for i = 2:n
        f = f * i;
    end
    disp(['The factorial of ' num2str(n) ' is ' num2str(f)]);
else
    disp(['Cannot compute factorial of negative number']);
end
```

This code reads in a number from the user and computes its factorial. (Note that MATLAB actually has a built-in function to compute factorials: it's called factorial.) The code first checks to see if the number is greater than or equal to zero. If not, program execution jumps to after the else statement, and an error message is displayed. If it is greater than or equal to zero, the program executes the statement f = f * i once for each value in the array 2:n. During each pass through the loop (an *iteration*), the variable i takes on the value of the current element of the array. So, for the first iteration i=2, for the second iteration i=3, and so on, until i takes on the value of n.

Note that the programmer does not need to explicitly update the loop variable. MATLAB will do this automatically; all the programmer needs to do is specify the array of values (e.g., 2:n) that the loop variable will take.

Note also how the *assignment* statement inside the for loop (f = f * i) has the variable f appearing on both sides. This might seem a little strange, and it is important to understand here that the part of the statement on the right hand side of the assignment operator (=) is evaluated first using the current value of f. The result of multiplying f by the current value of the loop variable i is found, and it is then assigned to f. In other words, the value of f is updated. After this, the for loop proceeds to the next iteration.

After the loop has finished all of its iterations (i.e., i has taken on all values of the array 2:n, and f = f * i has been executed for each one), it exits, and program execution continues after the end statement. Since f was initialized to 1, the effect of iterating through the loop is to multiply 1 by every integer between 2 and n. In other words, it computes the factorial of n.

Try typing in this code, saving it as a script *m*-file and running it. Read through it carefully to make sure you understand how it works. You can step through it using the debugger to help you. ∎

Now let us return to the flow chart in Fig. 2.4. First, in our MATLAB example, the *initialize array and loop variable* box corresponds to setting up the array 2:n and initializing i to its first element. The *condition* is to test if we still have elements to process in the array. When we reach the end, the condition becomes false, and the loop will exit. The *statements* correspond in our example to a single statement: f = f * i. The *update loop variable* box will modify i so that it takes on the value of the next element in the array.

SYNTAX: FOR LOOP

Formally, the syntax of a for loop in MATLAB can be defined as follows:

```
for loop_variable = loop_array
    statements
end
```

We can see that we need to use two MATLAB keywords: `for` and `end`. The `loop_array` represents the array of values, which will be iterated through (i.e., `2:n` in Example 2.5), and `loop_variable` is the variable that will take on values from this array in different iterations (i.e., `i` in Example 2.5).

■ Activity 2.7

02.C

Write a MATLAB script *m*-file that uses a `for` loop to compute the value of the expression:

$$\sum_{a=1}^{5} a^3 + a^2 + a$$

■

■ Activity 2.8

02.A, 02.C

In a *phase I* clinical trial for a new drug, one of the main goals is to determine the recommended dose of the drug for phase II of the trial. Typically, the drug is tested on a small cohort of patients with a low dose, and efficacy and toxicity are estimated. Then, the dose is *escalated* (i.e., increased) and efficacy/toxicity are estimated on a new cohort. This process is continued until toxicity becomes too high. The dose/toxicity results inform the choice of drug dose for phase II of the trial [3].

The increasing dose levels used in a phase I trial are known as the *dose escalation pattern*. A common way of calculating the dose escalation pattern is to increase the dose by 67% after the first cohort, by 50% after the second, by 40% after the third, and by 33% after each subsequent cohort.

Write a MATLAB script *m*-file to determine the dose escalation pattern for a new cytotoxic drug that is being tested for cancer treatment. The starting dose should be 0.2 mg/kg. The script should determine the first ten drug doses in the escalation pattern. These should be (in mg/kg):

```
0.2
0.334
0.501
0.7014
0.932862
1.240706
1.65014
2.194686
2.918932
3.882179
```

■

Generally, `for` loops are appropriate when it is known, before starting, exactly how many times the loop should be executed. For instance, in Example 2.5,

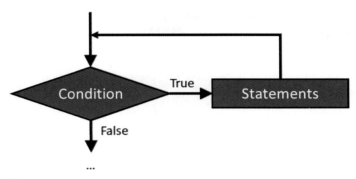

FIGURE 2.5
The control flow of a `while` loop.

the value of n was entered by the user, so the number of loop iterations could be determined from this value. If it is not possible to know how many times the loop should be executed; an alternative iteration construct is necessary, as will be described below.

2.6   Iteration: `while` loops

The second type of iteration statement available in MATLAB is the `while` loop. The control flow of a `while` loop is shown in Fig. 2.5. First, a *condition* is tested. If this condition evaluates to false, then control immediately passes beyond the `while` statement. If it evaluates to true, then some *statements* are executed, and the condition is retested. This *iteration* continues until the condition evaluates to false.

Therefore a `while` loop will execute continually until some condition is no longer met. This type of behavior can be useful when continued loop execution depends upon some computation or input that cannot be known before the loop starts.

■ Example 2.6

Consider the following simple example: *02.C*

```
i=randi(10) − 1; % random integer between 0 and 9
guess = −1;
while (guess ~= i)
    guess = input('Guess a number:');
    if (guess == i)
        disp('Correct!');
    else
        disp('Wrong, try again ...');
    end
end
```

This program first generates a random integer between 0 and 9. The MATLAB `randi` statement, used in this way, generates a random integer from a uniform distribution between 1 and its argument (in this case, 10), so by subtracting 1 from this, we can get a random integer between 0 and 9. The following `while` loop will iterate whilst the condition `guess ~= i` is true. In other words, the loop will execute, so long as `guess` is not equal to the random integer `i`. Since `guess` is initialized to −1 before the loop starts, the loop will always execute at least once. Inside the loop, a number is read from the user, and an `if` statement is used to display an appropriate message, depending on whether or not the guessed number is correct. The program will continue asking for guesses until the correct answer is entered. ■

SYNTAX: WHILE LOOP

Formally, the syntax of a `while` loop can be defined as follows:

```
while condition
    statements
end
```

Again, the loop is enclosed within two MATLAB keywords: `while` and `end`. The condition should be a logical expression (i.e., it should result in a true or false value). Both `condition` and `statements` correspond to the boxes shown in Fig. 2.5.

One word of warning regarding `while` loops is that we should always be careful that the `condition` should eventually evaluate to false. If this does not happen, we will end up in an infinite loop, and the program will not terminate. For example, if the `condition` involves testing the value of a variable (e.g., `x>0`), we should ensure that the value of the variable can change in each iteration so that progress toward this "stopping condition" can be made.

Generally, `while` loops are useful when it is not known in advance how many times the loop should be executed. Both `for` loops and `while` loops can be very useful, but in different types of situation. In time you will learn to recognize when to use each of these types of iteration programming construct.

■ Activity 2.9

02.B, 02.C

Write a MATLAB script *m*-file that continually displays random real numbers between 0 and 1, each time asking the user if they want to continue. If a "y" is entered, the program should display another random number, if not, it should terminate. Use a `while` loop in your implementation.

Once you have done this, similar to Activity 2.4, try to make your code more robust by also accepting an upper case "Y" and a lower case "y." ■

■ **Activity 2.10**

Modify your solution to Activity 2.9 so that rather than displaying a single   *02.C, 02.D*
random value, it asks the user *how many* random values they would like. If
they enter a number greater than zero, the program displays many random
values. Otherwise it should repeat the request for the number of random
values. The program should still terminate when any character other than a
"y" (or "Y") is entered when the user is asked if they want to continue.   ■

2.7   A note about efficiency

When we start to program using loops, it raises some issues about the efficiency
of code, both with respect to execution speed and use of memory. We will
deal with the topic of code efficiency in more detail in Chapter 10, but, for
now, we will introduce one common mistake that can result in quite inefficient
programs and a simple solution that can be used to avoid this mistake.

■ **Example 2.7**

There are many programming situations in which an array of values needs   *02.C, 02.E*
to be built up inside a loop. For example, the following piece of code uses a
for loop to build up an array of 1,000,000 random real numbers between
0 and 1:

```
clear
tic
nRands = 1000000;
for i=1:nRands
    rand_array(i) = rand(1);
end
toc
```

The code also uses some built-in MATLAB functions (tic and toc) to measure how long it takes to build the array. However, the code is not particularly efficient with regard to execution speed.   ■

The reason for the inefficiency of this code is that in MATLAB, arrays are *dynamically* allocated. This means that memory is set aside (or *allocated*) for the
array, based on the highest indexed element that we have created so far. If we
attempt to place a value into the array at an index that is greater than its current
size, MATLAB will recognize that we need a larger array, and it will automatically resize the original array to allow the assignment. This is a nice feature
that makes programming easier at times. However, it comes with a cost: our
program may not execute as quickly. This is because sometimes MATLAB may
need to allocate a new, larger block of memory and copy across all the values
from the block where the array is currently stored, and this takes some time. In
summary, if we know (or have a good guess) at how big the array needs to be

in the first place, we can "pre-allocate" it, and our program is likely to be more efficient.

■ Example 2.8

02.C, 02.E   The same piece of code from the previous example is shown below, now with a pre-allocation of the array.

```
clear
tic
nRands = 1000000;
rand_array = zeros(1,nRands);
for i=1:nRands
    rand_array(i) = rand(1);
end
toc
```

Here, the `zeros` function is used to pre-allocate a large array of zeros. Therefore MATLAB knows straight away how large a block of memory to set aside and will never have to allocate a new block and copy data across. Thus the code should run quicker. You can verify this yourself by typing in the two versions of the code and running them. The code should measure how long it took to execute each version and display this information.   ■

2.8   `break` **and** `continue`

Another way of altering the control flow of a program is to use jump statements. The effect of a jump statement is to unconditionally transfer control to another part of the program. MATLAB provides two different jump statements through the keywords `break` and `continue`. Both can only be used inside `for` or `while` loops.

The effect of a `break` statement is to transfer control to the statement immediately following the enclosing control structure.

■ Example 2.9

02.A, 02.C, 02.F   As an example, consider the following code:

```
total = 0;
while (true)
    n = input('Enter number: ');
    if (n < 0)
        disp('Finished!');
        break;
    end
    total = total + n;
end
disp(['Total = ' num2str(total)]);
```

This piece of code reads in a sequence of numbers from the user. When the user types a negative number, the loop is terminated by the break statement. Otherwise, the current number is added to the total variable. When the loop terminates (i.e., a negative number is entered), the value of the total variable, which is the sum of all of the numbers entered, is displayed.   ■

The continue statement is similar to the break statement, but instead of transferring control to the statement following the enclosing control structure, it only terminates the current iteration of the loop. Program execution resumes with the next iteration of the loop.

■ Example 2.10

For example, examine the following piece of code: *O2.A, O2.C, O2.F*

```
total = 0;
for i = 1:10
    n = input('Enter number: ');
    if (n < 0)
        disp('Ignoring!');
        continue;
    end
    total = total + n;
end
disp(['Total = ' num2str(total)]);
```

A sequence of numbers is again read in and summed. However, this time there is a limit of 10 numbers, and negative numbers are ignored. If a negative number is entered, the continue statement causes execution to resume with the next iteration of the for loop.   ■

■ Activity 2.11

The listing below shows a MATLAB program that reads 20 numbers from the   *O2.A, O2.C, O2.F*
user and displays the sum of their square roots.

```
sum = 0;
for x=1:20
    n = input('Enter a number:');
    sum = sum + sqrt(n);
end
disp(['Sum of square roots = ' num2str(sum)]);
```

Enter this script into a MATLAB *m*-file and modify it so that

- negative numbers are ignored, and

- entering the number 0 causes the program to display the sum so far, and then terminate. ∎

2.9   Nesting control structures

Any of the control structure statements we have covered in this chapter can be *nested*. This simply means putting one statement inside another one. We have already seen nested statements in some of the examples and activities of this chapter.

∎ Example 2.11

02.A, 02.C, 02.D

As another example, examine the following piece of code and see if you can work out what it does:

```
n = input('Enter number: ');
while (n >= 0)
    f = 1;
    for i = 2:n
        f = f * i;
    end
    disp(f);
    n = input('Enter number: ');
end
```

Here, we have a for loop nested inside a while loop. The while loop reads in a sequence of numbers from the user, terminated by a negative number. For each positive number entered, the for loop computes its factorial (similar to Example 2.5), which is then displayed. ∎

Notice how we have used *indentation* of code within control structures, i.e., the code between the while/end keywords is indented once, and the code between the nested for/end keywords is indented twice. This does not change the functionality of the code, but it does make the code easier to read and understand, and is generally accepted to be a good thing in computer programming.

Finally, although nesting is allowed in MATLAB and other programming languages, and is often necessary to achieve the behavior that we want, note that excessive nesting can make code difficult to read. As a general rule, if you have 4 or more levels of nesting, you should probably think about writing your code in a different way. Usually, it is possible to reduce or avoid excessive nesting, for example, by using the elseif keyword (see Section 2.2) or by using functions (see Chapter 4).

2.10 Summary

Control structures allow us to alter the natural sequential flow of execution of program statements. Two fundamental types of control structure are *conditional* statements and *iteration* statements. MATLAB implements conditional control flow using the if and switch statements. Iterative control flow can be implemented with for loops and while loops. All of these control structures can be *nested* inside each other to produce more complex forms of control flow. The break and continue statements allow "jumping" of control from inside loop statements.

2.11 Further resources

- MATLAB documentation on control flow statements: http://mathworks.com/help/matlab/control-flow.html.

2.12 Exercises

■ Exercise 2.1

Write a script that asks the user to enter a number, and then uses an if statement to display either the text "odd" or "even," depending on whether the number is odd or even.

02.A

(Hint: see the documentation for the mod function.)   ■

■ Exercise 2.2

Cardiac resynchronization therapy (CRT) is a treatment for heart failure that involves implantation of a pacemaker to control the beating of the heart. A number of indicators are used to assess if a patient is suitable for CRT. These can include the following:

02.A, 02.B

- New York Heart Association (NYHA) class 3 or 4; this is a number representing the severity of symptoms, ranging from 1 (mild) to 4 (severe);
- 6-minute walk distance (6MWD) less than 225 meters; and
- Left ventricular ejection fraction (EF) less than 35%.

Write a MATLAB script *m*-file containing a single if statement that decides if a patient is suitable for CRT, based on the values of three numerical variables: nyha, sixmwd, and ef, which represent the indicators listed above. All three conditions must be met for a patient to be considered suitable. You can test your program by assigning values to these variables from the test data shown in the table below.

Patient	NYHA Class	6MWD	EF	Suitable for CRT?
1	3	250	30	No
2	4	170	25	Yes
3	2	210	40	No
4	3	200	33	Yes

∎

∎ Exercise 2.3

02.A, 02.B

Not all patients respond positively to CRT treatment. Common indicators of success include:

- Decrease in NYHA class of at least 1;
- Increase in 6MWD of at least 10% (e.g., from 200 m to 220 m); and
- Increase in EF of at least 10% (e.g., from 30% to 40%).

The table below shows post-CRT data for patients 2 and 4 from Exercise 2.2 (who were considered suitable for CRT). Write a MATLAB script *m*-file that determines if a patient has responded positively to CRT treatment. All three conditions must be met for a patient to be considered a positive responder. The program should use a single `if` statement, which tests conditions based on the values of six numerical variables: `nyha_pre`, `sixmwd_pre`, `ef_pre`, `nyha_post`, `sixmwd_post`, and `ef_post`, which represent the pre- and post-treatment indicators.

Patient	NYHA Class	6MWD	EF	Responder?
2	3	220	30	No
4	2	230	45	Yes

∎

∎ Exercise 2.4

02.A

Write a MATLAB script *m*-file that reads in two numbers from the keyboard followed by a character. Depending on the value of the character, the program should output the result of the corresponding arithmetic operation ("+," "−," "*," or "/"). If none of these characters was entered, a suitable error message should be displayed. Use a `switch` statement in your program.

∎

∎ Exercise 2.5

02.A, 02.B

The figure below shows how blood pressure can be classified, based on the diastolic and systolic pressures. Write a MATLAB script *m*-file to display a message indicating the classification, based on the values of two variables representing the diastolic and systolic pressures. The two blood pressure values should be read in from the keyboard.

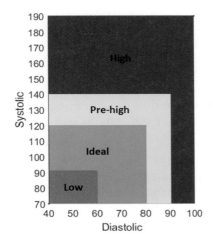

■ Exercise 2.6

Write a `switch` statement that tests the value of a character variable, called grade. Depending on the value of grade, a message should be displayed, as shown in the table below. If none of the listed values are matched, the text "Unknown grade" should be displayed.

02.A

Grade	Message
"A" or "a"	"Excellent"
"B" or "b"	"Good"
"C" or "c"	"OK"
"D" or "d"	"Below average"
"F" or "f"	"Fail"

■

■ Exercise 2.7

Write a MATLAB script *m*-file to read in an integer N from the keyboard, and then use an appropriate iteration programming construct to compute the value of the expression

02.C, 02.D

$$\sum_{n=1}^{N} \frac{n+1}{\sqrt{n}} + n^2.$$

■

■ Exercise 2.8

Write a MATLAB script *m*-file to compute the value of the expression

02.C, 02.D, 02.E

$$\sin(x) + \frac{1}{2}\cos(x)$$

for values of x between 0 and 2π in steps of 0.0001, and produce a plot of the results.

Write two separate MATLAB script *m*-files to implement this task: one that produces the result using MATLAB array processing operations, as we learned about in Chapter 1; and one that builds up the array result element-by-element using an iteration programming construct. Add code to time your two implementations. Which is quicker and why? ■

■ **Exercise 2.9**

02.C, 02.D

In Activity 1.6, we introduced the field of gait analysis. The file *LAnkle_tracking.mat* contains data acquired using an optical tracking system from the same patient mentioned in Activity 1.6, whilst they were walking. The file contains a single variable, called `LAnkle`, which is a matrix containing four columns. The first column represents timings in seconds, and the next three columns represent the x, y, and z coordinates of a marker attached to the patient's left ankle.

In this exercise you will write code to automatically identify a particular gait event: the *foot strike* of the left leg, which refers to the instant that the left heel touches the ground. The foot strike event can be determined by looking for *troughs* in the series of z coordinates of the ankle marker. The z coordinates represent the *inferior-superior* (i.e., foot-head) motion of the marker.

Write a MATLAB script *m*-file that loads in the *LAnkle_tracking.mat* file and first produces a plot of the z-coordinates against time. The script should then search the list of z-coordinates and report any troughs. To begin with, consider a trough to be any value that is less than both the preceding and following values.

Test your code. By looking at the plot and the identified foot strike events, do you think it is working correctly? What can you do to rectify any problems? ■

■ **Exercise 2.10**

02.C, 02.D

The Taylor series for the exponential function e^x is

$$1 + \frac{x^1}{1!} + \frac{x^2}{2!} + \frac{x^3}{3!} \cdots$$

Write a MATLAB script *m*-file to read a number x from the keyboard and compute its exponential using the Taylor series. Also compute the exponential using the built-in MATLAB function `exp`. Run your program several times with different numbers of terms in the Taylor series to see the effect this has on the closeness of the approximation. ■

■ Exercise 2.11

Modify your solution to Exercise 2.10 so that, as well as computing and displaying the result of the Taylor series expansion, it also computes and displays the number of terms required for the approximation to be accurate to within 0.01. ■

02.A, 02.C, 02.D

■ Exercise 2.12

The code below is intended to use the built-in MATLAB function `isprime` to find and display the first 20 prime numbers.

02.A, 02.C, 02.F

```
x = 0;
nPrimes = 0;
while (nPrimes < 20)
    if (isprime(x))
        disp(x);
        nPrimes = nPrimes + 1;
    end
end
```

1. Try entering this code and running it. What is the problem with the code as it stands and how would you fix it?
2. Starting with a corrected version of the code, add code that uses a `break` statement so that the `while` loop terminates as soon as the prime number found is greater than 50.
3. Again starting with the corrected version of the code from part 1, add code that uses a `continue` statement so that prime numbers less than 10 are not reported. The code should still report 20 prime numbers. ■

■ Exercise 2.13

Modify your solution to Activity 2.4 so that, rather than reading a single character/number, the program continually reads character/number pairs and displays the trigonometry result. The program should terminate when the character "x" is entered. In this case, no number should be read, and the program should exit immediately. If a number outside the range [−100, 100] is entered, the program should not display the trigonometry result and should ask for another character/number pair to be entered. ■

02.A, 02.C, 02.D, 02.F

■ Exercise 2.14

In Activity 1.11, we introduced the concept of the injury severity score (ISS). The ISS is a medical score to assess trauma severity and is calculated as follows: Each injury is assigned an abbreviated injury scale (AIS) score in the range [0–5] which is allocated to one of six body regions (head, face, chest, abdomen, extremities, and external). Only the highest AIS score in

02.C, 02.D

each body region is used. The three most severely injured body regions (i.e., with the highest scores) have their AIS scores squared and added together to produce the ISS. Therefore the value of the ISS is always between 0 and 75.

Write a MATLAB script *m*-file to compute the ISS, given an array of 6 AIS scores, which represent the most severe injuries to each of the six body regions. Test your code using the array [3 0 4 5 3 0], for which the ISS should be 50. ∎

■ Exercise 2.15

02.A, 02.C, 02.D

In Exercise 1.6, you wrote MATLAB code to check if an integer was divisible by 9 by summing its digits: if the digits add up to 9 the number is divisible by 9. For integers up to 90, only a single application of the rule is required. For larger integers, multiple applications may be required. For example, for 99, the digit sum is 18. Then, because 18 still has multiple digits, we sum them again to get 9, confirming that 99 is divisible by 9 (after two steps). In general, for a given starting number, you should repeatedly sum the digits until the result is a single digit number.

Write a MATLAB script *m*-file to read in a single number from the keyboard, and use the rule to determine if it is divisible by 9. An appropriate message should be displayed, depending on the result. You can assume that the number will always be less than 1000. Recall from Exercise 1.6 that the `int16`, `idivide`, and `mod` functions can be used to find the digits of a decimal number. ∎

FAMOUS COMPUTER PROGRAMMERS: ALAN TURING

Alan Turing was an English mathematician and computer scientist who was born in 1912. At school he was criticized by his teachers for not focusing on his work. His headmaster wrote of him: "If he is to stay at Public School, he must aim at becoming educated. If he is to be solely a Scientific Specialist, he is wasting his time at a Public School."

Despite these criticisms, Turing went on to study mathematics at King's College Cambridge. He became interested in algorithms and in 1936 he published a now-famous paper, in which he introduced the idea of a "Turing machine." This was similar in concept to Charles Babbage's analytical engine (see Chapter 1's Famous Computer Programmer), but a fundamental contribution of Turing's proposal was that the data and instructions would be stored in the same computer memory. This makes program development much easier, and is the way in which today's digital computers are designed.

In 1939, after the outbreak of the Second World War, Turing started to work at the UK Government Code and Cypher School at Bletchley Park. He was brilliant at cracking codes and developing electro-mechanical machines to assist him in code decyphering. In 1945 he was awarded an OBE for his contributions to the war effort (he kept his OBE in his tool box). After the war, Turing performed much of the pioneering work in developing the early digital computers, and designing algorithms and writing code for them.

However, several years later, Turing's life took a more tragic turn. In 1952 he was arrested for violation of Britain's laws against homosexuality. At the trial, he offered no defence except that he saw no wrong in what he had done. Alan Turing died of cyanide poisoning in 1954. Cynanide was found on a half eaten apple beside his body. An inquest returned a verdict of suicide, but his mother always maintained that it had been an accident.

In September 2009, the UK Prime Minister made an official apology for "the appalling way [Turing] was treated." In December 2013, Alan Turing was granted a posthumous royal pardon. In granting the pardon, the UK Justice Minister described Turing as "an exceptional man with a brilliant mind."

"How can one expect a machine to do all this multitudinous variety of things? The answer is that we should consider the machine as doing something quite simple, namely carrying out orders given to it in a standard form which it is able to understand."

Alan Turing, 1946

CHAPTER 3

Basic data types

LEARNING OBJECTIVES

At the end of this chapter you should be able to:

O3.A Explain how a data type acts to help interpret the sequences of 1s and 0s stored in the computer's memory

O3.B Identify the fundamental data types used in MATLAB®

O3.C Make appropriate use of integer and floating point numeric types and describe the differences between them

O3.D Make appropriate use of logical types, character types, character arrays and strings

O3.E Use MATLAB commands to identify the type of a variable

O3.F Convert between types and explain when this is done automatically by MATLAB

3.1 Introduction

As we have already seen, MATLAB easily handles numeric data and array manipulation is one of its strengths. As discussed in Section 1.8, MATLAB can also process other, non-numeric data types, and these can also be collected in arrays.

Data types in MATLAB can be broadly divided into fundamental types for everyday use and more advanced types. The fundamental types include the basic data types discussed in Section 1.8 (i.e., numeric values, characters, and Boolean values), as well as arrays constructed of these types. These will be our focus in this chapter. More advanced data types will be discussed in Chapter 6.

3.2 What is a data type?

Ultimately, any data in a computer's memory is stored as a sequence of 1s and 0s. These are the *bits* used to represent the data. A *byte* is a sequence of eight bits. Data are normally stored using a number of bytes, i.e., in multiples of 8 bits.

63

MATLAB® Programming for Biomedical Engineers and Scientists. https://doi.org/10.1016/B978-0-32-385773-4.00012-5

A *data type* represents a way of interpreting a sequence of bits in terms of some higher-level representation. A pattern of 1s and 0s arranged over a sequence of bytes can represent different things, *depending on how they are interpreted,* and this underlies what it means to have different data types.

■ Example 3.1

03.A   Any sequence of bits (in a byte) can be interpreted as an integer. For example, the sequence 01000001 can be viewed as the number 65, because 01000001 is the *binary* representation of 65 (i.e., $2^6 + 2^0 = 64 + 1$).

We can, however, also interpret the same pattern as the character "A." This is because of the definitions given in the ASCII[1] code, which defines a correspondence between the number 65 and the character "A." The correspondence continues in the alphabet, for example, "B" matches with the same bit pattern that would be used for 66, "C" matches with 67, etc. ■

■ Activity 3.1

03.A   How would the letter "D" be represented as a binary number in the computer's memory? ■

3.3   Numeric types

We briefly discussed the fundamental data types offered by MATLAB, including numeric types, in Section 1.8. To recap, we distinguish between two categories of numeric types: those used to represent integers and those used to represent floating point numbers (i.e., numbers with a fractional part). Types used for integers can be *signed* (these are allowed to be positive or negative, as well as zero) or *unsigned* (for non-negative numbers only).

Different integer data types can use different numbers of bytes to store their values. Using more bytes, it is possible to represent larger integers. The number of bytes used to store a floating point number can also vary. If more bytes are used, as well as being able to represent larger floating point numbers, we can also represent the fractional part with greater *precision*.

The names of the MATLAB types that are used for integers are given below.

- `uint8, uint16, uint32, uint64`
- `int8, int16, int32, int64`

The different types are prefixed with a `u` if they are unsigned. The number at the end indicates how many bits each one uses: 8, 16, 32, or 64, corresponding to 1, 2, 4, or 8 bytes, respectively.

---

[1]American standard code for information interchange: https://en.wikipedia.org/wiki/ASCII.

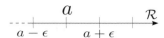

FIGURE 3.1
Illustration of numeric value precision.

The types that MATLAB uses to represent non-integer numbers are called `single` and `double`, meaning *single precision floating point numbers* and *double precision floating point numbers*, respectively. Again, these use different numbers of bytes. See below for further details on the precision of floating point numbers.

Precision for non-integer (floating point) numeric types

The precision of a floating point number relates to the smallest distinguishable decimal place that can be used within it. In a computer, precision can be represented by the difference between a number that it is possible to represent, and the *next nearest* representable number (above or below). For example, in Fig. 3.1, a value a that is stored would have two adjacent representable numbers $a - \epsilon$ and $a + \epsilon$ on the real line, $\mathcal{R}$. For high precision, the value of ϵ is very small, i.e., adjacent representable floating point numbers are very close. Because of the way floating point numbers are represented, *the size of ϵ depends on the size of the number a*. For a large value of a, there will be larger gap to the nearest representable numbers than there would be if the value of a were small.

The built-in MATLAB function `eps` (short for epsilon) can be used to find out the value of ϵ for a given number and data type. If we call `eps` with the name of the data type as an argument, it will return the value of ϵ for a value of $a = 1.0$ when represented in that data type.

```
>> eps('double')
ans =
   2.2204e-16

>> eps('single')
ans =
   1.1921e-07
```

In each case, we obtain the gap between $a = 1.0$ and the adjacent representable numbers when we use `double` or `single` data types.

If we call the `eps` function with a number as its argument, it will return the value of ϵ for that number when using the `double` type. For example,

```
>> eps(1000)
ans =
   1.1369e-13
```

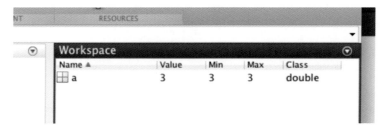

FIGURE 3.2
The workspace browser shows information about the type of a variable.

shows that the gap between 1000 (stored as a `double`) and the next representable number is 1.1×10^{-13}, which is significantly larger than the gap for 1.0 (2.2×10^{-16}, as calculated earlier).

MATLAB defaults to double precision for numbers

An important point to remember when using numeric data is that MATLAB always uses double precision for numeric data by default. For example, typing `a = 3` at the command window, followed by a call to the `whos` function, gives the following output:

```
>> whos
   Name       Size           Bytes  Class      Attributes

   a          1x1                8  double
```

We can see that there is only one variable, `a`, in our workspace, and the `whos` function tells us that

- MATLAB has used the `double` data type to represent our variable (under the heading `Class`).
- MATLAB views our variable implicitly as an array, with a size of 1×1. Even a scalar value is considered in MATLAB to be a (small) array.
- The amount of memory used to represent the variable is 8 bytes (i.e., 64 bits).

Similar information is shown by the workspace browser in the MATLAB environment, as shown in Fig. 3.2. Note that we can choose the columns displayed in the workspace browser by clicking the small triangle in its top bar to open its context menu.

How does MATLAB display numeric values by default?

We need to take some care when inspecting how MATLAB displays numbers in the command window, as this can vary, depending on the size of the number. By default, MATLAB tries to display numbers as compactly as possible in what is known as a "short" format. For example, asking for the value of π in the command window gives

```
>> pi
ans =
    3.1416
```

This does not mean that MATLAB stores only the first four decimal places of π; it is simply only *showing* the value rounded to 4 decimal places.

We can request that MATLAB uses a longer format to display numbers by typing `format('long')` in the command window.

```
>> format('long')
>> pi
ans =
    3.141592653589793
```

We can revert to the default format by typing `format('short')`.

For large numbers, MATLAB uses a compact scientific notation. In scientific notation, a number is written in the form $A \times 10^B$, where A is called the *mantissa* and B is called the *exponent*.

For example, the value of the factorial of 15 is around 130 billion (to be exact $15! = 130,767,436,800$). However, the default way in which this number is displayed in the MATLAB command window is as follows:

```
>> factorial(15)
ans =
    1.3077e+12
```

This corresponds to 1.3077×10^{12}, which means that MATLAB has again rounded to 4 decimal places (this time for the mantissa) when displaying the result. The exact value in scientific notation would be $1.307674368 \times 10^{12}$.

This way of displaying numeric values can require some care when looking at arrays, especially if an array contains both small and large values. For example, let us initialize an array using three integers with very different sizes.

```
>> x = [25 357346 2013826793]
x =
    1.0e+09 *
    0.0000    0.0004    2.0138
```

We can see that MATLAB tries to show all three numbers in scientific notation using *a single exponent*. In this case, the exponent is 9, and this is determined by the largest number in the array. If we were to write all three numbers *exactly* using this exponent, then the three numbers would appear as

$$0.000000025 \times 10^9 \qquad 0.000357346 \times 10^9 \qquad 2.013826793 \times 10^9$$

MATLAB shortens this to a single array with a common exponent:

$$10^9 \times [0.000000025 \quad 0.000357346 \quad 2.013826793]$$

Finally, applying short notation to each of the mantissa values in the array means rounding them to four decimal places. This is why the array appears in MATLAB as

```
x =
   1.0e+09 *
     0.0000    0.0004    2.0138
```

This does not mean that the first number in the array is zero. It just indicates that the non-zero decimals in the mantissa have "gone beyond" the 4th decimal place for this large exponent.

Take care when working with numeric types other than double

MATLAB's default behavior of using the `double` data type for numeric data is fine most of the time, even if, for example, we only want to work with integers. *The best advice is to leave the default behavior.*

However, this default behavior can occasionally be a problem, for example, if we want to work with *very* large amounts of data. In such cases, one possible way to address this is to force MATLAB to use one of the integer types above, or the `single` type if we still need to work with floating point data. This is because these types use fewer bytes, so our data will require less computer memory.

■ Example 3.2

03.B, 03.C

Here, we create array variables of the same length, but we force MATLAB to create them using a number of different data types, as well as the default `double` type.

```
>> a = zeros(100000,1);
>> b = int8(zeros(100000,1));
>> c = int16(zeros(100000,1));
>> whos
  Name        Size        Bytes     Class      Attributes
  a         100000x1      800000     double
  b         100000x1      100000     int8
  c         100000x1      200000     int16
```

Note the different amounts of memory required to store each variable.   ■

One warning here is that variables that have a type other than `double` will no longer work properly with functions that expect a `double` as their argument(s), and there are quite a few of these. For example, a call to the square root function, such as the following, will now generate an error:

```
>> a = int16(3);
>> b = sqrt(a);
```

```
Check for incorrect argument data type or missing
argument in call to function ''sqrt.''
```

If we had assigned to a by simply typing a=3, the call to sqrt would have worked, because the variable a would have been, by default, a double precision floating point number, and therefore suitable for passing to the sqrt function.

Ranges of numeric types

Another difficulty that can be encountered when not using MATLAB's default double type for numbers relates to the range of values that a numeric type can store.

We start by looking at the ranges of some of the integer data types.

■ Example 3.3

For example, an unsigned single byte integer (uint8) can only store values in the range 0–255. We can check this using the built-in functions intmax and intmin, which take as an argument a character array specifying the data type.

03.C

```
>> intmax('uint8')
ans =
  255

>> intmin('uint8')
ans =
    0
```

This tells us that if we use eight bits to represent an unsigned integer, then we can represent the numbers from 0 to 255 inclusive. The number 255 is the decimal value that corresponds to the binary number 11111111, i.e., the one where all of the eight bits are set to 1.

If we use 16 bits, then we can represent a larger range of unsigned integers, for example,

```
>> [ intmin('uint16') , intmax('uint16') ]
      0   65535
```

Note how the minimum and maximum values change if we want to represent *signed* integers, i.e., both positive and negative numbers:

```
>> [ intmin('int16') , intmax('int16') ]
  −32768  32767
```

■

We need to take care when working with mixed numeric types, because we can have unexpected behavior if we try to represent a number beyond the range of a data type we are using. This is shown by the following command window statements:

```
>> a = 250 + 10
a =   260

>> b = uint8(250 + 10)

uint16

b =   255

>> whos
  Name        Size            Bytes  Class     Attributes
  a           1x1                 8  double
  b           1x1                 1  uint8
```

Both commands aim to set their variables to a value of 260, but this failed in the case of variable b, which was forced to be a one byte unsigned integer. Therefore, the assignment to b gave an incorrect result due to what is called an *overflow* error.

To reiterate, if we leave all numeric values in MATLAB's default data type of `double`, then we reduce the chance of such unexpected behavior.

■ Example 3.4

03.C   We can check the range of *floating point* numbers using the `realmax` and `realmin` functions.

```
>> realmax('single')
ans =
   3.4028e+38

>> realmax('double')
ans =
   1.7977e+308
```

This shows that the biggest `single` precision number we can represent is 3.4028×10^{38}, whilst the biggest `double` precision number is 1.7977×10^{308}.   ■

■ Activity 3.2

03.A, 03.B, 03.C   Find the problems with the following pieces of code and rectify them.

1. Code to compute and display the value of e^8:

```
% Compute value of e^8
x = int16(8);
y = exp(x)
```

2. Code to compute and display the exponential sequence 1, 10, 100, 1000, etc.:

```
% Generate exponential sequence 1,10,100,1000, ...
a = 1:10;
b = 10^a;
disp(b);
```

3. Code to compute squares of an array of numbers:

```
% Compute squares of array
w = uint8([1 16 24 19 21 12]);
q = w.^2
```

■

Infinity and NaN

Two special MATLAB identifiers are used to deal with cases where we encounter infinite numbers or numbers that cannot be determined.

The first of these is Inf. MATLAB uses Inf to represent infinite numbers. For example,

```
>> 1/0
ans =
   Inf
```

The function isinf can be used to test if a number is infinite. It can be applied to an array of numbers and will return 1 (true) at locations with an infinite value and 0 (false) otherwise.

■ Example 3.5

For example, consider the following element-wise division for two arrays: *03.C*

```
>> a = [1 2 3];
>> b = [2 0 1];
>> a ./ b
ans =
    0.5000       Inf    3.0000

>> isinf( a ./ b )
ans =
     0     1     0
```

The 1 in the second element of the result indicates that isinf returned a true value for the result of the operation 2/0.   ■

Another special MATLAB identifier is NaN. This is used to represent indeterminate values or values that cannot be interpreted as a number. For example, the expression 0 / 0 cannot be interpreted as a number, so is represented by NaN. NaN is also returned by functions that attempt to interpret character arrays as numbers when the given sequence of characters cannot be interpreted properly. The function `isnan` can be used to test for NaN values in a similar way to the `isinf` function, and it can also be applied to arrays.

■ Example 3.6

03.C For example, consider the following calls to the built-in `str2double` function, which converts a character array into a floating point number:

```
>> str2double('2.3')
ans =
    2.3000

>> str2double('i')
ans =
   0.0000 + 1.0000i

>> str2double('pi')
ans =
   NaN

>> n = str2double('blah')
n =
   NaN

>> isnan(n)
ans =
   1
```

■

■ Activity 3.3

03.C The *end diastolic volume* (EDV) and *end systolic volume* (ESV) of the left ventricle of the heart refer to the volume (in mL) of blood just prior to the heart beat, and at the end of the heart beat, respectively. The *ejection fraction* (EF) is the fraction of blood pumped out from the heart in each beat, and can be computed from the EDV and ESV as follows:

$$EF = \frac{EDV - ESV}{EDV}.$$

The file *edv_esv.mat* contains EDV and ESV values for a number of patients. Both are integer values. Write a MATLAB script *m*-file to read in these data, and compute EF values for each patient. The script should then display the EDV, ESV, and EF values for all patients to the command window. Note that the data may contain some erroneous values, so you should write your code in such a way that it can detect such cases and display a message indicating that there is a problem with the data.

■

3.4 The Boolean data type

In programming, we use the term "Boolean" to refer to data and operations that work with the two values "true" and "false." As described in Section 1.8, MATLAB uses the `logical` data type for such data. For this type, MATLAB can also use the identifiers `true` and `false`, representing them internally with the numbers 1 and 0, respectively.

For example, here we assign a value of `true` to a variable, display it, and ask MATLAB what class it is:

```
>> v = true;
>> disp(v)
    1

>> class(v)
ans =
logical
```

Note that when the value is displayed, it is shown as the number 1, even though the data type is `logical` (i.e., Boolean).

A common way to generate logical *arrays* is with comparison operators. Here, we generate a list of random numbers between 0 and 1 and test which elements are greater than 0.5:

```
>> a = rand([1,5])
a =
    0.1576    0.9706    0.9572    0.4854    0.8003

>> b = a > 0.5
b =

  1x5 logical array

   0   1   1   0   1
```

The `whos` command can give a description of these variables, where we see that the data type of variable `b` is logical:

```
>> whos
  Name      Size            Bytes  Class      Attributes

  a         1x5                40  double
  b         1x5                 5  logical
```

■ Activity 3.4

Consider the code listing shown below, which defines a number of different variables to store data about a hospital patient.

03.B, 03.D, 03.E

```
name = 'Joe Bloggs';
initial1 = name(1);
```

```
initial2 = name(strfind(name, ' ')+1);

heart_rates = [76 80 95 110 84 70];
cholesterol = 4.3;

bradycardia = heart_rates < 60
tachycardia = heart_rates > 100

class(name)
ischar(initial1)
isnumeric(initial2)
isa(heart_rates, 'int16')
class(cholesterol)
class(bradycardia)
class(tachycardia)
```

Predict what you think the output of the code would be. Enter the code into a MATLAB script *m*-file, run it, and check if your predictions are correct. ■

3.4.1 Logical indexing

One very powerful MATLAB feature that arrays of logical values facilitate is *logical indexing*. The idea of logical indexing is quite simple: we use an array of logical values to index into another array, *and the values returned are only those where there was a true value in the logical array.*

■ Example 3.7

03.D

Let us consider an example. The code below defines a two-dimensional (2-D) array, called `patients`, in which the first column contains an ID number for each patient, and the second column contains their age.

```
>> patients = [1234 45; 1235 37; 2345 60; 2346 77; ...
              3456 21; 3457 58]
patients =
         1234            45
         1235            37
         2345            60
         2346            77
         3456            21
         3457            58
```

Suppose that we wished to find the ID numbers of all patients over the age of 40. In MATLAB, we can do this quickly and easily using logical indexing. For illustration purposes here, we will do it in steps. First, we can produce a logical array, indicating which patients are over 40 by using a simple comparison operator.

```
>> patients(:,2) > 40
ans =
```

```
6x1 logical array
   1
   0
   1
   1
   0
   1
```

Here, we have asked MATLAB to extract every row of the `patients` array (by using the `:` symbol), and the second column. The resulting array was compared with the scalar value 40 to give an array of true (1) or false (0) values.

Next, we use this array to index into the original `patients` array to select only rows where the logical array has true values. We also specify that we want every column, again by using the `:` operator.

```
>> patients(patients(:,2) > 40, :)
ans =
        1234        45
        2345        60
        2346        77
        3457        58
```

■

■ Activity 3.5

The table below shows some data recorded from a group of patients. The columns represent an ID number, age, and systolic/diastolic blood pressure.

03.D

ID	Age	Systolic b.p. (mmHg)	Diastolic b.p. (mmHg)
1234	45	125	85
1235	37	100	70
2345	60	121	79
2346	77	75	62
3456	21	95	66
3457	58	122	77

Write MATLAB code to enter these data into a 2-D array. Next, use logical indexing to select only those patients with a systolic blood pressure greater than 120 mmHg (which indicates high blood pressure). The data for all four columns should be displayed for these patients. ■

3.5 Characters and character arrays

It is often necessary to work with text values, for example, to report output or communicate with a user. In MATLAB, characters and character arrays can be used to represent text, and we use single quotes to enclose such data. We can assign a single character to a variable with a command such as s='A'. The command whos allows us to inspect how MATLAB represents this variable:

```
>> whos
  Name        Size            Bytes  Class     Attributes
  s           1x1                 2  char
```

This shows us that MATLAB views the type as a size 1×1 array with a single element of the character type (i.e., its Class is char). It also shows us that two bytes are needed to store this single character.

We can also assign an array of characters in exactly the same way. For example, typing s='ABC' followed by whos gives the following:

```
>> whos
  Name        Size            Bytes  Class     Attributes
  s           1x3                 6  char
```

As we can see, we still have an array with elements of type char, but now its size is 1×3, and it uses 6 bytes of memory (2 for each element).

The above illustrates that a sequence of characters is represented by MATLAB as an array of character elements. We can make this explicit during assignment by using the command s=['A', 'B', 'C'], which achieves exactly the same result as s='ABC'.

Generally, character arrays have a size $1 \times N$, where N is the number of characters in the array. Therefore they can be viewed as row vectors, and we can concatenate two or more such vectors to generate a longer array.

■ Example 3.8

03.B, 03.D

For example, the following commands create and manipulate some character array variables.

```
>> s1 = 'hello';
>> s2 = ' ';
>> s3 = 'world!';
>> s = [s1, s2, s3]
s =
hello world!

>> size(s)
ans =
     1    12
```

Concatenating character arrays in this way produces a single long array with twelve elements (characters). ■

If we want an array that contains three *separate* character arrays (as opposed to a single long one) then this is more difficult. One way to do it is to combine the three one-dimensional (1-D) arrays into a single 2-D array (using the semicolon to separate entries). Doing this can, however, lead to an error. For example, using s1, s2, and s3 from the previous example, we would get the following:

```
>> s = [s1 ; s2 ; s3]

Error using vertcat
Dimensions of matrices being concatenated are not consistent.
```

This is because the lengths of the arrays s1, s2, and s3 are 5, 1, and 6, respectively. MATLAB is trying to place the three row vectors, one beneath the other, to create a 2-D array. When using 2-D character arrays to store multiple sequences of characters like this, it is only possible if all of the row vectors have the same length (i.e., they must be consistent). We can make the arrays consistent by modifying them.

■ Example 3.9

The following commands address this issue by *padding* some of the arrays with spaces so that they all end up having the same length.   *03.B, 03.D*

```
>> s1 = 'hello ';
>> s2 = '      ';
>> s3 = 'world!';
>> s = [s1 ; s2 ; s3]
s =
hello

world!

>> size(s)
ans =
     3     6
```

The extra spaces in s1 and s2 mean that they now each have a length of 6, the same as s3. We combine the three row vectors vertically, to create a 3 × 6 array of characters.   ■

■ Activity 3.6

The file *patient_names.mat* contains a 2-D array of characters. Each row of the array is a character array of length 20 characters representing a patient name. Each patient name consists of a first name, followed by a space, then a second name, and then more spaces to make its length equal to 20. Write a program to read in this array; separate each name into a first and second name, and then build up two new 2-D arrays of the first and second names, respectively. These two new arrays should be written to new *MAT* files. The new 2-D arrays of first and second names that you produce should also be padded with spaces so that each name is an array of 20 characters.   *03.B, 03.D*

(Hint: The MATLAB functions `strfind` and `blanks` might be useful to you. Look at the MATLAB documentation for details of how to use them.) ■

3.5.1 The string type

The approach of padding out character arrays with spaces to enable them to be combined into a 2-D array is quite cumbersome. The built-in MATLAB string type offers a more elegant way to handle such situations. Like character arrays, strings can also be used to represent sequences of characters, and we specify that we want to form a string, rather than a character array, by using double quotes (i.e., `"`), rather than single quotes.

■ Example 3.10

For example, let us revisit Example 3.9, but this time we will use the string type, rather than character arrays.

```
>> s1 = "hello";
>> s2 = " ";
>> s3 = "world!";
>> s = [s1 ; s2 ; s3]
s =
  3x1 string array
    "hello"
    " "
    "world!"

>> whos
  Name      Size            Bytes  Class      Attributes

  s         3x1               258  string
  s1        1x1               134  string
  s2        1x1               134  string
  s3        1x1               134  string
```

Note that the types of `s1`, `s2`, and `s3` are now `string`. With the string type, it is possible to create an array `s` of such strings, even though they have different lengths. ■

Accessing and manipulating string variables is slightly different to the way in which we use character arrays. As we saw in Example 3.10, when we define a string variable, MATLAB actually creates a 1×1 *array* of strings. Therefore if we try to index into a string in the same way that we index into a character array, we will get an error.

```
>> xchar = 'MATLAB is great';
>> xstr = "MATLAB is great";
>> xchar(1:6)
ans =
    'MATLAB'
```

```
>> xstr(1:6)
```

```
Index exceeds the number of array elements (1).
```

Because xstr is a 1 × 1 array of strings, we cannot access its first six elements. We can confirm this by using the size function.

```
>> size(xstr)
ans =
     1     1
```

If we want to access the first six characters of the string, we can use the extractBetween function. Furthermore, we can find out the length of the string by using the strlength function.

```
>> strlength(xstr)
ans =
    15

>> extractBetween(xstr,1,6)
ans =
    "MATLAB"
```

MATLAB provides two ways to concatenate string variables: either through using the + operator or using the append function (which will also work with character arrays).

■ Example 3.11

For example, let us revisit Example 3.8, but this time use strings, rather than character arrays. We demonstrate two equivalent ways to concatenate string variables.

03.B, 03.D

```
>> s1 = "hello";
>> s2 = " ";
>> s3 = "world!";
>> s = s1 + s2 + s3
s =
"hello world!"

>> size(s)
ans =
     1     1

>> t=append(s1,s2,s3)
t =
    "hello world!"

>> size(t)
ans =
     1     1
```

Table 3.1 MATLAB built-in string and character array functions.

`newS=erase(s,match)`	Creates a copy of s with all occurrences of `match` removed
`yn=contains(s,match)`	Returns a logical result, indicating whether or not `match` is contained within s
`p=strfind(s,match)`	Returns an array p of the indices of all occurrences of `match` in s
`strcmp(s1, s2)`	Compare strings or character arrays, return `true` if they are. N.B.: Strings can be compared with the `==` comparison operator.
`newS=replace(s, old, new)`	Creates a copy of s with all occurrences of `old` replaced by `new`
`newS=lower(s)`	Creates a copy of s with all characters converted to lower case
`newS=upper(s)`	Creates a copy of s with all characters converted to upper case
`sArr=strsplit(s, del)`	Creates an array of strings/character arrays formed from the words in s separated by `del`. If s is a character array, then `sArr` will be a cell array (see Section 6.2), whereas if s is a string, then `sArr` will be a string array.
`newS=extractBetween(s,p1,p2)`	Creates a copy of s containing only characters between indices `p1` and `p2`
`newS=extractBefore(s,p)`	Creates a copy of s containing only characters before index `p`
`newS=extractAfter(s,p)`	Creates a copy of s containing only characters after index `p`
`newS=append(s1, .., sN)`	Creates `newS` which is a concatenation of `s1` ... `sN` (see also `strcat`)

Note that the result of concatenating multiple 1×1 strings is another 1×1 string, in contrast to the use of character arrays, where we saw the size of the array increase as we concatenated more arrays. ∎

3.5.2 Built-in MATLAB functions

MATLAB features a number of built-in functions for manipulating strings and character arrays, and most of them work for both types. Some of the more useful ones are summarized in Table 3.1.

■ Activity 3.7

03.B, 03.D   Recall Activity 3.6, in which we split up a character array of patient names into first name and second name arrays. Now we will repeat this activity using strings.

The file *patient_strings.mat* contains an array of strings containing the same patient name data, as in Activity 3.6. Modify your code from Activity 3.6 to split these data into first name and second name arrays (which should now both be string arrays), and save them to new *MAT* files. You may need to make use of some of the built-in functions listed in Table 3.1. ∎

3.6   Identifying the type of a variable

As we have seen in Section 1.8, we can use the `whos` command to display a summary description of the variables that are in the workspace. This is useful, but sometimes we may have many variables in the workspace, and we may not need to know about some of them. Also, the `whos` command gives information on the size of variables (in bytes), which we may not want.

The built-in function `class` can also be used to identify the type of a variable. It produces a text output that describes the variable type.[2]

∎ Example 3.12

For example, consider the following sequence of commands: *03.B, 03.D, 03.E*

```
>> a = 1:5
a =
     1    2    3    4    5

>> b = 'Some text'
b =
Some text

>> class(a)
ans =
double

>> class(b)
ans =
char
```

Here, we can see that the `class` command has identified the types of the variables a and b as `double` and `char`, respectively. ∎

Another way to identify the type of a variable is to use the built-in `isa` function. This function takes a variable and a character array (or string) as its arguments (where the character array describes a type) and returns a Boolean result (see Section 3.4), with a 1 indicating `true` (the variable *is* of the specified type) and a 0 indicating `false` (it is not).

---

[2]Note that the use of the word "class" in MATLAB to describe the data type of a variable is at odds with the use of the same word in other programming languages, such as C++ or Python, where the word "class" is reserved for something much more specific.

Further functions for testing data types are specific to each type and do not need a description of the type to be given. For example, we can test if a variable is a character or a character array with the function `ischar`, and the function `isnumeric` can be used to test if a variable's data type is one of the numeric types (i.e., `uint8`, `int64`, etc.).

■ Example 3.13

03.B, 03.E This example uses the same variables defined in the previous example.

```
>> isa(a, 'double')
ans =
     1

>> isa(a, 'char')
ans =
     0

>> ischar(b)
ans =
     1

>> isnumeric(b)
ans =
     0
```

■

3.7 Converting between types

Now that we have considered some of the fundamental data types and how to find out the types of variables, we can look at converting between data types. It can be useful to do this on occasion; for example, we may need to read a text file as a set of character arrays, and then reinterpret them as numeric values. The simplest way to convert data types is to use the functions associated with each type.

Common conversions include converting between numbers and characters and converting between numbers and logicals.

3.7.1 Converting between a number and a character

We stick to the default numeric type `double` in this example. The `char` function can convert a number to a character:

```
>> char(66)
ans =
B
```

To go the other way, i.e., convert a character to a number, we can use the function named `double`.[3]

```
>> double('B')
ans =
    66
```

This shows that conversion between numbers and characters is based on the ASCII character encoding.

3.7.2 Converting between a number and a logical type

We can convert a number to a logical type using the `logical` function. Consider the following commands:

```
>> logical(1)
ans =
    1

>> logical(3.5)
ans =
    1

>> logical(-12)
ans =
    1

>> logical(0)
ans =
    0
```

These show that the conversion of any non-zero number to a Boolean (`logical`) type will always return `true` (i.e., 1). It will only produce a `false` result if the number is zero (as in the last example above).

Converting a `logical` type to a number can again be carried out using the `double` function. There are of course only two cases to check, and the results are what we would expect. The following commands illustrate the conversion of `logical` values to numbers:

```
>> double(true)
ans =
    1

>> double(false)
ans =
    0
```

---

[3]Take care to distinguish between the *function* named `double` and the data type `double` that it returns.

3.7.3 Converting arrays

Converting arrays of data is also possible. For example, the function `double` will attempt to convert any argument it is given to a double precision floating point number. If the argument is an array, then it should give an array of `double` values. Let us initialize an array `arr` with a list of unsigned 8-bit integers.

```
>> arr = uint8(5:9);
>> class(arr)
ans =
  uint8
```

Passing `arr` as an argument to the `double` function performs a conversion to the `double` type.

```
>> arrTwo = double(arr)
arrTwo =
     5    6    7    8    9

>> class(arrTwo)
ans =
  double
```

Note that we cannot distinguish the arrays above by examining their contents, but we can by using the `class` function.

We can convert from numbers to characters using the `char` function. For example,

```
>> arr = [104    101    108    108    111];
>> char(arr)
ans =
  hello
```

In other words, the list of numbers initially put into the array, when converted to characters individually, produces a `char` array containing the characters `'h'`, `'e'`, `'l'`, `'l'`, `'o'`.

MATLAB will sometimes convert values automatically. This can happen, for example, when a mixture of data types are put into an array. In this case, a set of rules determine how the data types contained in the array are converted so that the resulting array has a uniform type.

■ Example 3.14

03.B, 03.D, 03.F

This example illustrates automatic type conversion and a possible problem it can cause. We initialize an array as follows:

```
>> x = 70;
>> str = ['This is number seventy: ' x];
```

Here, we place a character array and a numeric type together in the array. MATLAB will automatically try to convert the numeric variable x to characters and concatenate the result with the first character array. We need to take care, as the result may not be what we expected. If the above character array is displayed, we get the following:

```
>> disp(str)
This is number seventy: F
```

showing that the conversion was defined by the ASCII codes. The programmer may not have intended this and might instead have wanted to print a character array representing the numeric value 70.

MATLAB has a specific function for this kind of conversion, num2str, which can be used as follows:

```
>> y = num2str(x)
y = 70

>> disp(class(x))
double

>> disp(class(y))
char
```

This shows that the variable y contains the characters required to represent the number 70. In particular, it contains two characters: a '7' and a '0'. We can now modify the displayed text above to give a more sensible message:

```
>> str = ['This is number seventy: ' num2str(x)];

>> disp(str)
This is number seventy: 70
```
■

Going the other way, the built-in function str2double takes a character array representation of a number and converts it into a numeric value. For example, compare the different behaviors in the following two command window calls, which lead to very different results:

```
>> '3.142' + 1
ans =
    52    47    50    53    51

>> str2double('3.142') + 1
ans =
    4.1420
```

In the first command, the char array ['3','.','1','4','2'] was automatically converted to numeric values, based on the ASCII codes of each character, then a value of 1 was added to each. In the second command, the character array '3.142' was reinterpreted (converted) directly as the numeric value that is close to pi (π), and then a value of 1 was added. The second command is perhaps the more sensible one to use in general.

Conversion between strings and character arrays is also possible using the `string` and `char` functions, as the following code illustrates:

```
>> xstr = "hello";
>> xchar = char(xstr);
>> xchar_conv = string(xchar);
>> whos
  Name              Size            Bytes  Class      Attributes

  xchar             1x5                10  char
  xchar_conv        1x1               150  string
  xstr              1x1               150  string
```

Terminology - Casting: The process of converting from one data type to another is often called *casting*. We say that a variable is cast from one type to another. This cast can be *explicit*, using a function, or it can be *implicit* (i.e., automatic).

Terminology - Strong and Weak Types: Different programming languages vary in how strict they are regarding the type of a variable. Some languages, such as C++, require the type of a variable to be explicitly stated when it is defined. Also, some languages allow little or no conversion of the type of a variable. In this case, we say that the language is *strongly typed*.

In the examples we have looked at, we can see that MATLAB is fairly permissive about converting types of variables, with many conversions taking place implicitly (automatically). Therefore we say that MATLAB is a *weakly typed language*.

Good Advice: In general, it is always safer to convert between data types using explicit function calls (e.g., `double`, `char`, `str2double`, etc.), rather than relying on automatic conversions. This leads to clearer code that is easier to debug.

■ Activity 3.8

03.B, 03.F

In Example 1.3, we wrote code to visualize the radial displacement of a single segment of a patient's left ventricular myocardium.

The files *radial1.mat …radial17.mat* contain the displacement data for all 17 segments, based on the standard delineation of the myocardium into segments produced by the American Heart Association (AHA). The files each contain two variables: `radial` represents the radial displacement measurements (in mm), and `t` represents the time (in milliseconds) of each measurement. *Both of these variables are character array representations of the arrays of numeric time and displacement values.*

Write a script *m*-file that does the following:

1. Prompts the user to enter a segment number at the command window.
2. Checks that the number entered is between 1 and 17 (the number of AHA segments). If not, the user should be continually prompted to reenter the number until a valid segment is entered.
3. Reads in the data file corresponding to the segment entered and performs any required type conversions.
4. Produces a plot of the numeric radial displacement against time. The plot should be appropriately annotated and the title should indicate the segment number.
5. Asks the user if they want to display another plot.
6. If they type "y," repeat the above steps. Otherwise, exits.

Note that to read in the correct file and to display the segment number of the myocardium in the title of the plot, you will need to consider the types of the data being used and convert them if necessary. ∎

3.8 Summary

In this chapter we have looked at how a data type is used to interpret sequences of ones and zeros in the computer's memory. The same pattern of ones and zeros will be interpreted differently, depending on the data type, e.g., whether it is numeric, a character, Boolean, or a string.

We have considered the range of values it is possible to represent with numeric types and how this varies. We have also looked at the precision that floating point numbers can achieve. We have introduced the special NaN and Inf identifiers to represent "Not a Number" and "Infinity," respectively.

We have looked at Boolean (or logical) arrays and how these enable *logical indexing*, a powerful feature that allows quick and easy extraction of rows or columns from an array that match a specified condition.

We have reviewed the use of character arrays and strings and how these two types differ. We have also introduced some built-in MATLAB functions for manipulating text data using these types.

Finally, we have looked at ways of converting a variable between data types and how, for numeric types, MATLAB defaults to using doubles. Accepting this default behavior is generally the best advice, as many functions in MATLAB expect to use doubles.

3.9 Further resources

The MATLAB documentation can be a useful resource for discovering more about data types:

- General documentation on data types: https://mathworks.com/help/matlab/data-types.html.
- Explanation of logical indexing: https://mathworks.com/help/matlab/math/array-indexing.html#MatrixIndexingExample-3.
- Specific information on character arrays and strings: https://mathworks.com/help/matlab/characters-and-strings.html.
- A full list of the specific functions for testing the type of a variable: https://mathworks.com/help/matlab/data-type-identification.html.
- Further details on conversions between data types: https://mathworks.com/help/matlab/data-type-conversion.html.

For further description of the ASCII encoding for characters, the Wikipedia entry is a good place to start: https://en.wikipedia.org/wiki/ASCII.

3.10 Exercises

■ Exercise 3.1

03.A, 03.B, 03.D, 03.E, 03.F

1. Explain why the binary pattern 01100011 can be viewed as the decimal number 99 or the character `'c'`.
2. Initialize your own 1×3 array of numbers. Try and choose them so that there is at least one integer that matches the ASCII code for an alphanumeric character.
3. Make an explicit conversion of the array to a character array. If some of the characters appear to be missing, why should this occur?
4. Make an explicit conversion of the array to the `logical` type. Use a MATLAB function to check that the type of the result is correct.
5. Make an *implicit* conversion of the array's contents to a character. *(Hint: We can use concatenation.)* ■

■ Exercise 3.2

03.B, 03.D, 03.E, 03.F

1. Create a variable containing a one row character array containing `'some string'`. Do this using square brackets and individual characters. Use `class` to confirm the type of the variable.
2. Repeat the last part, but create the character array variable using a single set of single quotes.
3. Make a two row character array containing `'some'` and `'string'` on separate rows. What do we need to do to make sure these two words can be fitted into a two row array?
4. Make an explicit conversion of the character array from the last part to a numeric type.
5. Make an *implicit* conversion of the character array to a numeric type.

6. For each of the following character arrays, try to generate a corresponding numeric value using the `str2double` function. Give the reason why some of these conversions can fail and how we can test for it.

```
'2.3'
'e'
'0.7'
'X.3'
'1.3e+02'
```

■

■ Exercise 3.3

03.C

1. A variable is set by the command `a=10`. The default MATLAB behavior is to set `a` to double precision. Use the `eps` command to determine the precision of the variable.
   How would the precision change if the initial command had been `a=1000000`? Confirm your prediction, again with the `eps` command.
2. The `intmax` function can be used to show that the largest unsigned sixteen-bit integer is 65535. Show the numeric calculations that prove this is the largest possible unsigned sixteen-bit integer. ■

■ Exercise 3.4

Consider the following definitions of numeric variables:

03.B, 03.D, 03.F

```
a = pi
b = uint8(250)
c = int32(100000)
d = single(17.32)
```

1. Give the full name of the type for each of the variables above.
2. An array is created by concatenating the variables above:

   ```
   X = [a b c d]
   ```

   Identify the type of the resulting array.
3. Create arrays by concatenating different subsets of the elements in different orders:

   ```
   X = [a d]
   X = [d a]
   X = [a d b]
   X = [a d c]
   X = [a d c b]
   X = [a d b c]
   ```

   Check the type of the result each time.
   Identify the rule that MATLAB uses for deciding on the type of the output:

a. When the array contains one or more integer types.
b. When the array contains a mixture of floating point types and no integer types.
4. A fifth variable is assigned a character: e = 'M'. What is the data type of the resulting array if variable e is included in any of the concatenations above? ∎

■ Exercise 3.5

03.D   The file *patient_temps.mat* contains a variable called age_temps. The first column of this variable contains patients' ages, and the second column contains their body temperatures. Each row represents a different patient. By mistake, some of the temperatures were recorded in Fahrenheit and some in Celsius. Write MATLAB code to load in the data, and convert all Fahrenheit values to Celsius using the following formula:

$$\text{Celsius} = (\text{Fahrenheit} - 32) \times \frac{5}{9}.$$

You can easily find which values are in Fahrenheit by comparing them to some threshold (e.g., 50 degrees). You should use logical indexing when choosing which values to convert. ∎

■ Exercise 3.6

03.D, 03.F   The file *diseases.mat* contains a variable called diseases, which is a string array containing names of different diseases. Currently some of the names are in upper case and some in lower case. Write a MATLAB script to read in these data, and convert all names so that they are in *title case*, i.e., the first letter of each word should be upper case and the rest lower case. Save the resulting string array in a new file called *newdiseases.mat*.

(Hint: You may find it easier to convert the strings to character arrays to perform the case conversions.) ∎

FAMOUS COMPUTER PROGRAMMERS: JOHN VON NEUMANN

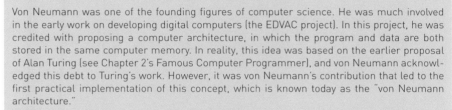

John von Neumann was a mathematician who was born in 1903 in Hungary. Von Neumann was a child prodigy. As a 6-year-old, he could divide two 8-digit numbers in his head, and was fluent in both Latin and Ancient Greek. By the age of 8, he was familiar with differential and integral calculus. By the time he was 19, von Neumann had published two renowned mathematical papers. He received his PhD in mathematics at the age of 22.

In 1933, at the age of 30, von Neumann left Hungary and joined Princeton University in the USA. He remained there for his entire career. He was a remarkable academic figure, who made major contributions to a wide range of subjects, including mathematics, physics, economics, statistics, and computer science. He was also a principal member of the Manhattan project that developed the world's first atomic weapon.

Von Neumann was one of the founding figures of computer science. He was much involved in the early work on developing digital computers (the EDVAC project). In this project, he was credited with proposing a computer architecture, in which the program and data are both stored in the same computer memory. In reality, this idea was based on the earlier proposal of Alan Turing (see Chapter 2's Famous Computer Programmer), and von Neumann acknowledged this debt to Turing's work. However, it was von Neumann's contribution that led to the first practical implementation of this concept, which is known today as the "von Neumann architecture."

In his personal life, John von Neumann was a sociable figure and was known for throwing large parties while at Princeton. However, he was a notoriously bad driver. When he was hired as a consultant at IBM, they had to quietly pay the fines for his multiple traffic offences. He enjoyed music, and regularly received complaints from colleagues at Princeton (including Albert Einstein) for playing his German marching music loudly at the office.

John von Neumann died of cancer in 1957. Because of his involvement in the USA's atomic weapons program, his final days were spent in a military hospital, in case he revealed any military secrets on his death-bed. There has been speculation that his cancer was the result of his attendance at some of the early atomic weapons tests.

"You wake me up early in the morning to tell me that I'm right? Please wait until I'm wrong."

John von Neumann

Functions

4.1 Introduction

In this chapter, we will consider the programming construct known as the *function*. We have already seen how MATLAB provides a large number of built-in functions for our use, so you should be aware that *a function is a piece of code that can take some arguments as input and carry out some tasks that can produce a result*. Functions can be particularly useful if we want to perform exactly the same (or a similar) operation many times with different input values. They also offer a way to promote *code reuse*, which refers to the (re)use of existing software when developing a new application. Code reuse is generally considered to be a good thing in computer programming, as it saves time and reduces the chance of introducing errors into new code. By learning to write our own functions, we can enable others (and ourselves) to reuse our code and take advantage of a lot of the power available in MATLAB (or indeed in many other programming languages).

4.2 Functions

MATLAB has its own built-in functions, such as `sin`, `max`, `exp`, etc., but it also allows us to define our own functions. We can save any function we write in an *m*-file in the same way as we can save any set of commands to form a script. In both cases this helps us to share our work with others.

So it is important to note that there *two kinds of* m-*file*. There are *function* m-files and *script* m-files. We have already seen script m-files in Section 1.12. Later, we will discuss them and how they differ from function m-files in Section 4.4.

The following examples illustrate some functions:

■ Example 4.1

04.A   Here is a very simple (and quite silly) function.

```
function out = add_two(in)
% Usage:
%   out = add_two(in)
%   in  : Argument passed in
%   out : Output, result of adding two to input.

out = in + 2;
end
```

This very simple function has a single input argument, `in`, and a single output argument, `out`, which is the result of adding two to the input.

Note how the code begins with the `function` keyword. This is then followed by the name of the output argument `out`, then an equals sign, then the name of the function `add_two`. The first line is then ended by the input argument, `in`, inside a pair of brackets. This first line is known as the function's *prototype*. It is also sometimes described as the function *definition* or *signature*.

After the first line, some comments describe how the function should be used, and then the "work" of the function is actually carried out in the line `out=in+2;`. We indicate the end of the function with the `end` keyword.

The code for a function can be saved in a file.   ■

Note that, for the previous example, the function is saved in a file with a name matching the name of the function (i.e., *add_two.m*). This is in line with the naming rules on function *m*-files, which are described in more detail in Section 4.9.

■ Activity 4.1

04.A   Use the *Editor* window (see Fig. 1.1) to type the code for the function from Example 4.1 into a new file. Then save the file with the name *add_two.m*. This kind of file is called a *function m*-file.

Once the file containing the function is saved, go to the command window and call the function as follows:

```
>> a = 5
a =
     5

>> b = add_two(a)
b =
     7
```

In the above, we assign a value of 5 to the variable with the name a, and then pass it to the function as an input argument. The return value from the function is assigned to the variable named b.

Experiment with different input arguments, and assign the return value to variables with different names.   ■

Note that a function can have any number of input arguments (including none) and any number of outputs (including none).

■ Example 4.2

The next function takes two input arguments. It does not return any output   *04.A*
arguments. It simply prints the sum of the inputs to the command window
when it runs.

```
function report_sum(x, y)
% Usage:
%   report_sum(x, y)
%   x, y : numbers to be added

fprintf('The sum is %u\n', x + y)
end
```

Here, we have used the built-in `fprintf` function to print output to the screen. We will discuss this function in more detail in Section 7.5.   ■

■ Activity 4.2

Type the code from Example 4.2 in the *Editor* window, and save it in a func-   *04.A*
tion *m*-file with the name *report_sum.m*.

Once the file containing the function is saved, go to the command window, and call the function to make sure it works, for example, as follows:

```
>> report_sum(3, 25)
The sum is 28
```

Note that there was no need to assign any return value to a variable, because the function `report_sum` did not have any output arguments.   ■

■ Example 4.3

04.A   The next function takes no inputs and gives no return value! It simply prints one of the words "low" or "high" to the screen at random.

```
function say_low_or_high()
% Usage:
%    say_low_or_high

a = rand();
if a < 0.5
  fprintf( 'low\n' )
else
  fprintf( 'high\n' )
end

end
```

Here, we have again used the built-in MATLAB function `fprintf` to print output to the screen. We have also used the built-in function `rand()` to generate a random number in the interval [0, 1] (type `help rand` at the command window to get an idea of how it works).   ■

■ Example 4.4

04.A   This example is slightly more complex than the previous ones.

```
function [l,m] = compare_to_mean(x)
% Usage:
%    [l,m] = compare_to_mean(x)
% Input:
%     x — A list of numbers
% Outputs:
%     l — Number of values in x less than mean
%     m — Number of values in x greater than mean

m = mean(x);

less_array = find(x < m);
l = length(less_array);

more_array = find(x > m);
m = length(more_array);

end
```

We will use and discuss this function in more detail in the next few pages.   ■

■ Activity 4.3

04.A   Create a new file in the *Editor* window and type in the code for the function `compare_to_mean`. Save it with the file name *compare_to_mean.m*.

Now switch to the command window and call your function by typing the code below. First, we create an array with some numbers of our choosing, e.g.,

```
>> a = [1 2 4 9 16 25 36 49 64 81 100];
```

Now, we check the function works by passing our array as input:

```
>> [countLess, countMore] = compare_to_mean(a)
countLess =
      6
countMore =
      5
```

We find that six elements of the array were below the mean value, and five were above the mean value in this example. Check for other arrays of your own.   ■

Note that the call to the function in Activity 4.3 captures *both* return values and assigns them to the two variables called `countLess` and `countMore`. Inside the function code, these are the values of the output arguments that are returned (with the names `l` and `m`).

Whereas the function `compare_to_mean` will always return two output arguments, when we call the function, we do not have to assign all of them to variables in the command window call. For example, the following call

```
x = compare_to_mean(a)
```

will only capture the value of the *first* of the return values (`l` in the original function), assigning it to a command window variable called `x`. In this case, we collect fewer values from the function than it returns, and the second return value is simply lost or discarded.

However, if we try to collect *more* values than the function can provide, we will obtain an error:

```
>> [a, b, c] = compare_to_mean(a)

 Error using compare_to_mean
 Too many output arguments.
```

Similarly, we can encounter errors if we do not pass the correct number of input arguments to a function. Using the function `report_sum` from Example 4.2, MATLAB will give an error message in the following examples:

```
   >> report_sum(3,4,5)

 Error using report_sum
 Too many input arguments.
```

```
>> report_sum(3)

Not enough input arguments.
Error in report_sum (line 6)
fprintf('The sum is %u\n', x + y)
```

We give too many inputs in the first case, and not enough inputs in the second.

As we have seen in all of these examples, we can create our own functions, and we can call them from the command window in the same way that we can call built-in MATLAB functions.

The function in Example 4.4 takes an array as its argument and returns the counts of the array elements that are less than or greater than the mean value of the array elements. Therefore we can assign the return values of the function to *two* variables, enclosed by square brackets.

Let's look more closely at the compare_to_mean function code. The first line is the function *prototype* (a.k.a. *definition* or *signature*). The prototype line looks like this:

```
function [l,m] = compare_to_mean(x)
```

This important line specifies

- That the code represents a function, using the function keyword.
- The name of the function: compare_to_mean.
- The input argument(s) that the function expects: in this case, just one input, x.
- The output values that it will return (l and m).

We do *not* need to state the *types* of the values (e.g., number, character, array; see Section 1.8). This is a feature of programming using MATLAB and is in contrast to many other programming languages. The reason for this difference is that MATLAB programming uses *weak typing*, i.e., we can assign values to variables without specifying their type, and the type of a variable can also change during the running of a MATLAB program.

The next set of lines in Example 4.4 are all comments, as they start with the % symbol:

```
% Usage:
% ...
%     m - Number of values in x greater than mean
```

As pointed out in Section 1.13, it is good practice to include comments like this to help other people (and ourselves) to understand and make use of our functions. The comments at the start of a function are also used by MATLAB when the built-in help command is called for our function. If we type help compare_to_mean in the command window, the comments at the top of the function are printed to the screen. As noted in Section 1.5, this works for all built-in MATLAB functions; e.g., type `help sin` in the command window.

After the comments at the start of `compare_to_mean`, a series of steps are carried out to do the work of the function:

- The mean value of the input array, `x`, is computed and assigned to the variable `m`.
- The built-in `find` command is used to find the elements of the array that are less than `m`. Note that the expression `x < m` will return an array containing 1s (true), where the original element was less than `m` and 0s (false) where they weren't. The `find` command will return a new array containing the indices (i.e., the array element numbers) of the non-zero elements of this array.
- Therefore the subsequent call to obtain the `length` of the array returned by the `find` command will give the number of the original array elements that were less than the mean. This is assigned to the variable `l`, one of the return variables specified in the function prototype.
- An almost identical pair of steps are then taken to assign a value to the other return variable, `m`.

Read through this code carefully, and ensure you understand how it works.

SYNTAX: FUNCTIONS

Formally, we can define the syntax rule for a MATLAB function as follows:

```
function [y1, ..., yN] = myfun(x1, ..., xM)
    ...
end
```

In this definition, we can see that we are allowed to have multiple input arguments (`x1,...,xM`), although, as we have already seen, we can have a single input argument or no input arguments at all. Likewise, the output arguments (`y1,...,yN`) can consist of multiple values, a single value, or no values. The end keyword is advised, but MATLAB will not raise a syntax error if it is omitted.

4.3 Checking for errors

We often need to check for errors during the running of a function and, if we find one, we can stop execution before it completes all of its tasks. One common check is to ensure that the user has called the function with the correct arguments. For instance, in Example 4.4, the user might give an empty array as the input argument from the command window, i.e.,

```
>> a = [];
>> [lo, hi] = compare_to_mean(a);
```

In this case, the function should "raise an error" before attempting any calculation. This can be achieved using the built-in `error` function. Example 4.4 can be adjusted slightly to address this by using the built-in `isempty` function as shown below:

```
function [l,m] = compare_to_mean(x)
% [Usage and help section goes here ....]

if isempty(x)
  error('compare_to_mean: %s', 'The input array is empty.');
end

% [... function continues as before]
```

The `error` call has two arguments: the first is a character array, which contains a *format specifier*: `%s`. This format specifier is substituted with the second argument to make a single character array, which is used as the message displayed by the `error` function.

When the `if` condition is true and the user has given an empty array, the full formatted message "`compare_to_mean: The input array is empty.`" is printed in the command window, and the function exits immediately without giving any return value. It is important that the error message contains some text to say which function was being executed when the error occurred, especially in complex code, when lots of functions are being called.

As well as serious errors, which should cause execution of a function to stop, it is sometimes the case that we may encounter less serious situations, which should be reported, but should not stop execution. For example, suppose the `compare_to_mean` function received an array argument containing a single value. This will not prevent the function from running, since the mean of a single value *can* be calculated, but it is possible that this input argument was not intended. In such cases, rather than terminating execution with the `error` statement, we can use the `warning` statement. The `warning` statement will display a warning message to the command window, but then continue with execution. Consider the following added lines to `compare_to_mean`:

```
function [l,m] = compare_to_mean(x)
% [Usage and help section goes here ....]

if length(x) == 1
  warning('compare_to_mean: %s', 'Input array of length 1.');
end

% [... function continues as before]
```

Now the function will spot the possibly unintended input array and report it, but continue with execution of the rest of the function.

In the above examples, we have seen the use of the `%s` format specifier when using the `error` and `warning` commands. Other format specifiers are possible, for example, %u can be substituted with a subsequent integer argument. For more information on the `error` and `warning` functions, use `doc` or `help` at the command window. We will return to the topic of format strings in Section 7.4.

■ Activity 4.4

Use the code for the `compare_to_mean` function in Example 4.4, and include a check to see if the correct number of input arguments has been provided. There should be exactly one argument, and the function should give an `error` call if not.   *04.D*

(Hint: Use the built-in function `nargin` to check how many arguments have been passed to the function. Type `doc nargin` or `help nargin` at the command window to see how it works.)   ■

4.4   Function *m*-files and script *m*-files

Recall that we described script *m*-files in Section 1.12. To recap, a script *m*-file simply contains a sequence of MATLAB commands. It is different from a function *m*-file, because it does not take any input argument(s) nor return any value(s). Also, it does *not* have a function prototype line at the top of the file.

The following table lists the main differences between function *m*-files and script *m*-files:

Script *m*-file	Function *m*-file
A simple list of commands.	A list of commands with clearly defined inputs and outputs.
No `function` keyword at start.	`function` keyword at start of file.
No input arguments.	Prototype line at start lists input arguments.
No output arguments.	Prototype line lists output arguments.
No `end` keyword at the finish.	`end` keyword to indicate we have reached the end of the function.

■ Example 4.5

This example is a script *m*-file. It is just a list of commands, and no function keyword or prototype is used.   *04.B*

```
disp('Hello world')

total = 0;

for i = 1:1000
        total = total + i;
end

fprintf('Sum of first 1000 integers: %u\n' , total)
```
■

■ Activity 4.5

04.B   Type the commands from Example 4.5 into a script *m-file*, and save it with the name *run_some_commands.m*.

Run the script at the command window by simply typing the name of the file (without the .m suffix). Check that it gives the correct output as shown below.

```
>> run_some_commands
Hello world
The sum of the first 1000 integers is 500500
```

What would you need to do to make the script produce a different output, such as the following?

```
>> run_some_commands
Hello world
The sum of the first 10 integers is 55
```

■

Important: The script *run_some_commands.m* will *always do the same thing* unless we modify it. If we want to sum the first 100 or 1,000,000 numbers, we need to edit the file and change it to give us what we want. This is one reason why it is generally better to write a function that can take an input argument.

■ Example 4.6

04.A   We can create a function to do something similar, but with input and output arguments as follows:

```
function total = sum_function(N)

total = 0

for i = 1:N
        total = total + i;
end

end
```

■

In this case, the function does not specify the number N; it is an input argument that the user can pass in and any integer will suffice.

■ Activity 4.6

04.B   Use the *Editor* window to type in the code for the function sum_function, and save it in a function *m-file* called *sum_function.m*.

We can call the function in a similar way to the script, but now we must pass an input argument using brackets. Check that the function returns the correct output for the sum of the first 1000 integers, i.e.,

```
>> value = sum_function(1000)
value =
   500500
```

Show how you can find the sum of the first 20, 50, and 300 integers by simply modifying how you call the function at the command window (you do not need to adjust the function's *m*-file!). ∎

4.5   The `return` keyword

We have now seen that functions can be called from the command window, and that they will return control when they have no more code to execute. However, it is sometimes desirable to return from a function before we have reached the end of its code, and for such cases MATLAB provides a special keyword: `return`.

∎ Example 4.7

Consider the following function, which will search an array for the first pair of consecutive identical values and return the index of the first value of this pair:

04.A

```
function ind = spot_pair (x)
% Usage:
%   ind = spot_pair (x)
%   x: input array
%   ind: position of first pair of identical values

ind = −1;
for n = 1:(length(x)−1)
    if (x(n) == x(n+1))
        ind = n;
        return;
    end
end

end
```

This code uses a `for` loop to search for adjacent values that are the same. Once a pair of values is found, the output argument `ind` is assigned with the current value of the loop variable `n`. We then use the `return` statement to immediately return control to where the function was called from. If the `return` statement was not there, the code would continue to search the array and possibly find another pair of values, overwriting the `ind` variable, and hence giving the wrong result. Note that, in general, when using the `return` statement in a function that has output arguments, it is important to make sure that these arguments are correctly set before returning control. ∎

4.6   A function *m*-file can contain more than one function

If we are working on a *m*-file that contains a function that gets long and complicated, we can break it up into a "main" function and pass on some of its work to be carried out in "sub-functions" or "helper" functions. These can also be described as local functions and can appear later within the same file after the "main" function. The following example illustrates how to do this:

■ Example 4.8

04.A   The code below represents three functions, and all of them can be saved in a single file called *get_basic_stats.m*. The "main" function is called `get_basic_stats`, and there are two helper functions that follow it.

```
function [xMean, xSD] = get_basic_stats(x)
% Usage :
%    [xMean, xSD] = get_basic_stats(x)
%
% Return the basic stats for the array x
% Input: x, array of numbers
% Outputs: mean, SD of x.

xMean = get_mean(x);
xSD   = get_SD(x);

end

% Helper functions below:
function m = get_mean(x)
m = mean(x);
end

function s = get_SD(x)
s = std(x);
end
```

■

This example illustrates the following rules that must be followed:

- The main function appears first (`get_basic_stats`).
- The name of the main function and the *m*-file must match.
- The helper functions are typed in after the `end` of the main function. (In the above example the helper functions are `get_mean` and `get_SD`.)
- A helper function can be called by the main function (and by any of the helper functions).

Note that these helper functions *can only be called by a function from within this file.* They cannot be called, for example, from the command window or from a function in another *m*-file.

Good advice: When typing multiple functions into a single *m*-file, make sure to end one function before starting another. Otherwise, there can be an error or unpredictable behavior. In other words, the structure of a file with more than one function in it should be the following:

```
function val = f1(input1)

% ... f1 code here

end % for f1

function val = f2(input2)

% ... f2 code here

end % for f2

% etc..
```

It is even better to put a comment line in between functions to make the code readable, as in the following:

```
function val = f1(input1)

% ... f1 code here

end % for f1

%%%%%%%%%%%%%%%%%%%%%

function val = f2(input2)

% ... f2 code here

end % for f2
```

(It is actually possible to nest one function inside another, but you are *strongly* advised not to do this. It is rarely used, and only in special circumstances.)

4.7 Script *m*-files and functions

As we saw above, a script *m*-file is just a list of commands. We might also consider adding some functions into a script file so that we can take advantage of their benefits such as code reuse. Whether this is possible or not depends on the MATLAB version.

The following is a modified version of the code in Example 4.5 that includes a function:

```
% Start of script m—file with a function
disp('Hello world')

total = find_sum_illegal(1000);

fprintf('Sum of first 1000 integers: %u\n' , total)

% Add a function into the script m—file:

function tot = find_sum_illegal(N)
    tot = 0;
    for i = 1:1000
        tot = tot + i;
    end
end
```

This code will work in recent versions of MATLAB, but will fail for older versions. Before MATLAB version 2016b, it was illegal to start an *m*-file as a script and to later introduce a function. In such older versions of MATLAB, to make the above work, we would need to separate the script part and function part, and put them in separate files: a script *m*-file and a function *m*-file.

■ Example 4.9

04.A, 04.B

This example shows the correct separation of the code above in versions of MATLAB older than 2016b.

Script *m*-file:

```
% In a script m—file run_some_commands_2.m
disp('Hello world')

total = find_sum_legal(1000);

fprintf('Sum of first 1000 integers: %u\n' , total)
```

Function *m*-file:

```
% In a function m—file find_sum_legal.m
function tot = find_sum_legal(N)
tot = 0;
for i = 1:N
        tot = tot + i;
end

end
```

■

4.8   *m*-files and the MATLAB search path

When using *m*-files to save functions, we may need to save them in different locations in the file system. This relates to the set of directories in which

MATLAB will search for *m*-files. Collectively, the set of all locations where MAT-LAB searches is called the *path*, and we can inspect the current set of locations by typing the built-in function `path` in the command window.

Recall that the current working directory is shown near the top of the MATLAB environment (see Fig. 1.1). If a function *m*-file that we have written is in the current working directory, then we can call it from the command window directly.

If, on the other hand, the function that we want to call has its *m*-file in a different location from the current working directory, there are two options:

- Change the working directory to where the file is located.
- Add the location of the file to the *path*.

We can take the second option, and add the location to the path list, by using the `addpath` command, giving the name of the required file's location in single quotes:

```
>> addpath('/path/to/some/directory')
```

Once this is done, any script or function *m*-files that are contained in the directory can be called from the command window, no matter what the current working directory is.

■ Activity 4.7

Make a note of where you saved the function *m*-file for the code in Activity 4.6. Let's say it is called `'/path/to/file'`.

04.E

Now change the working directory to a different location. You can use the address bar at the top of the MATLAB environment (see Fig. 1.1), or you can type `cd /to/some/other/location` in the command window.

Try to run the instruction at the command window as before.

```
>> value = sum_function(1000)
```

You should get the following error:

```
Unrecognized function or variable 'sum_function'.
```

This is because the function *m*-file containing the code is not in the path.

Now add the location of the function *m*-file to the path by typing `addpath('/path/to/file')` in the command window.

Now check to confirm that the call to the function runs without an error. ■

4.9   Naming rules

When saving an *m*-file for a script or a function, we need to abide by the following rules on how it can be named. The name of a *m*-file

- *Must* start with a letter ...
- The letter with which the name starts can only be followed by
  - letters or
  - numbers or
  - underscores.
- Must have a maximum of 64 characters (excluding the .m extension).
- Should not have the same name as a MATLAB reserved word or built-in function (e.g., for or sin).
- Must match the name of the function for *function m*-files.

For example, names that start with a number, or contain dashes or spaces, cannot be used.

For *function m*-files, the name of the *m*-file should match the name of the function it contains (the main one if it contains more than one). If they do not match, then MATLAB may give a warning (depending on the version used). For example, if the function that follows is saved to a file called *give_random.m*, then there will be a mismatch between the function definition and the filename.

```
function result = get_random()
% Usage get_random: Return a random number
result = rand;
end
```

4.10   Scope of variables

The *scope* of a variable refers to the parts of a program or the MATLAB environment where the variable is 'visible', i.e. where it can be assigned to, read, or modified. Scope is an important concept in most programming languages.

The scope of a variable that is inside a MATLAB function is described as *local*. This means that it can only be accessed and modified by commands inside the same function. It cannot be affected by, for example, instructions that are typed at the command window or instructions inside different functions or scripts.

■ Activity 4.8

04.A, 04.C

Using the *Editor* window, save the following into a function *m*-file called *my_cube.m*:

```
function y = my_cube(x)
% my_cube(x) : returns the cube of x
```

```
y = x * x * x;
fprintf('The cube of %u is %u\n', x, y);
end
```

Now, in the command window, assign values to variables called x and y, and then call the function my_cube. For example, use the following commands:

```
>> x = 4;
>> y = 7;
>> my_cube(12)
>> disp([x, y])
```

What output do you expect to be produced by the call to my_cube(12) above?

What output do you expect from the subsequent call to disp([x,y]) that follows? ■

The important point about the code in Activity 4.8 is that there are two variables called x, but that they are in different scopes. The same applies to the variables called y. We assign values of 4 and 7 to the variables x and y in the command window. We then make a call to the function with an argument of 12. The fprintf command in the function will produce the following output:

```
>> my_cube(12)
The cube of 12 is 1728,
```

which means that the variables x and y in the scope of the *function* have values of 12 and 1728.

After the function is complete, the call to disp([x,y]) will produce the following output for the variables in the scope of the *command window*:

```
>> disp([x y])
    4     7
```

We can see that these variables were unchanged by the call to my_cube.

In summary, for this example, the variables x and y that we created in the command window are not considered to be the same as the variables x and y in the my_cube function, because they have different *scopes*.

The scope of a variable in a function is also local when we use more than one function, as shown below.

■ Activity 4.9

Save the code that follows into a function *m*-file called *log_of_cube.m*. This function contains a call to the my_cube function from Activity 4.8.

04.A, 04.C

```
function log_of_cube(y)
% log_of_cube(x) : return the log of the cube of x
```

```
z = log(my_cube(y))
fprintf('Input: %f\n', y);
fprintf('Output: %f\n', z);
end
```

Use the command window to test this function by calling it with various inputs. For each call, determine what the value is for the variable called y in the scope of the function log_of_cube. Do the same for the variable called y in the scope of the function my_cube. ■

Calling the log_of_cube function will create a variable y that is local to the log_of_cube function, and pass this as an argument to the function my_cube. The function my_cube also has a local variable named y (as an output variable), but this is distinct from the one in the function log_of_cube. The value it has in the *calling* function does not affect the value it has in the *called* function.

Be careful about script *m*-files and scope

We have seen some of the rules for the scope of variables in functions. Note that these rules *do not apply to script m-files*.

■ Activity 4.10

04.B, 04.C Confirm this by saving the command that follows into a script *m*-file called *my_script.m*.

```
a = sqrt(50);
```

Now, return to the command window, set the value of a variable called a, and then call my_script, i.e.,

```
>> a = 4;
>> disp(a)
     4

>> my_script
>> disp(a)
     7.0711
```

Try this activity again, but starting with different assigned values to a, and following up with a call to my_script. ■

For the above, you should see that the variable we set in the scope of the command window *is altered by running the script*. This would not have happened if we called a *function*, even if that function also had a local variable with the name a within it.

Altering a variable in a different scope can be an unintended consequence of using scripts, which generally makes using functions (rather than scripts) a safer option.

4.11   Recursion: a function calling itself

We can define some functions *recursively*. This means that, while the function is running, it can call itself again, which in turn calls itself again, and so on.

A good illustration of a recursive definition for a function can be obtained using the *factorial* function. The factorial of an integer is denoted with a ! sign and is equal to the product of all the numbers between 1 and the integer. The factorial of zero is defined as 1, and the factorials of the first few integers are the following:

$0! = 1$

$1! = 1$

$2! = 2 \times 1 = 2$

$3! = 3 \times 2 \times 1 = 6$

$4! = 4 \times 3 \times 2 \times 1 = 24$

$$\vdots$$

This shows that when we evaluate the factorial for a number, we carry out nearly all the same operations that are needed for evaluating the factorial of the previous number. For example $4! = 4 \times 3!$. This fact is what we will use to write a recursive definition for the factorial function.

■ Activity 4.11

Use the *Editor* window to save the code that follows in a function *m*-file called *my_factorial.m*.   *04.A, 04.F*

```
function result = my_factorial(n)
% my_factorial(n) : return the factorial of n

if (n<=0)
  result = 1;
else
  result = n * my_factorial(n-1);
end

end
```

This code shows that *the function calls itself,* and this is what makes it a recursive function. If it is called, for example, with n = 4, then the result will be the value of n (which is 4), multiplied by the result that is obtained from calling the same function with a value of n−1, which is 3.

Test the function by calling it from the command window to check that it is working correctly.

Modify the function code by inserting a line to display what the value of n is at the start of the function's body, just before the `if` clause. What output do you obtain when calling the function from the command window?

Move the display line from the start of the function to the point just before the end of the function. How does the output change when it is called? ∎

When the `my_factorial` function is called with an input argument greater than one, e.g.,

```
>> my_factorial(3),
```

this will lead to the second part of the `if-else` clause being executed:

```
...
   result = n * my_factorial(n-1);
...
```

with a value of $n = 3$. In other words, a call is made *to the same function* passing an argument of $n - 1 = 2$. The return value from that call is multiplied by 3 to give the final result.

Each call for a value of $n > 0$ will involve a call to the same function with a value $n - 1$. Thus the value passed as input to the function will decrease by one each time. Eventually, we will have $n = 0$. In this case, and only in this case, the function will actually run the first part of the `if-else` clause:

```
...
   result = 1;
...
```

and a value of 1 will be returned, and *no further recursive calls will be made*. Then the return value will be passed to the calling function, which will multiply the result by *its* local value of n and return the result to its calling function, which will multiply the result by *its* local value of n, and so on.

The first part of the `if` clause above is important; it runs if $n \leq 0$ and prevents further recursive calls. This is known as a *stopping condition* for the recursive function. Stopping conditions are needed to prevent a recursive function from calling itself infinitely. A stopping condition is also known as a *ground case*.

4.12 Summary

Functions can be defined by the user in MATLAB, and they can be very powerful. They allow programmers to expand the basic capabilities of MATLAB for their own specific purposes. They can be saved in files in the same way as scripts, but one needs to be careful regarding the differences between scripts and functions.

Functions should be saved in a file with the same name as the function (with an added *.m* suffix). If the file is saved in a different location from the current working directory, then that location must be added to the *path* that MATLAB uses to search for *m*-files.

A function *m*-file can also contain *sub-functions* or *helper* functions, which are called by the "main" function in the file.

Functions take input values (arguments) and return values. Variables defined inside functions have a *scope* that is *local* to the function, and this means that they cannot be accessed from outside the function, for example, from the command window or from other functions.

MATLAB allows the definition of *recursive* functions. Recursive functions are functions that make calls to themselves. Recursive functions need a stopping condition to prevent them from calling themselves infinitely.

The `error` and `warning` commands can be used to report errors within functions, for example, by checking that the input arguments have the correct size or data type.

4.13   Further resources

Look at the *File Exchange* section of the MathWorks web site (the company that develops MATLAB). You will find a large number of useful MATLAB functions that have been written and shared by other people. This can be an extremely valuable time-saving resource. The URL is: http://mathworks.com/matlabcentral/fileexchange/.

4.14   Exercises

■ Exercise 4.1

Write MATLAB code to solve the following problems: *04.A, 04.B*

1. Write a function called `mycube` to compute and return the cube of a number provided as an argument.
2. Call this function several times from the command window using different input argument values.
3. Write a script to call the function twice with different input arguments, and assign the results to two different variables.
4. Extend the script to read a number from the user, and supply this as an argument to the function. The script should display the result of the calculation to the command window.   ■

■ Exercise 4.2

04.A   Identify problems and errors in the following functions that relate to their input and output arguments.

1. `func1`:

```
function f = func1(x, y)
x = 3 * y;
end
```

2. `func2`:

```
function f = func2(a, b)
f = sqrt(3 * a);
end
```

3. `foo`:

```
function [a, b] = foo(x, y)
b = a + x + y;
end
```

4. `findSquareRoot`:

```
function findSquareRoot(x)
% Usage : findSquareRoot(x)
%    Provide the square root of the input
returnValue = sqrt(x)
end
```

■

■ Exercise 4.3

04.A   In Exercise 2.14, you wrote a script *m*-file to compute the injury severity score (ISS), given an array of 6 abbreviated injury scale (AIS) values. Modify your solution so that it uses a *function m*-file, rather than a *script m*-file. Specifically, the function should take an array of 6 AIS values as its only input argument, and produce the ISS value as its output. As described in Exercise 2.14, the ISS is computed from the AIS values as follows:

- Find the three highest AIS values in the array.
- Square their values.
- Sum the squares.

Also, write a short *script m*-file that defines the AIS array `[3 0 4 5 3 0]` and contains a call to your function using this array as an argument. The correct ISS value for the array should be 50.

■

■ Exercise 4.4

04.E   Modify your solution to Exercise 4.3 so that the function *m*-file is contained in a different location in the file system. For example, make a subdirectory of the directory containing your script *m*-file, and move the function *m*-file into it. Then, modify your script so that it is still able to find the function *m*-file. ∎

■ Exercise 4.5

Examine the code that follows which is saved in the file *fn1.m*.   *04.A, 04.C*

```
function x = fn1(y)

x = fn2(y + 2);
disp([x,y]);

end

function x = fn2(y)

x = y^2;
disp([x,y]);

end
```

The file is available to you through the book's web site.

Without using MATLAB, predict what the output of the function would be if the following call was typed in the command window:

```
>> fn1(10);
```

Verify your prediction by downloading the function *m*-file and making the call. ∎

■ Exercise 4.6

A function takes three numeric arguments representing the coefficients of a   *04.A, 04.D*
quadratic equation and returns two values, which are the two roots of the equation. The name of the function is quadRoots.

1. Write a suitable prototype or definition for the function. Add an appropriate help and usage section at the top of the function, and save it in a suitably named file.
2. Add a test to check if the quadratic roots are complex and to return with an error if this is true (see Section 4.3, doc error or help error).
3. Complete the function quadRoots and test that it gives the correct results.
4. Modify the quadRoots function so that it accepts a single array containing all three coefficients and returns a single array containing the roots. ∎

■ **Exercise 4.7**

04.A, 04.B, 04.D

In Exercise 2.9, you wrote a script *m*-file to automatically identify troughs in gait tracking data. As well as identifying troughs in such data, it can be useful to identify peaks too. Recall that a trough can be defined as a value that is less than the value(s) before and after it. Similarly, a peak can be defined as a value that is greater than the value(s) before and after it.

First, convert your code from Exercise 2.9 to use a function *m*-file. The function should take a single argument, representing the array of values to be searched, and return two values: an array of indices of peaks and an array of indices of troughs. The main script *m*-file should load in the same data file that you used in Exercise 2.9 (*LAnkle_tracking.mat*). Display the plot of *z*-coordinates over time as in Exercise 2.9, and then call the new function to determine the peaks and troughs. Finally, the script should use the indices returned to determine the timings of all peaks and troughs.   ■

■ **Exercise 4.8**

04.A, 04.B, 04.D

X-ray computed tomography (CT) scanners produce images whose intensities are in *Hounsfield units* (HU), named after the inventor of the CT scanner, Sir Godfrey Hounsfield. The HU scale is an approximately linear transformation of attenuation coefficient (AC) measurements into one in which the radiodensity of distilled water at standard pressure and temperature is defined as zero HU, while the radiodensity of air is defined as −1000HU. When reconstructing positron emission tomography (PET) images the reconstructions need to be corrected for the effects of attenuation using the attenuation coefficients. Therefore there is a need to convert from HU to AC. The figure below illustrates such a relationship, reported in [4].

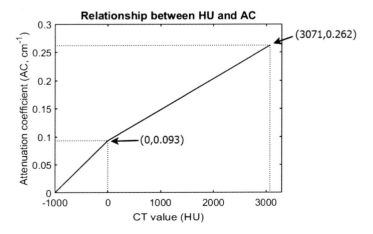

Using the relationship illustrated in the figure, write a MATLAB function to convert HU to AC. Your function should check if the HU value is between −1000 and 3071, and generate an error if it isn't. Write a script *m*-file to read a HU value from the user, convert it into AC, and display the result. ■

■ Exercise 4.9

A company is trialing a portable personal device that monitors a range of clinical indicators. One of these is heart rate, and the company would like to embed some software in the device to raise an alert if a sudden change occurs in the heart rate.

04.A, 04.B, 04.D

The file *heart.txt* contains sample heart rate data, acquired once per minute from a patient who was wearing the new device. Your task is to write a MATLAB script *m*-file that reads these data into an array, and searches for any sudden changes in heart rate.

A sudden change is defined as a heart rate that is more than 20% different (i.e., larger or smaller) than the average of the previous five heart rate values. Therefore you should only start to look for sudden changes from the 6th measurement onwards. When your program identifies a sudden change it should display a warning such as:

```
Warning - abnormal increase in heart rate at 123 minutes
```

As the heart rate measurements are made once per minute you can simply use the array index of the abnormal value as the time of the sudden change. Note that the program should indicate whether the heart rate change was an increase or a decrease. All sudden changes within the array should be reported. If no sudden changes are detected the following message should be displayed:

```
No abnormal changes detected
```

As part of your solution you should define a function *m*-file that takes an array of heart rate values as its only input argument, returns as its output an array of the indices of all sudden changes, and also values indicating whether each change was an increase or decrease. If the function is provided with an empty array it should exit with an appropriate error message.

A separate script should call your function and should also take responsibility for reading the data (see Section 1.9) and printing the output message(s). ■

■ Exercise 4.10

04.A, 04.B

Many medical imaging modalities acquire imaging data in a number of acquisitions over time. *Gating* is a common technique that aims to reduce artifacts in the resulting images that are due to cyclic motion (such as heart beats). The approach involves accepting only data that were acquired at approximately the same motion position and discarding the rest. One way to decide whether to accept data or not is to use a simple signal that is related to the motion of interest (for example an electrocardiographic signal, or ECG). The percentage of data accepted, out of all those acquired, is known as the *scan efficiency*.

A *bellows* can be used to generate a signal for *respiratory gating* of magnetic resonance (MR) data. It is an air-filled bag that is strapped around the chest. As the subject breathes, the air moving into and out of the bag is measured by a sensor. The file *bellows.mat* contains a set of respiratory bellows data.

For respiratory gating using the bellows signal, we need to determine a pair of signal values: a low value and a high value. Only data with a signal between these values will be accepted. The low and high values define the *gating window*.

The bounds of the gating window are typically determined using the signals acquired during a short "preparation phase." One way to set the high value is simply to use the maximum value observed during the preparation phase. The low value then needs to be set so that the percentage of data accepted during the preparation phase is equal to the desired scan efficiency.

Write a MATLAB function *m*-file to determine the bounds of a gating window, given an array of bellows data acquired during a preparation phase. Write a short script *m*-file to read in the data in *bellows.mat*, and call the function treating the first 5000 measurements as the signals from the preparation phase.

Once the gating window has been determined, your script should compute the indices of all elements in the full bellows array that are within the gating window. The script should then compute and display the *final* scan efficiency, i.e., the percentage of the full set of bellows signal values that lie within the gating window. ■

■ Exercise 4.11

04.A, 04.F

The Fibonacci sequence is $1, 1, 2, 3, 5, 8, \ldots$. It is defined by the starting values $f_1 = 1$, $f_2 = 1$, and after that by the relation $f_n = f_{n-1} + f_{n-2}$. The Fibonacci sequence has many applications in a wide range of fields, including in biology [5].

Write a *recursive* function to find the nth term of the Fibonacci sequence. Use the function prototype that follows, and ensure that the code has a help section and good comments.

```
function f = fibonacci(n)
```

■

■ Exercise 4.12

A good example of working recursively is given by the evaluation of a polynomial.

04.A, 04.F

A polynomial is defined by its coefficients. As an illustration, assume we have a cubic polynomial with the coefficients 2, 3, 7, 1, i.e., the polynomial $2x^3 + 3x^2 + 7x + 1$. If we evaluate this expression directly it requires 3, 2, and 1 multiplications for the first three terms, and three additions to give a total of nine operations. We can, however, write the evaluation in a recursive way that requires fewer operations: If we take out successively more factors of x from the terms that contain them we get the following:

$$2x^3 + 3x^2 + 7x + 1 = x(2x^2 + 3x + 7) + 1$$
$$= x(x(2x + 3) + 7) + 1$$
$$= x(x(x(2) + 3) + 7) + 1$$

The last line shows how the polynomial may be evaluated with three additions and three multiplications, giving a total of six operations, three fewer than the direct approach. This illustrates one of the benefits of working recursively.

The above suggests how a function to evaluate a polynomial may be written recursively. An evaluation of the above polynomial could therefore be written as a sequence of calls to a function f that starts off by taking an array containing the coefficients.

$$2x^3 + 3x^2 + 7x + 1 = f([2, 3, 7, 1], x)$$
$$f([2, 3, 7, 1], x) = x\, f([2, 3, 7], x) + 1$$
$$f([2, 3, 7], x) = x\, f([2, 3], x) + 7$$
$$f([2, 3], x) = x\, f([2], x) + 3$$
$$f([2], x) = 2$$

Write a *recursive* MATLAB function to evaluate a polynomial with a given set of coefficients and a given x value. The prototype of your function should be

```
function y = my_poly_value(coeffs, x)
```

Make sure that the code is well-commented, and that there is a help section at the top. Also ensure that you include a stopping condition for when the array of coefficients contains a single element.

Use your function to find the value of y when $x = -2$ for the polynomial $x^4 - 2x^3 - 2x^2 - x + 3$. You might also want to check that your function agrees with the built-in function `polyval`. Type `doc polyval` for details of how to use this function. ∎

FAMOUS COMPUTER PROGRAMMERS: BETTY JEAN BARTIK

Betty Jean Bartik (née Jennings) was a mathematician who worked at the University of Pennsylvania in the 1940s. She was one of a team of six women who were responsible for programming the ENIAC, which was the world's first electronic digital computer. Along with her colleagues Kay McNulty, Betty Snyder, Marlyn Wescoff, Fran Bilas, and Ruth Lichterman, Betty Jean Bartik programmed the ENIAC by physically modifying the machine, i.e., moving switches and rerouting cables, rather than using a symbolic language, such as those that we are used to today.

Although the way they programmed the ENIAC was very different to what we understand today by computer programming, these "ENIAC girls" pioneered some of the features of the programming languages we use now, such as subroutines (i.e., functions) and nesting. Arguably they invented the discipline of programming modern digital computers.

However, for a long time these women did not receive the recognition they deserved. Greater credit was given to the team that built the computer's hardware, rather than the team that programmed it. It was only many years later that historians of computing began to appreciate the contributions of these women. In 2008, two years before her death, Betty Jean Bartik received a Pioneer Award from the IEEE Computer Society and became a fellow of the Computer History Museum in California, USA.

"I was told I'd never make it to VP rank because I was too outspoken. Maybe so, but I think men will always find an excuse for keeping women in their 'place.' So, let's make that place the executive suite and start more of our own companies."

Betty Jean Bartik

CHAPTER 5

Program development and testing

LEARNING OBJECTIVES

At the end of this chapter you should be able to:

O5.A Incrementally develop a MATLAB® function or program
O5.B Test each stage of the incrementally developing function or program
O5.C Identify common types of error that can occur in MATLAB functions and scripts
O5.D Debug and fix errors in scripts and functions
O5.E Use MATLAB's built-in error handling functions

5.1 Introduction

Most tasks in realistic computer programming settings involve writing substantial amounts of code. Such programs typically consist of functions or libraries that can call each other in complex ways. Furthermore, tackling large computer programming tasks can often involve more than one person collaborating and working on the same code. Therefore there is a need to work methodically and break a large project down into manageable and testable portions.

We already saw some of the debugging and code analysis tools that MATLAB has to offer (Section 1.14), and other languages such as C++, Python, and Java also have similar debugging and analysis tools. In this chapter, we will revisit these tools and discuss a general approach for developing and implementing a significant programming project.

5.2 Incremental development

Working on the principle of breaking a large task down into smaller, more manageable, and (importantly) testable sub-tasks, the process of *incremental development* can be useful when trying to solve a difficult task or model a complex system using a computer program.

When programming something substantial, it is better to avoid trying to take a "big bang" approach to coding. We will generally fail if we try to solve it all at once.

123

General tips for incremental development are:

- Write code in small pieces.
- The pieces may well be incomplete, but should work without error.
- Don't try and write everything in one massive function in one go.

For example, if the program needs to perform some user interaction, a typical incremental development process might involve the following:

- Start off with just enough code to display output.
- Run that part, and check it.
- Then write a bit more code to ask the user for input.
- Check the new part by printing to the screen what the user provides as input.
- Add a bit more code to actually do something (simple) with the user's input.
- Check the "doing something simple part," perhaps by again printing out its result.
- Add a little more code to do something a bit more than the simple stuff already done.
- Continue these steps of adding a bit more code and checking it works until we have a final working version of the program.

This kind of iterative way of working with frequent checking is important. Adding a large amount of code before checking can be difficult. First, the program itself may not run. Second, it may give incorrect output. In either case, if a large amount of code was introduced since the last check, it will be more difficult to find where the bug or error is.

The bottom line is that we should aim to always have a version of our program that runs without error.

This chapter tries to illustrate this principle by example. We first describe the task in the next example, and, after that, one way to take the incremental steps of building and testing some code to carry out the task is given.

■ Example 5.1

05.A, 05.B, 05.C, 05.D

This example will look at data representing measurements of heart contraction (see Example 1.3). In a healthy heart, all parts of the cardiac muscle (the *myocardium*) will contract in time with each other. *Cardiac dyssynchrony* occurs when there are differences between the timings of the contractions of different parts of the myocardium. Such differences may indicate the presence of heart disease.

One form of dyssynchrony is *intraventricular dyssynchrony*, which occurs when the timing differences are between the contractions of different segments of the myocardium of the left ventricle of the heart. A common way of defining the different parts of the left ventricle, to assess for dyssynchrony, is to use the American Heart Association's (AHA) 17-segment model.

Using echocardiography or a magnetic resonance (MR) scanner, it is possible to measure the radial displacement of each segment over time, and these measurements form the starting data for the analysis. There are a number of ways of measuring intraventricular dyssynchrony, and one method is to use the standard deviation of the *time-to-peak* radial displacement of the AHA segments. Using the radial measurements, the time between the start of the cardiac cycle and the peak radial displacement is determined for each segment. The standard deviation of these times is the indicator of dyssynchrony.

The task is to write a MATLAB function to compute the standard deviation of time-to-peak displacements. The input to the function should be an array containing radial displacement data broken down by AHA segment. The array for a patient contains 17 columns, one for each segment. Each column contains the displacement measurements for the corresponding segment. Sample input arrays for three patients are provided in the file *radial_displacements.mat*, which can be used to test the function. Loading this file will create arrays with names `patient1`, `patient2`, and `patient3`. The *temporal resolution* (i.e., the time between displacement measurements) of each segment's data is 30 ms.

The function should first determine the index of the maximum radial displacement for each segment. As the temporal resolution is known, this index can be converted to give the time-to-peak value for the segment. When the times to peak are calculated for all segments, their standard deviation can be computed and returned as the output of the function.

A script should load the test data and, for each patient, it should call the function, display the standard deviation value, and display whether or not the patient has "significant" dyssynchrony. We consider a patient to have significant dyssynchrony if their time-to-peak standard deviation is greater than 60 ms. ∎

Step 1

In keeping with the principles of incremental programming, we start by writing a very small and basic script:

```
clear; close all;

% load data
load('radial_displacements.mat');

% display results
disp('Inside the main script!');
```

Note that this script does very little of what we ultimately aim to do. We can save this in the file *main_script.m*.

Test 1

It does not matter that the script does very little, we should still run this script from the command window to test it:

```
>> main_script
Inside the main script!
```

All our script does so far is to clear the workspace, close any open figures, load in the measurement data, and give a dummy message. Once we are happy that it does this, we can move on to the next step.

Step 2

Now we start writing the function that will do the work. We can use the editor to create a file *std_dev_ttp.m*. We begin with the following:

```
function ttp_SD = std_dev_ttp(rad)

% Standard deviation of time-to-peak, return value.
ttp_SD = 100;

end
```

It is okay that our function does not yet do what we want; the function still has some important features. It has the correct *prototype*, i.e., it expects an argument of radial displacements as input, and says that it will return an output value indicating the required standard deviation (even though this is currently hard-coded to a "dummy" value of 100).

Now that we have written a (very limited) function; we can adjust the main script to call it. We can pass one of the arrays loaded from the data file to the function and collect the result that is returned:

```
clear; close all;

% Load data
load('radial_displacements.mat');

% Call the function.
sttp1 = std_dev_ttp(patient1);

% Display results
disp(['  Std dev of time-to-peak displacement = ' num2str(sttp1) ...
    'ms']);
```

Test 2

Now we can check the output of the code by once again calling the main script at the command window:

```
>> main_script
   Std dev of time—to—peak displacement = 100ms
```

The test gives the output we expect: the function is called by the main script, and it is currently hard-coded to return a value of 100 ms. This returned value is then displayed as was required.

Step 3

Now we can add a little more functionality to our main script. We could include a check to see if the dyssynchrony measure is safe or not, and display an appropriate message. We do this by modifying *main_script.m* as follows:

```
clear; close all;

% load data
load('radial_displacements.mat');

% Call the function.
sttp1 = std_dev_ttp(patient1);

% display results
disp(['  Std dev of time—to—peak displacement = ' num2str(sttp1) ...
     'ms']);

if (sttp1 < 60)
    disp('  Patient is dyssynchronous');
else
    disp('  Patient is not dyssynchronous');
end
```

Here we have added a few lines of code at the end (containing an `if-else` clause) to check if the value is above 60 or not.

Test 3

We test the code as before:

```
>> main_script
   Std dev of time—to—peak displacement = 100ms
   Patient is not dyssynchronous
```

Here we can see that the code does *not* work as expected; we have found a bug! The function currently is hard-coded to return a value of 100 ms, which is above the limit of 60 ms for determining dyssynchrony, but the message displayed says that the patient is not dyssynchronous. To fix this, we only need to look back over the (small) amount of code we just added. On reviewing the if-else, it is clear that the test if (sttp1<60) is the cause of the problem, and the bug can be fixed easily by re-writing it as if (sttp1>60) instead.

A second quick and easy test at this stage is to modify the hard-coded value that the function returns, for example, to a value of 30 ms. We modify *std_dev_ttp.m* as follows:

```
function ttp_SD = std_dev_ttp(rad)

% Standard deviation of time—to—peak, return value.
ttp_SD = 30;

end
```

and then re-run the code:

```
>> main_script
  Std dev of time—to—peak displacement = 30ms
  Patient is not dyssynchronous
```

This quick test also confirms what we expect (having fixed the bug in the main script as described above).

Step 4

In the steps so far, we have only really built a skeleton of the program we want. Now we can begin to think about the "core" functionality that is required. We can modify the function *m*-file so that we find the peak radial displacement for each segment. To do this, we will need to loop over the array of radial displacement measurements rad. Recall that there are 17 segments in the AHA definition, and the array has a column of measurements for each segment. We need to find the index of the largest value (peak) in each column. We add code as follows:

```
function ttp_SD = std_dev_ttp(rad)

nSegments = 17;

% Loop over the 17 segments and compute time—to—peak
% displacements for each one.
for seg = 1:nSegments
   [peak, peak_index] = max(rad(:,seg));

end

% Standard deviation of time—to—peak, return value.
ttp_SD = 30;

end
```

Here, we have added a loop to find the index of the maximum value for each segment (column). We are not actually using this information yet, but it is a small incremental addition of code that will help us later.

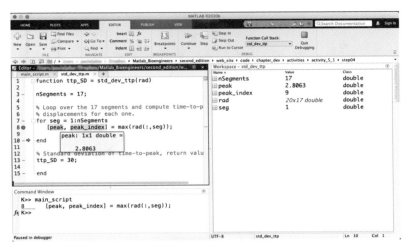

FIGURE 5.1
Using a breakpoint to inspect program variables.

Test 4

One way to check that the new code is working is to place a breakpoint in the loop that was added. Then we can call the main script and inspect variables while the code is running, as shown in Fig. 5.1.

Here, we have added a breakpoint at the end of the loop, and the code has stopped in the first iteration (see the value of the loop variable seg in the workspace window on the right). We can check variables in the code by hovering the mouse over them. In the above, we can see that the variable peak_index has a value of 9. How can we test if this index for the peak is sensible? One way is to look at the contents of the current column in the radial measurements array, for example, by simply typing it at the command window:

```
>> rad(:,1)'
ans =
  Columns 1 through 6
        0    0.1031    0.7285    1.3087    1.7075    2.2302
  Columns 7 through 12
   2.5135    2.7005    2.8063    2.7488    2.4591    2.1285
% .... etc.
```

Here we can see that, at index 9, the value is 2.8063, which is indeed the highest. If the array we want to inspect is very large, displaying it in the command window is inconvenient, and an alternative would be to use the plot command to visualize the data, i.e.,

```
>> plot(rad(:,1))
```

The resulting plot is shown in Fig. 5.2; it confirms that the index we have found (9) corresponds to a peak.

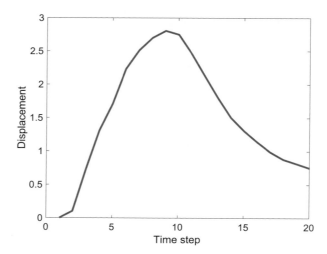

FIGURE 5.2

A plot of radial displacements for a segment of the left ventricular myocardium for the first patient (see Test 4 in the main text).

The remaining steps

To complete the task, we have a few more items to add to our function's code:

- Usage and help text for the function. This can be done at any stage and could have been done right at the start.
- Code to calculate the *time* of a peak from the index that we have found so far.
- A variable to store the times for *all* the segments.
- A line or two of code to find the standard deviation of these times.

All of these items can be added incrementally in the same way as we have been working above. After each step, a quick test can be carried out to check (a) if it runs and (b) gives sensible results/values.

■ Activity 5.1

05.A, 05.B

Complete the task above so that the code includes the items remaining. Use the same incremental add-code-then-test approach described above.   ■

5.3   Are we finished? Validating user input

Once the code for the dyssynchrony task is finished, we could decide that the task is complete, as we have a working function that achieves its initial aim. However, in another sense, we may still have work to do, because the function might need to be made more *robust*. One way of making code robust is to protect it against possible misuse.

Validating user input is an important part of programming; one of the main reasons for program failure is when a user (or another part of the program) gives input to a function or program that does not match what the programmer expected. We have already seen in Section 4.3 that MATLAB provides the built-in `error` and `warning` commands for reporting such situations once they are detected.

In the case of Example 5.1, we could still check to see if a sensible array has been provided as input, for example, it should not be an empty array perhaps. The array should also be consistent with the AHA heart segment model; in other words, it should have exactly 17 columns.

For example, if we want to ensure that the number of columns is correct, then we can introduce the following code near the top of the function:

```
function ttp_SD = std_dev_ttp(rad)
% ... Usage and help text here

nSegments = 17;

if (size(rad,2) ~= nSegments)
  error('std_dev_ttp: Expecting exactly 17 columns in the input ...
    array.');
end

% ... rest of function as before.
```

If the function receives an array that does not have 17 columns, it now prints out an error message and returns without continuing to the rest of the code.

Notice that the error message begins by giving the name of the function where the error is raised. As we noted in Section 4.3, this is good practice, as programs can use multiple functions, and it makes identifying the locations of errors easier.

5.4   Debugging a function

The basics of the MATLAB debugger were described in Section 1.14. Here, we consider some of the types of errors that the debugger can be used to help identify and fix.

When writing a program or function that is non-trivial, the chances of it working perfectly the first time are quite low, even for a skilled MATLAB user. Debugging tools are there to help identify why a program does not work correctly or as expected.

When debugging, we need to work systematically, and run our program with various sets of test data to find errors, and correct them when we find them. If we work incrementally, as we did in Example 5.1 above, then the repeated nature of the testing allows errors to be caught earlier, which reduces the chance of later errors occurring or minimizes their severity.

There are three main types of error that a function or program can contain:

- Syntax errors.
- Run-time errors.
- Logic errors.

The term "syntax" relates to the *grammar* of the programming language. Each language has a specific set of rules for how commands can be written in that language, and these rules are collectively known as the language's syntax.

When a line in a MATLAB function contains a syntax error, the built-in *Code Analyzer* (see Section 1.14.2) should highlight in red the corresponding line of code. Hovering the mouse over the line should result in a message being displayed to the user that describes what MATLAB has decided is the particular syntax error in the line. In other words, MATLAB can detect an error with the script or function before it is run.

■ Example 5.2 (A syntax error)

05.C, 05.D   If the code that follows is entered into a script, say *my_script.m*, using the editor, then the line containing the `if` statement should be highlighted in red to indicate a problem with the syntax.

```
clear, close all

randVal = rand;

if randVal > 0.5
  disp('Greater than 0.5')
```

This is because the syntax rules of MATLAB require a closing `end` statement for each `if` statement.

If we fail to notice this error, and attempt to run the script from the command window, we will get an error when MATLAB reaches the relevant part of the code, and a more dramatic message will be displayed:

```
>> my_script

Error: File: my_script.m Line: 5 Column: 1
At least one END is missing: the statement may begin here.
```

In this case, clicking on the part of the error report with the line number will take us directly to the corresponding place in the script file so that we can figure out how to correct the error. In this case, adding an end statement after the disp call should be enough. ∎

■ Activity 5.2

Type the code from the previous example into a script. Confirm that the code analyzer highlights the error. Fix the error and confirm that the highlighting disappears. ∎

05.C, 05.D

■ Example 5.3 (Another syntax error)

The code that follows contains a syntax error. The MATLAB code analyzer should highlight it in red. The code begins by generating a single random number between 1 and 10, and then it seeks to decide if the random number equals three or not, adjusting its output in each case.

05.C

```
clear, close all

a = randi(10, 1);

if a = 3
  disp('Found a three!')
else
  disp('Not a three')
end
```
∎

■ Activity 5.3

Type the code from the previous example into a script. Confirm that the code analyzer highlights the error. Fix the error, and confirm that the highlighting disappears. ∎

05.C, 05.D

The reason for this syntax error is that the = sign used in the if statement is an *assignment operator* and should not be used to test for *equality*. The correct symbol to use here is the *comparison operator*, which is == in MATLAB, i.e., two equals signs, not one (see Example 2.2). This is also the case in quite a few other languages, and it is very common to mix up these operators.

■ Example 5.4 (A run-time error)

Some errors cannot be determined by MATLAB while we are editing the code and can only be found when the code is run. This is because a specific sequence of commands needs to be run to bring about the conditions for the error. Consider the code that follows in a script called *someScript.m*. It aims to generate three sequences of numbers, and assign them to array variables a, b, and c, then it tries to find the sums of each of these sequences (line numbers have been shown here to help the description).

05.C

```
1    clear, close all
2
3    a = 1:10;
4    b = 2:2:20;
5    c = 3:3:30;
6
7    sumA = sum(a);
8    disp(sumA)
9
10   sum = sum(b);
11   disp(sum)
12
13   sumC = sum(c);
14   disp(sumC);
```

In this case, there are no syntax errors, but running the script at the command window leads to a run-time error as follows:

```
>> someScript
   55

   110

Index exceeds matrix dimensions.

Error in someScript.m (line 13)
sumC = sum(c);
```

■

■ Activity 5.4

05.C   Type the code from the previous example into a script. Run the script from the command window to reproduce the error.   ■

The problem with the code in Example 5.4 is related to the choice of variable names. The sums of the arrays a and c have been given sensible names (sumA and sumC). The sum of the array variable b has been stored in a variable called sum. This is a problem, because the local script has used the name of a *built-in MATLAB function* as a name for a local variable.

Using the name of a built-in function for a variable that we use locally is called *shadowing*. We can check if a name is already used by MATLAB or in the workspace using the exist command. In this case, we can trivially check

```
>> exist('sum')
ans =
     5
```

The return value of 5 indicates that the name sum exists and is attached to a function (as opposed, say, to a file). Type help exist at the command window for more details. We can check a more sensible name as follows:

```
>> exist('sumB')
ans =
     0
```

The value of 0 returned indicates that the name `sumB` is not being used, and we can safely use it in our code.

Returning to the code in Example 5.4, when the execution reaches line 13, the programmer's intention was to use the built-in MATLAB function called `sum`, but this now refers to a local variable that contains a single number. The use of the brackets in line 13 is now interpreted as a request for an element or elements taken from an array called `sum`. The elements requested are at the indices specified by another variable, `c`. That is, the indices requested start at 3 and go up to 30 in steps of 3. The problem however, is that the variable called `sum` is just a single scalar value, i.e., it as an array of length one. It is not possible to obtain an element from it at index 3 or higher. This is the reason for the code crashing at run-time.

Apart from giving an example of a run-time error, this example shows why it is very important *not* to give names to variables that are already used to refer to built-in functions. MATLAB will not complain when you do this, so be careful!

■ Example 5.5 (A run-time error)

Here is some more code that gives a run-time error: *05.C*

```
clear, close all

a = 60 * pi / 180;

b = sin(a + x)
```

In this case, when the code is run, the following output is given:

```
>> someScript

Unrecognized function or variable 'x'.

Error in someScript.m (line 5)
b = sin(a + x);
```
■

■ Activity 5.5

Type the code from Example 5.5 into a script. Run the script from the com-   *05.C*
mand window to reproduce the error. Can you see why this error occurs? ■

The script in Example 5.5 calls `clear` at the start. This means that even if the workspace contained a global variable, called `x`, before calling the script, then it would have been cleared and would no longer be available. Hence, the request to calculate `sin(a + x)` will always fail, as there is no variable called x. This would need to be defined before the point where the `sin` function is called to make the code work (but after the `clear`).

■ Activity 5.6

05.D   Fix the code from Example 5.5 by introducing an appropriate local variable and assigning a value to it. Run the script to check it works.   ■

Logic errors can be some of the hardest to find, and identifying them often requires using the debugger to carry out careful line-by-line inspection of the code as it is running. This is because no error is reported by MATLAB, and the code analyzer does not highlight anything either. However, when the code is run, an incorrect output or behavior results, and MATLAB will generally not provide a clue as to why this happened.

■ Example 5.6 (A logic error)

05.C   In this example, the code that follows attempts to show that $\sin\theta = \frac{1}{2}$ when $\theta = 30°$.

```
clear, close all

theta = 30;
value1 = sin(theta);
value2 = 1/2;

fprintf('These should be equal: ')
fprintf(' %0.2f and %0.2f\n', value1, value2)
```

When we actually run the code, we get the following output with no error reported:

```
>> someScript
These should be equal: -0.99 and 0.50
```
■

This shows two numbers that are clearly not equal, so what has gone wrong?

■ Activity 5.7

05.C, 05.D   Type the code from Example 5.6 into a script, and run it to reproduce the same output.

Can you see why the output is incorrect? Place a breakpoint in the code, and run the script again. Check the values of the variables to see if there is a clue.   ■

To track down the reason for the behavior in Example 5.6, we need to carry out some further investigation. In the command window, we can look at the help on the `sin` function, which we might suspect of being broken:

```
>> help sin
 sin    Sine of argument in radians.
    sin(X) is the sine of the elements of X.
```

This shows that the built-in MATLAB trigonometric function `sin` expects its argument in *radians,* so that is what we should give it. The function when used above gave the *correct result* for the sine of 30 radians, not degrees. That is, there was a mismatch between our expectation of the function and what it was designed to do. This error is readily fixed in a number of ways, for example, by replacing `theta = 30;` with `theta = 30 * pi / 180;` where we have made use of the built-in MATLAB constant `pi`.

■ Activity 5.8

Fix the code in Example 5.6, and confirm that $\sin 30°$ does indeed equal 1/2!   ■   *05.C, 05.D*

Because no error message is given, logic errors can be difficult to detect. Sometimes we might not even notice that the program or function is behaving wrongly. Even when we suspect a logic error exists, by noticing something incorrect in the program's behavior, it can still be difficult to pinpoint where the error is. There is no error message helpfully pointing us to a specific line in the code, as in the case of syntax errors. Worse still, the error will not necessarily be restricted to just one line; it can arise from the way different sections of code interact.

The error in the previous example was due to the intention to use degrees where radians should have been used instead. This is only noticed by observing the actual values of the output and seeing that the numeric value is not correct; in other words, by careful inspection of the output.

One way to identify logic errors in a function is to use the MATLAB interactive debugger to *step through the function line-by-line,* perhaps with a calculator or pencil and paper in hand (see Section 1.14).

Another way is to take each line of the code and copy-and-paste it into the command window, executing each line at a time, and assessing its result to see if it is right.

Putting lots of statements in the code that display output and the values of the most relevant variables using the `disp` or `fprintf` commands can also help. Values of variables can be displayed before and after critical sections of the code, e.g., calls to other functions or important control sections, such as `if` clauses and `for` or `while` loops. Once we have checked and are satisfied that the function or program is working correctly, the extra output statements should be removed.

5.5 Common reasons for errors when running a script or a function

There are a few common reasons for errors when running a script or a function. These include the following:

- We use matrix multiplication when we meant to use *element-wise* multiplication. For example, the following code:

```
x = [1 2 3];
y = [6 7 8];
z = x .* y;
```

multiplies x and y *element-wise* to give a result of [6, 14, 24]. It is a common mistake to write `z = x * y`, instead of the above, which would lead to an error. In this case, it would lead to a run-time error, because it is not possible to matrix-multiply the values of x and y, as defined above (because the sizes don't match).
- The location of the code that we want to run is not in the MATLAB path, or some other code that it relies on is not in the path (see Section 4.8).
- We forget that MATLAB is *case-sensitive*, e.g., the variables x and X are different, as are `myVar`, `myvar`, and `MyVar`.
- We accidentally set off an infinite loop, for example,

```
count = 0; sum = 0;
while(count < 100)
  sum = sum + count;
end
```

will start, but never end, because the variable `count` never gets updated within the loop. The following is one way to fix this:

```
count = 0; sum = 0;
while(count < 100)
  sum = sum + count;
  count = count + 1;
end
```

- We have exceeded the memory available for our computer. For example, the command `rand(N)` returns a matrix of random numbers in an array of size $N \times N$. So, if we run the following code:

```
N = 100000;
R = rand(N);
```

then the computer could get very "tired" as it runs out of memory trying to create a $10^6 \times 10^6$ array. This corresponds to around 3 or 4 terabytes if four bytes are needed for each array element. This can be viewed as a run-time error.

5.6   Error handling

In the preceding sections, we discussed some of the kinds of errors that may occur when writing code. Some of these can be hard to predict, but it is still a good idea to write code in a way that anticipates the kinds of errors that might occur and tries to deal with them in advance to reduce any disastrous consequences. This can be done through *error handling*; we give some illustrations in this section of how this can be done.

In general, if there is a run-time error that is encountered when the program is run, then MATLAB itself will stop running the program and return. It will give an error message (see Example 5.5), but sometimes the error message that MATLAB gives can be uninformative, and we might want to make the error reporting more useful and specific.

5.6.1   The `error` and `warning` functions

We first introduced the `error` function in Section 4.3, and in Section 5.3, we saw a way of using it while validating user input. This function, and the closely related `warning` function, can be useful when testing for correct behavior when code is running. Once an error is detected, we can use them to give tailored messages to the user.

■ **Example 5.7**

For example, we may have a function that is expected to carry out some processing and save a record of its activity to a given file. For this to work, the file may be required to already exist:

05.E

```
function processFile(logFileName)

if ( ~ exist(logFileName, 'file') )
  msg = sprintf('No file: %s. Bailing out!\n', logFileName);
  error('MATLAB:processFile:fileMissing', msg)
end

% further code below
```

This code first tests to see if the log file with the given name exists. If it does not, a specific error message saying this is generated using the `sprintf` function. The message is contained in a character array variable called `msg`, which is passed as the second argument to the `error` function. This is similar to the use of `error` we saw in Section 4.3. However, here we also have an extra argument. The first argument is a unique character array, which identifies the location and nature of the error call: although not required, it is a good idea for each call to `error` to have a unique identifier. In the code above, if a file with the given name does not exist, then execution will transfer to the error function, and MATLAB will return and halt processing the script and any that have called it. Any further code below this point will not be run. ■

■ Example 5.8

05.E

In the above example, we could be a little less strict and decide that if the log file is not present, we should still carry out the processing, but warn the user that there will be no logging carried out. This is where the warning function is useful: the above code can be modified slightly to do this.

```
function processFile(logFileName)

if ( ~ exist(logFileName, 'file') )
  msg = sprintf('No such file: %s. No logging.\n', ...
                logFileName);
  warning('MATLAB:processFile:fileMissing', msg)
end

% further code below
```

In this case, if the log file does not exist, a warning message is displayed, but the code below the warning is still executed (although in this example, without logging). ■

In other words, both the error and warning functions can be used to report a problem to the user, but the error function then returns and stops running the script or function file.

5.6.2   The try and catch method

The error and warning functions described above can be used when we anticipate a particular type of problem that can lead to an error. Sometimes we may want to prepare for a possible error, even if we do not know in advance what its type will be.

One phrase to describe a program encountering an error is to say that the program has *raised an exception*. The exception itself can be raised for a number of reasons, which include

- Trying to access an element of an array that is beyond the last element it contains.
- Trying to open a file that does not exist.
- Performing a type error, such as trying to perform an arithmetic operation on a variable that is not numeric.

If we have a section of code that might raise an exception, we can wrap this code inside a *try-catch clause*. This a little like an *if-then* clause, where a section of code is "tried" and, if it raises an exception, this exception is then "caught" by the second part of the clause.

■ Example 5.9

As a simple and artificial example of the use of a try-catch clause, the follow-   *05.E*
ing is adapted from the MATLAB documentation:

```
A = rand(3);
B = ones(5);

try
    C = [A; B];
catch

    % handle the error
    msg = ['Dimension mismatch: First argument has ', ...
            num2str(size(A,2)), ' columns, second has ', ...
            num2str(size(B,2)), ' columns.'];
    error('MATLAB:try_catch_ex:dimensions', msg);

end  % end try/catch
```

The code within the `try` section tries to concatenate two arrays vertically, placing A above B, and then store the result in a third array, C. If there is an error when doing this, then the execution of the program will immediately switch to the code inside the `catch` section.   ■

It is important to note that *any* error in the `try` section will lead to the code in the `catch` clause being run. This allows different possible reasons for the error to be tested inside the `catch` part of the clause.

5.7 Summary

This chapter has considered how we can incrementally develop a function or set of functions that can be used to solve a programming problem. We have looked at how it is important to make small steps, ensuring that the code runs at each stage, and testing its output to see if it behaves as we expect. We have also looked at common types of error, such as run-time errors, syntax errors, and logic errors, and what we can do to identify and fix them, i.e., to *debug* them. We have also briefly looked at how to handle errors in our code; testing for them explicitly or using try-catch clauses to wrap sections of code that might raise an error.

5.8 Further resources

- The MATLAB help is always a good place to start. From the main help menu for MATLAB, go to the section on *Programming* (see screenshot in Fig. 5.3) and browse the subsections.
- There is a very good MATLAB coding style set of guidelines by Richard Johnson at: https://mathworks.com/matlabcentral/fileexchange/46056-matlab-style-guidelines-2-0.

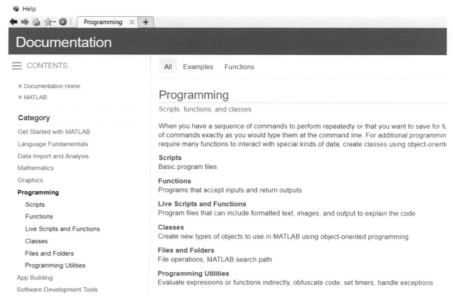

FIGURE 5.3
Accessing the MATLAB help on Programming.

- An informative paper on good practice when programming, with a more scientific perspective, can be found at the link: http://journals.plos.org/plosbiology/article?id=10.1371/journal.pbio.1001745.

5.9   Exercises

■ Exercise 5.1

O5.C, O5.D

The code that follows is intended to loop over an array and display each number it contains one at a time. It is quite badly written and needs to be corrected.

```
1   % Display elements of an array one by one.
2
3   % Array of values
4   myArray = [12, −1, 17, 0.5]
5
6   for myArray
7     disp(myArray)
8   end
```

1. Type the code into the MATLAB editor, and use the hints from the editor to locate any syntax errors. While there are syntax errors in the code, the MATLAB editor will not allow you to use breakpoints for

debugging. Once you have corrected any syntax errors, insert a break-point at the start, and run it. Step through the code to identify further possible errors, and fix them as you go along.

2. Identify and fix any style violations in the code, i.e., things that do not prevent the code from running, but can be changed to improve the appearance or behavior of the code. ∎

■ Exercise 5.2

Type the following code into the MATLAB editor (note that it contains errors):

05.C, 05.D

```
clear, close all

% Get 30 random integers between −10 and 10
vals = randi(21, 1, 30) − 11;

for k = 1:1:numel(vals)
  if (vals(n) > 0)
    sumOfPositives = sumOfPositives + vals(n)
  end
end

disp('The sum of the positive values is:')
disp(sumOfPositives)
```

1. Identify and explain what the different parts of this piece of code are trying to do. What seems to be the main aim of the code?
2. Find the errors and style violations in the code, and fix them using the debugger. ∎

■ Exercise 5.3

A function with errors in it is given below. It aims to evaluate a quadratic polynomial for a given x value or for an array of given x values. Type this function into a file, and save it with the correct name.

05.C, 05.D, 05.E

```
function [ ys ] = evaluate_quadratic(coeffs, xs)
% Usage: evaluate_quadratic(coeffs, xs)
%
% Find the result of calculating a quadratic
% polynomial with coefficients given at the x
% value(s) given. Return the result in the
% variable ys which must contain the same
% number of elements as xs.

clear
close all

ys = zeros(1, xs);
```

```
ys = a * xs^2 + b * xs + c;

end
```

Now type what follows into a script, and call it *testScript.m*. This will be used to test the function above.

```
% Test script for evaluate_quadratic.m
clear, close all

x = 3;
coeffs = [1 2 1];
y = evaluate_quadratic(coeffs, x);
fprintf('f(%0.1f) = %0.1f\n', x, y)

xs = -1:0.5:3;
ys = evaluate_quadratic(coeffs, xs);
disp(xs), disp(ys)
```

1. Try to run the script with the debugger, placing breakpoints in the main function. List the errors in the function, and suggest how they should be fixed.
2. Add code to the corrected function to validate the input arguments `coeffs` and `xs` to make sure that they can be correctly processed by the function. ■

The exercises that follow contain longer, more open-ended problem solving activities. There are many ways that they can be tackled, and you are encouraged to adopt the incremental programming style described in the chapter. Try to break the tasks of each problem down into multiple functions; these could perhaps be called by a single main function or script. Use the debugger to check for errors as you go along, and try to make use of the MATLAB error handling functions, such as `error` and `warning`, to guard against misuse of the code.

■ Exercise 5.4

O5.A, O5.B, O5.C, O5.D, O5.E

An *oscillatory* monitoring device automatically measures systolic and diastolic blood pressure, as well as heart rate. The device works by placing a cuff on the arm and inflating it to a pressure that blocks off blood flow through the artery. The pressure is gradually reduced until blood starts to flow again; the pressure at which this happens is the patient's systolic blood pressure. After the flow restarts, the pressure exerted by the cuff causes a vibration in the vessel walls, which can be detected by the device. The vibrations stop when the pressure falls below the patient's diastolic blood pressure. The vibrations are *oscillatory* (hence the name of the device), and the frequency of these oscillations gives an estimate of the patient's heart rate.

The file *blood_pressure.mat* contains a set of the device's measurements. Specifically the following:

- p: an array of pressure measurements (without the oscillations) in mmHg, which gradually decreases throughout the monitoring.
- peaks: an array of ones and zeros, in which a one indicates a peak in the oscillatory vibrations.

Both arrays are measured with a temporal resolution of 0.05 seconds (i.e., 20 measurements per second).

Write a MATLAB function *m*-file that takes these two arrays as input arguments and returns three values as the output of the function: the systolic and diastolic blood pressures and the heart rate. Write a separate short script *m*-file that reads in the data provided, calls the function, and displays the results.

Summary of the key points:

- The systolic blood pressure is the value of p at the index where the first non-zero value of peaks occurs.
- The diastolic blood pressure is the value of p at the index where the last non-zero value of peaks occurs.
- The heart rate is the number of heart beats per minute. One way this can be calculated is by first estimating the average time of a single heart cycle, i.e., the average time between non-zero values in the array peaks. ■

■ Exercise 5.5

Exercise 4.10 made use of data acquired with a respiratory bellows. In this exercise, we analyze these data further.

O5.A, O5.B, O5.C, O5.D, O5.E

The file *bellows_sub.mat* contains a subset of the original bellows signal. It contains the data acquired when the subject was taking deep breaths. The goal of this exercise is to automatically determine the average length of a respiratory cycle from this subset of the data.

The program should perform the following tasks:

- Load in the bellows signal data from the file.
- Determine the measurement times (in seconds) of the bellows data. To do this, use the fact that the bellows signals were acquired at a rate of 500 Hz. You can assume that the measurements start at a time of zero seconds.
- Smooth the signal data to eliminate peaks that are caused by noise. You can use the MATLAB smooth function for this purpose (part of

the curve fitting toolbox), and you will need to experiment with different levels of smoothing, which you can specify by changing the value of the second argument to `smooth` (the *span* of the averaging that is performed). See the MATLAB documentation for `smooth` for details.

- Produce a plot with the smoothed bellows signal on the *y*-axis and the measurement time on the *x*-axis. This plot will help you judge whether the smoothing level you chose in the previous step was sufficient.
- Find the measurement times of all *peaks* in the smoothed bellows signal. Recall that you wrote a function to perform this task in Exercise 4.7, so it would make sense to reuse the code you wrote then.
- Find the lengths of all *respiratory cycles* (i.e., the times between successive peaks), and compute the average of these times.

Your final program should display the plot of the smoothed bellows data, and print to the command window the times of all peaks and the average length of the respiratory cycles.   ■

■ Exercise 5.6

05.A, 05.B, 05.C, 05.D, 05.E

A research project wants to produce a portable monitoring device for measuring blood sugar levels. Normal blood sugar levels for adults are 4.0–5.9 mmol/L but may rise to 7.8 mmol/L after meals. A level that is higher than 7.8 mmol/L may indicate diabetes.

The data files called *sugar1.txt*, *sugar2.txt*, ... *sugar5.txt* contain measurements made by a prototype of the device. A program is required that reads in a single file and does the following:

- Reports the times and blood sugar levels of any measurements that were above 7.8 mmol/L or below 4.0 mmol/L.
- Reports the times and average blood sugar levels for any given 2-hour period in which the average blood sugar was greater than 5.9 mmol/L.

The measurements in the files were taken every 10 minutes over a period of 24 hours, i.e., there are 144 measurements in each file.

Develop a program to meet these requirements.   ■

■ Exercise 5.7

05.A, 05.B, 05.C, 05.D, 05.E

Monte Carlo simulations are a class of algorithms that use repeated random sampling to obtain numerical results. They have found practical use in a range of biomedical applications [6].

This exercise is about using a Monte Carlo simulation to estimate the value of π. The method uses random points chosen inside the unit square in the two-dimensional (2-D) Cartesian axes, i.e., $0 \le x \le 1$ and $0 \le y \le 1$. This is illustrated in the figure below, which shows part of the unit circle and random points in the unit square.

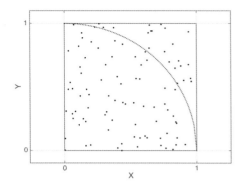

The method picks random points inside the unit square, and these are shown by dots in the diagram. Some of these points will lie inside the unit circle (which is also plotted), and some of the points will lie outside the unit circle.

We can estimate the area of the quarter circle by finding the fraction of points that are inside it. From this, we can estimate the area of the whole circle, and therefore the value of π. For example, if we pick 20 points inside the square, and 17 turn out to be also inside the unit circle, the estimate for the area of the circle (i.e., π) is $4 \times 17 \div 20 = 3.4$.

Write MATLAB code to do this. Break the problem down into a number of sub-tasks and use different functions to carry out each sub-task. Use a main function or main script to control the flow of the program. The user should be able to choose how many points are picked, and the program should print a suitable output message.

Run your program and experiment with different numbers of points to see the effect on the estimate of π. ∎

■ Exercise 5.8

The trapezoidal rule can be used to estimate the integral of a function be-tween a pair of limits. The formula for the trapezoidal rule is

05.A, 05.B, 05.C, 05.D, 05.E

$$\int_a^b f(x)dx \approx \frac{\Delta x}{2}\left(f(x_1) + 2f(x_2) + 2f(x_3) + \cdots + 2f(x_{n-1}) + f(x_n)\right),$$

where f is the function to integrate, a and b are the limits of the integration, and $\{x_1, x_2, \ldots, x_{n-1}, x_n\}$ are a sequence of points on the x-axis with $x_1 = a$ and $x_n = b$. Δx is the width of the interval between successive points. See the figure below for an illustration.

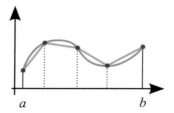

Write code that uses the trapezoidal rule to estimate the integral of the specific function $f(x) = \sin x$ between a pair of limits and for a given number of points. The user should be able to decide the limits and the number of points; for example, they might choose ten points between $x = 0$ and $x = \pi/4$. ■

FAMOUS COMPUTER PROGRAMMERS: ANNIE EASLEY

Annie Easley was brought up by her single mother in Birmingham, Alabama in the 1940s, an era when opportunities for African-Americans were severely limited. But her mother told her that she could do anything she wanted to; she would just have to work at it. She was committed to getting a good education and enrolled in classes at Xavier University, New Orleans.

In 1955, she heard about some women who worked as "computers" at the National Advisory Committee for Aeronautics (NACA, later to become NASA). In the early days of computing, "computers" were mathematicians who performed calculations by hand. She excelled at this and worked on projects ranging from nuclear powered rockets to the space mission that sent John Glenn into orbit in 1962. When human computers were replaced by computing machines, Annie adapted and became adept at programming in a number of languages, and developed code that made significant contributions to NASA's work on power technologies and its space program.

When she started work at NASA, Annie was one of only four African-Americans in a workforce of 2500 people. During her long career, she faced great discrimination. Her image was routinely cut out of NASA promotional material, and she was denied financial support to further her education, which was available to other staff. But she never gave up and always followed her mother's advice to keep working.

As well as her own achievements, throughout her life, Annie was committed to helping others. In the 1950s, she helped African-Americans prepare for the literacy test that they needed to pass to be able to vote. Later, she was enthusiastic in taking part in outreach programs to encourage children, particularly girls and those from minority backgrounds, to consider a career in technology.

Annie passed away in 2011. Reflecting on her life, she said, "I think of the poem, 'Mother to Son' [by Langston Hughes], Life for me ain't been no crystal stair, but you got to keep struggling."

"If I can't work with you, I will work around you."

Annie Easley

CHAPTER 6

Advanced data types

LEARNING OBJECTIVES

At the end of this chapter you should be able to:

O6.A Use cell arrays in MATLAB® to contain mixed data types
O6.B Make use of the `struct` data type in MATLAB
O6.C Define categorical arrays in MATLAB to contain data which can take values from a limited set of categories
O6.D Define MATLAB tables for storing mixed data in a tabular form
O6.E Make use of MATLAB maps to store data as key/value pairs

6.1 Introduction

In Chapter 3, we discussed in detail the basic data types provided by MATLAB. Often, these types are sufficient for our purposes when programming for biomedical applications. However, there will be occasions when these basic types are not flexible enough for what we want to do, and for this reason MATLAB provides a number of other, more advanced, data types. In this chapter, we review some of the more useful advanced types.

6.2 Cells and cell arrays

We have already seen that standard MATLAB arrays can only contain data of a single type. However, it is sometimes useful to be able to store data of mixed types in a single array. For example, we may wish to construct an array consisting of character, logical, and numeric types, representing patients' names, diagnoses, and ages. This can be done using an array containing a MATLAB-specific data type known as a *cell*.

We can construct a cell array using curly brackets: "{" and "}." Recall that we normally use the square brackets "[" and "]" to construct a standard MATLAB array. Here, we initialize a cell array to contain three variables of different types.

```
>> a = {'hello', true, 3.14};
```

151

MATLAB® Programming for Biomedical Engineers and Scientists. https://doi.org/10.1016/B978-0-32-385773-4.00015-0

A call to `whos` gives some details of what is stored:

```
>> whos
  Name        Size              Bytes  Class     Attributes

  a           1x3                 331  cell
```

We say that the array contains three *cells*, and that the cells contain a character array, a logical, and a double, respectively. Combining these values of different types together into a cell array, we see that it takes up 331 bytes in memory; much more than we would expect when storing each value in a separate variable. Cell arrays are therefore not efficient in terms of memory usage, but as long as we are not seeking to store too much data, the convenience of a cell array outweighs the extra memory needed.

There are two ways in which we can now access the data in the cell array. One is where we view its elements as cells or sub-arrays of cells. The other way is where we view the elements in terms of their "native" data type, which in the above example were character array, logical, and double.

■ Example 6.1

06.A   This example illustrates the different ways of accessing the elements of a cell array. We start by putting three values into a cell array called `test`, which is constructed using curly brackets.

```
>> test = {[1 2 3], "hello", false};
```

We can see that the cell array contains double array, string, and logical types. After constructing the array, the different ways of accessing the data in `test` are illustrated below.

```
>> first_cell = test(1);
>> rest_of_cells = test(2:3);
>> second_string = test{2};
```

In the above code, the cell array created can be visualized by the diagram shown in Fig. 6.1. This illustrates an array containing three cells, and each cell contains, respectively, a double array, a string, and a logical type.

After creating the cell array, the variable `first_cell` is assigned the value of a single cell (the first cell in the array `test`). The variable `rest_of_cells` is assigned with a sub-array of cells from `test` (the second and the third). Both variables `first_cell` and `rest_of_cells` are assigned using the round brackets that MATLAB normally uses for accessing array elements: "(" and ")." Therefore the types returned are *cells*, rather than the native types contained within the cells.

The variable `second_string` is different. This time, curly brackets, "{" and "}," are used to *access the contents of a cell* (the second one), rather than the

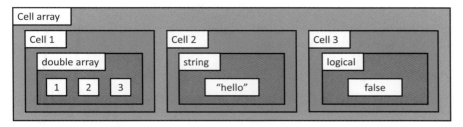

FIGURE 6.1
An illustration of a cell array.

cell itself. So second_string is assigned the string value "hello". We can check the types of our variables using the whos command.

```
>> whos
  Name            Size        Bytes  Class      Attributes

  first_cell      1x1           128  cell
  rest_of_cells   1x2           359  cell
  second_string   1x1           150  string
  test            1x3           487  cell
```

Note how first_cell and rest_of_cells are both cell arrays (with sizes 1 × 1 and 1 × 2, respectively), whereas second_string is a string type. ■

Another example of a cell array containing a mixture of different data is shown below.

```
>> a = {1, -2, [3,10]}
a =
   [1]    [-2]    [3,10]
```

In this case, all data are of the same type (double), but they are a mixture of single values and arrays. However, we have already seen that MATLAB considers a single value to actually be a 1 × 1 array, so the data consist of double arrays of different sizes. These could not be combined using a standard MATLAB array, but they can using a cell array.

The different kinds of bracket: recap

It is important to be clear about the use of the different types of brackets (round, square, and curly). When dealing with arrays in MATLAB, brackets can be used to do one of two things:

- Create an array.
- Access an element (or elements) in the array.

Not all combinations are possible. For example, in MATLAB, we cannot use square brackets to *access* elements in an array; they are only used for *constructing* arrays.

Depending on whether we have a cell array or a standard array of a basic data type (e.g., `char`), we can distinguish the different behavior of each of the types of bracket: square, curly, and round.

Creating an array

- Square brackets "[ ]": When used for array creation, square brackets are used to form a standard array, in which all entries have the same type, e.g., `double` or `char`. For example,

  ```
  >> numArray = [5, 62.7, 3];
  >> charArray = ['a', 'b', 'c'];
  ```

- Curly brackets "{ }": Curly brackets can be used to create a cell array, the cells of which can contain data of any type. For example,

  ```
  >> sameTypeCells = {'Once', 'upon', 'a', 'time'};
  >> diffTypeCells = {'Number', 17};
  ```

Accessing elements

- Round brackets "( )": These can be used to access one or more elements from any type of array. The elements returned always have the same type as the array, e.g., for an array of doubles, the use of round brackets will return doubles; for an array of cells they will return cells. For example, using the arrays created above

  ```
  >> numArray(2:3) % Second and third elements.
  ans =
        62.7000 3.000

  >> y = sameTypeCells([1, 3]) % first and third
  y =
      1x2 cell array

      {'Once'}    {'a'}

  >> class(sameTypeCells(2)) % Compare with below
  ans =
          'cell'
  ```

- Curly brackets "{ }": When used for accessing array elements, curly brackets are specific to cell arrays. They can be used to obtain the contents of a cell *in its native data type*, i.e., not in the form of a cell. For example,

  ```
  >> disp(sameTypeCells{2})
  upon

  >> class(sameTypeCells{2}) % Compare with above
  ans =

      'char'
  ```

Table 6.1 Data about hospital patients.

Name	Nationality	Born	ID Number	Gender	Admittance Reason
Joe Bloggs	UK	1960	123456	Male	Cardiac arrest
Zhang San	China	1977	234567	Male	Diabetes
Erika Mustermann	Germany	1986	345678	Female	Pregnancy

■ Activity 6.1

Examine the following code, which defines a number of arrays to store patient data: 06.A

```
name = 'Joe Bloggs';
heart_rates = [76 80 95 110 84 70];
cholesterol = 4.3;

patient1 = {name, heart_rates, cholesterol};

class(patient1{1})
class(patient1(1))
class(patient1{2})

name = 'Zhang San';
heart_rates = [60 61 58 59 60 66 64];
cholesterol = 7.0;

patient2 = {name, heart_rates, cholesterol};

patients = [patient1 patient2];

class(patients)
class(patients(2))
```

Predict what you think the output of the code would be. Enter the code into
a MATLAB script *m*-file, run it, and check if your predictions are correct.   ■

6.3   Structures

A "structure" is a commonly used data type in a number of programming languages. It is used to wrap up information, represented by different types, into
a single entity. One way of thinking about it is as a record that contains some
specific fields of data that describe different aspects of an object.

For example, we might want to represent some information relating to hospital
patients, as given in Table 6.1.

■ **Example 6.2**

We can build a structure to contain the details for the first patient in the list. We begin with a call to the built-in function `struct`.

```
>> patientA = struct
patientA =
  struct with no fields.
```

The function returns with a message that it has constructed an empty structure. Now we can add the different fields (name, nationality, born, etc.), and assign the details for our chosen patient.

```
>> patientA.name = 'Joe Bloggs';
>> patientA.nationality = 'UK';
>> patientA.born = 1960;
>> patientA.ID = 123456;
>> patientA.gender = 'M';
>> patientA.reason = 'Cardiac arrest';
```

Notice the use of the dot operator to access or set each field in the structure.

We can also perform the above with a single command by passing arguments in pairs to the `struct` function. This is done in the format `field,value,field,value` ... to give

```
>> patientA = struct('name', 'Joe Bloggs', ...
                     'nationality', 'UK', ...
                     'born', 1960, ...
                     'ID', 123456, ...
                     'gender', 'M', ...
                     'reason', 'Cardiac arrest');
```

After the assignment above, we can use the dot operator to *access* the different fields in the structure.

```
>> disp([patientA.name ' is from ' patientA.nationality])
Joe Bloggs is from UK
```

In the same manner, we can create structures for the other patients:

```
>> patientB = struct('name', 'Zhang San', ...
   'nationality', 'China', 'born', 1977, 'ID', 234567, ...
   'gender', 'M', 'reason', 'Diabetes');
>> patientC = struct('name', 'Erika Mustermann', ...
   'nationality', 'Germany', 'born', 1986, 'ID', 345678, ...
   'gender', 'F', 'reason', 'Pregnancy');
```

■

MATLAB also allows multiple structures to be collected into a single array. One way to do this is to simply concatenate them within square brackets.

```
>> patients = [patientA patientB patientC]
patients =
  1x3 struct array with fields:
```

```
name
nationality
born
ID
gender
reason
```

MATLAB reports that it has constructed a `struct` array with three elements and lists the available fields. Note that this is only possible if the fields of all structures in the array match exactly. If they do not, we need to use a cell array, rather than a standard array.

We can access the fields for the elements in the array using a combination of round brackets and the dot operator. For example, to get the names of the last two structures in the array, we can type

```
>> patients(2:3).name
   ans =
       'Zhang San'
   ans =
       'Erika Mustermann'
```

and MATLAB prints out the required names one after the other to the command window. Similar calls can be made for the other available fields.

■ Activity 6.2

In Exercise 4.7, you wrote code to visualize gait tracking data and automatically identify its peaks and troughs. In this exercise, you will repeat this task but using the same data supplied in two different formats. (It is not an uncommon problem in biomedical engineering to receive data in formats that are non-standard and not clearly specified!)

06.B

The files *LAnkle_tracking1.mat* and *LAnkle_tracking2.mat* contain the same data that you used in Exercise 4.7. The data types used are not specified, and you will need to use MATLAB commands to work out what format the data are stored in and access the parts that you want.

You can start with your solution to Exercise 4.7 (or download the provided solution from the book's web site). Make two copies of the script *m*-file, one to work with each of the two new data files. Modify the scripts so that they perform the same tasks as specified in Exercise 4.7, but work with the new data files. To recap, the tasks are

- Load in the data.
- Identify the data type used.
- Display a plot of *z*-coordinates against time.
- Automatically identify the times of all peaks and troughs of the *z*-coordinates.   ■

6.4 Categorical arrays

Cell arrays and structures are flexible ways of grouping together data of different types. But the types that we have been working with so far have been limited to the basic data types, introduced in Chapters 1 and 3, i.e., numerical, character (and string), and logical, as well as arrays of these types. However, some data that we might encounter in biomedical and other applications do not fall easily into these types.

For example, consider the example of blood types. Common blood types can only have one of eight possible values: A+, A-, B+, B-, O+, O-, AB+, and AB-. In statistics, data like these are known as *categorical data*, because the values represent categories, rather than, for example, numbers or arbitrary strings. How might we store categorical data in MATLAB? It would be possible to store them as strings or character arrays, but this would ignore the fundamental nature of the data, i.e., that only a limited number of different values, or categories, are permitted.

MATLAB provides a special type for such situations: the *categorical array*. With a categorical array, elements in the array can only take values from a limited number of categorical labels.

■ Example 6.3

06.C We can easily create categorical arrays in MATLAB by using the `categorical` function. In its simplest form, this function takes a single argument, which should be either a cell array of character vectors or a string array. Let us consider our blood type example.

```
>> types1 = {'A+', 'B-', 'O+', 'A+', 'B+', 'B+', 'AB-'}

types1 =
  1x7 cell array
    {'A+'}  {'B-'}  {'O+'}  {'A+'}  {'B+'}  {'B+'}  {'AB-'}
    >> blood_cat1 = categorical(types1)
    blood_cat1 =
      1x7 categorical array
        A+      B-      O+      A+      B+      B+      AB-

    >> types2=["A+", "B-", "O+", "A+", "B+", "B+", "AB-"]
    types2 =
      1x7 string array
        "A+"     "B-"     "O+"     "A+"     "B+"     "B+"     "AB-"

    >> blood_cat2 = categorical(types2)
    blood_cat2 =
      1x7 categorical array
        A+      B-      O+      A+      B+      B+      AB-
```

We can see that the result is the same, whether we provide a cell array of character vectors or a string array. Note that, when defining a categorical array, we are not listing all possible categories. Rather, we list the categories that we wish to store in a specific array. Thus not all categories need to appear, and some categories may appear more than once. For example, the variables `blood_cat1` and `blood_cat2` might represent the blood types of a list of patients in a research study.

We can also ask MATLAB for a list of the different categories that appear in a categorical array, and it provides its answer as a cell array of character vectors.

```
>> categories(blood_cat1)
ans =
  5x1 cell array
    {'A+' }
    {'AB-'}
    {'B+' }
    {'B-' }
    {'O+' }
```

Finally, comparison operators can be used to compare categorical data with hard-coded string or character array values. For example,

```
>> blood_cat1(2) == 'B-'
ans =
  logical
   1
```

∎

Now let us consider another, slightly different, example. Suppose that we wish to define categories for severities of burn injuries. In this case, our categories would be first degree, second degree, third degree, and fourth degree. Again, these are categorical data, as there is no numeric meaning to the values (i.e., it is not true that a second degree burn is twice as bad as a first degree burn). However, there is a clear and natural ordering between the categories, which was not the case with our blood type example. Categorical data with a natural ordering like these are known as *ordinal* or *ranked* data. In MATLAB, ordinal categorical data can be created using the same `categorical` function.

■ Example 6.4

The following code creates an ordinal categorical array to represent burn types:

06.C

```
>> burns = ["firstDegree", "firstDegree", "thirdDegree", ...
            "secondDegree", "thirdDegree", "fourthDegree"];
>> ordering = ["firstDegree", "secondDegree", ...
               "thirdDegree", "fourthDegree"];
>> burns_cat = categorical(burns, ordering, "Ordinal", true)
```

```
burns_cat =
  1x6 categorical array
  Columns 1 through 3
    firstDegree    firstDegree    thirdDegree
  Columns 4 through 6
    secondDegree    thirdDegree    fourthDegree
```

Note that this time we have used a second argument to the `categorical` function to indicate the ordering of the categories (smallest first). We also have to specify that the data are ordinal by adding the extra arguments `"Ordinal", true`.

Because our categorical array is ordinal, we can now perform comparisons between the values, just as we can with numerical data. This would not be possible with standard (i.e., non-ordinal) categorical data.

```
>> burns_cat(3) > burns_cat(1)
ans =
  logical
   1
```

Finally, note that we did this example with strings, but we could have done it all with character arrays, as in Example 6.3. ∎

■ Activity 6.3

06.A, 06.C

In oncology, tumors are classified into *grade 1*, *grade 2*, *grade 3*, or *grade 4*, depending on their severity. Tumor grades are ordinal categorical data, since there is no numerical meaning to the grades (i.e., grade 2 is not twice as bad as grade 1), but there is a natural ordering.

The data shown in the table below have been gathered from a number of cancer patients. For each patient, we have their ID number and their tumor grade. Write MATLAB code to store these data using appropriate data type(s). All data for all patients should be stored in a single variable. Next, write a `for` loop to display the ID numbers of all patients with a tumor, whose severity exceeds grade 2.

ID	Tumor grade
123	Grade 1
124	Grade 3
125	Grade 1
456	Grade 2
457	Grade 2
458	Grade 3
678	Grade 4
789	Grade 4

∎

Table 6.2 Table of data of mixed types about hospital patients.

Name	Gender	Vulnerable	Age
Joe Bloggs	Male	False	60
Zhang San	Male	True	43
Erika Mustermann	Female	False	34

6.5   Tables

We have now seen a range of different data types in MATLAB. In biomedical research, it is not uncommon to collect a large amount of data of different types from study participants, so that associations and causative links can be identified between different variables. It is natural to represent such data in a tabular form, and for this reason MATLAB provides a specific data type for *tabular data*. By tabular data, we mean data that can be visualized in a table with rows representing individuals and named columns indicating the variables collected. For example, Table 6.2 shows a tabular representation of data collected from a group of hospital patients. The table contains character data (name), categorical data (gender), Boolean data (vulnerable), and numerical data (age).

■ Example 6.5

We can create tables in MATLAB using the `table` function. Consider the following code, which creates a table containing the first three columns and the first row of Table 6.2:

O6.D

```
>> t = table; % create empty table
>> t.name = "Joe Bloggs"; % string column
>> t.gender = categorical({'male'}); % categorical column
>> t.vulnerable = false; % logical column
>> t % view table
t =
  1x3 table
       name         gender     vulnerable

    _____    _____    _____


    "Joe Bloggs"     male        false
```

Note the similarity of this code to the way in which we used the `struct` function in Section 6.3. Both can be used to group together fields of mixed data types and give these fields names. However, tables allow us to do more than this. For example, we can easily add new rows to our table.

```
>> newrow = {"Zhang San", categorical({'male'}), true};
>> t = [t; newrow];
>> newrow = {"Erika Mustermann", categorical({'female'}), ...
             false};
```

```
>> t = [t; newrow]
t =
  3x3 table
          name              gender      vulnerable
     _____    _____      _____

     "Joe Bloggs"           male          false
     "Zhang San"            male          true
     "Erika Mustermann"     female        false
```

Note that we specify the new row data by using a cell array, since the columns of the table contain mixed types.

We can also add new columns to our table by simply assigning a column vector of data to a new field in the table.

```
>> t.age = [60; 43; 34];
>> t
t =
  3x4 table
          name              gender    vulnerable    age
     _____    _____    _____    ____

     "Joe Bloggs"           male         false       60
     "Zhang San"            male         true        43
     "Erika Mustermann"     female       false       34
```

Accessing data in the table is also easy, and can be done either by numerical indexing, similar to standard two-dimensional (2-D) arrays, or by using the column names, as shown below.

```
>> t.gender % access entire column by name
ans =
  3x1 categorical array
     male
     male
     female

>> t(:,1:3) % access first three columns, all rows
ans =
  3x3 table
          name              gender      vulnerable
     _____    _____      _____

     "Joe Bloggs"           male          false
     "Zhang San"            male          true
     "Erika Mustermann"     female        false

>> t(1,4) % access first row, fourth column
ans =
  table
     age
     ___
```

```
      60

>> t(1,'age') % same but using column name
ans =
    table
      age
      ___

      60
```

■

■ Activity 6.4

The file *patient_table.mat* contains the patient data entered into the table in   *06.D*
Example 6.5. Write a MATLAB script to perform the following tasks with this
table:

1. Add a new row for `"Jane Doe"`, who is female, vulnerable, and aged
   77.
2. Add a new column for the patients' heart rate in beats per minute,
   the values of which should be 75, 69, 55, and 80 for `"Joe Bloggs"`,
   `"Zhang San"`, `"Erika Mustermann"`, and `"Jane Doe"` respectively.
3. Write a `for` loop to inspect all rows in the table, and display the names
   of all elderly and vulnerable patients. Patients are considered to be
   elderly if their age is greater than or equal to 70.
4. Logical indexing can be used with tables in the same way as it can be
   used with standard arrays. Perform the same operation as in part 3,
   but this time using a single statement and logical indexing.   ■

6.6   Maps

The final data type we will introduce in this chapter is the *map*. MATLAB maps,
which are also known as *associative arrays* or *dictionaries* in other programming
languages, are a useful generalization of the standard arrays that we have seen
so far.

We are already familiar with the idea of an array as a collection of data elements
that we can access via an index. We can construct a variety of arrays, e.g.,

```
>> a = [12, 2.4, 50];
>> b = {'red', 'blue', 'green'};
```

gives us two arrays, one numeric array and one cell array. When we want to
access an element, we use an integer (or an array of integers) as the index (or
indices). For example,

```
>> disp(a(2))
    2.4000
```

```
>> disp(b([1,3]))

{'red'}    {'green'}
```

The key point here is that *integers* are used as the indices when specifying which element(s) in an array we wish to access. More specifically, the integers used to index the data are ordered successive integers. As we introduced in Section 1.6, in MATLAB, the integers used for array indices start at 1, so, for example, the elements of the arrays a and b above can be indexed by the ordered list {1, 2, 3}.

The map data type in MATLAB is a generalization of the standard array type. It is still used to collect a number of variables in something similar to a standard array, but we do not need to use successive integers to index the elements.

Instead, a map relies on the use of *key-value* pairs to perform indexing. We say that each entry in the map has a unique *key*, and we can use the key to index an element in the map so that we can read or modify that element's *value*.

The keys that we use as indices can be more than just ordered integers. For example, we can use a character array to access an element, a floating point value, or a set of integers that are not ordered. The important thing is that each element in the map has a *unique* key; if we want to distinguish two different elements, then they *must* have different keys.

■ Example 6.6

06.E

Suppose that three people have the following ages: Joe (60), Zhang (43), and Erika (34). We would like to define a map to contain this information, and we would like to look up the age of a person by specifying their name. An empty map in MATLAB can be created by a call to the containers.Map function.

```
ages = containers.Map;
```

If we ask MATLAB to display the map, it gives a brief description:

```
>> disp(ages)
  Map with properties:
        Count: 0
      KeyType: char
    ValueType: any
```

This tells us that the count of stored elements is zero as expected. It also says that the map will use char arrays as the type for the keys. In other words, it will use character arrays for indexing. Character arrays are the default type for keys in MATLAB maps. We will see in the next example how to use a different type.

Finally, the description of the map variable `ages` says that it will accept any type as a value. In our case, we will store numeric values for the ages of the people.

Now we can actually assign our data to the map. We do this using round brackets containing the names of the people as character arrays:

```
ages('Joe') = 60;
ages('Zhang') = 43;
ages('Erika') = 34;
```

A call to `disp(ages)` will now give the same output as before, but the count will be 3.

Now that the map has data in it, we can access the elements in the map and read their values:

```
>> disp(ages('Erika'))
   34
```

We can also modify the elements in our map. For example, if Zhang has a birthday,

```
>> disp(ages('Zhang'))
   43

>> ages('Zhang') = ages('Zhang') + 1;

>> disp(ages('Zhang'))
   44
```

We can also add a new person to our map, for example,

```
ages('Ahmed') = 34;
```

and the map will now contain four elements.

The map object behaves in a similar way to a structure or a table, and we can ask for *all* of the keys or values for the map by using the dot operator with the fields `keys` and `values`:

```
>> ages.keys
ans =
  1x4 cell array
    {'Ahmed'}    {'Erika'}    {'Joe'}    {'Zhang'}

>> ages.values
ans =
  1x4 cell array
    {[34]}    {[34]}    {[60]}    {[44]}
```

Note that if we seek to index an element that is not present in the map, then we get an error:

```
>> ages('Elijah')

Error using containers.Map/subsref
The specified key is not present in this container.
```

This is because the character array that we have used (`'Elijah'`) is not present in the set of keys for the map. We can, of course, create an entry in the map with this key, in the same way that we did for `'Ahmed'` above.

It is also possible to create a map more efficiently by preparing the keys and corresponding values in advance.

```
>> keys = {'Joe', 'Zhang', 'Erika', 'Ahmed'};
>> values = [60, 43, 34, 34];
```

The above shows that the keys for the map are generated as a cell array of character arrays. The values for the elements are a numeric array. It is important that the order of the values array matches that of the keys array. Now we can create our map in one line as follows:

```
ages = containers.Map(keys, values);
```

which uses the same `containers.Map` function, but this time passes two arguments to assign the keys and values for all entries in a single call. ∎

■ Example 6.7

06.E

Now we consider a map that uses a different data type for its keys (i.e., not a character array). Suppose that we have the same people with the ages given in the previous example. Now we want to build a map that will give us all the people for a specific age. That is, we want to use the numeric age as the key for indexing elements and the names as the values. In this case, we can construct an empty map as before, but this time, we tell MATLAB specifically that we would like to use a numeric type for the key. We will use the `double` data type in case we ever need to use fractional ages.

When we specify the type of the key, we also need to specify the value type. Fortunately, we can be non-specific for the value type and can simply say "any."

```
>> people = containers.Map('KeyType', 'double', ...
                           'ValueType', 'any')
people =
  Map with properties:
        Count: 0
      KeyType: double
    ValueType: any
```

Now we can populate the elements of our map as follows:

```
>> people(34) = {'Erika', 'Ahmed'};
>> people(43) = {'Zhang'};
>> people(60) = {'Joe'};
```

Note that we have used a cell array of character arrays for each of the values. This allows us to store the name of more than one person if we need to, although some of the cell arrays contain only one character array.

We can access elements in the new map by using a numeric key. For example, people(34) will return all those with age 34.

```
>> disp(people(34))
    {'Erika'}    {'Ahmed'}
```

A missing key will still generate an error:

```
>> people(40)
```

```
Error using containers.Map/subsref
The specified key is not present in this container.
```

because there is nobody aged 40 in the map.

Using a key of the wrong type will also generate an error. Here, we attempt to access an element in the people map using a character array:

```
>> people('Fred')
```

```
Error using containers.Map/subsref
Specified key type does not match the type expected for
this container.
```

∎

■ Activity 6.5

This activity builds upon Example 6.7, so you can download the code for this to act as your starting point.   *06.E*

One problem with the way in which we have used maps so far is that we have not checked to see if a key already exists before overwriting its value. This may result in loss of information.

Write a new function that adds a new key/value pair to a map. Before assigning the value to the key, the function should check if the key already exists in the map. If it does, the new value should be appended to the value(s) already stored. If not, the new key/value pair should be assigned in the normal way. The prototype for the new function should be

```
function newmap = map_add (map, key, value)
```

Use the function to add two new entries into the people map: 'Elijah' who is 43 and 'Rahel' who is 25.

Table 6.3 MATLAB built-in conversion functions for advanced data types.

Conversions to standard arrays	
`table2array`	Convert table to standard array
`cell2mat`	Convert cell array to standard array
Conversions to cell arrays	
`cellstr`	Convert character array, categorical array, or string array to cell array of character vectors
`table2cell`	Convert table to cell array
`mat2cell`	Convert standard array to cell array
`struct2cell`	Convert structure array to cell array
Conversions to structure arrays	
`table2struct`	Convert table to structure array
`cell2struct`	Convert cell array to structure array
Conversions to tables	
`array2table`	Convert standard array to table
`cell2table`	Convert cell array to table
`struct2table`	Convert structure to table

(Hint: There is a built-in function that tells us if a particular key exists or not: it is called `isKey`. Check the MATLAB documentation to find out how it can be used.) ■

6.7   Conversion of advanced data types

A number of built-in functions are provided to convert data to and from the advanced types we have introduced in this chapter. Table 6.3 lists some of them and provides a brief summary of their purpose. For full details, please view the MATLAB documentation.

6.8   Summary

In this chapter, we have looked at more advanced MATLAB data types, specifically cell arrays, structures, categorical arrays, tables, and maps.

Cells can be used to enable arrays to contain data of mixed types, in contrast to standard MATLAB arrays, which have the restriction that all elements must be of the same basic type.

Structures also allow mixed data to be collected together as fields in a single entity, which can be accessed by the field names.

Categorical arrays can be used to store categorical data, which represent non-numerical labels. These categories can be ordinal (i.e., ordered) or non-ordinal.

Tables are used to store rows of mixed data with named columns in tabular form.

Maps are the way in which MATLAB implements associative arrays, in which data are stored as key/value pairs.

Many conversion functions are provided in MATLAB to convert data to and from these advanced types.

6.9 Further resources

As usual, the MATLAB documentation is a good place to start to learn more about these advanced types:

- Cells and cell arrays: https://mathworks.com/help/matlab/cell-arrays.html.
- Structures: https://mathworks.com/help/matlab/structures.html.
- Categorical arrays: https://mathworks.com/help/matlab/categorical-arrays.html.
- Tables: https://mathworks.com/help/matlab/tables.html.
- Maps: https://mathworks.com/help/matlab/map-containers.html.

The MathWorks web site also contains some informative and accessible short videos to introduce some of the advanced data types:

- Structures and cell arrays: https://mathworks.com/videos/introducing-structures-and-cell-arrays-68992.html.
- Tables and categorical arrays: https://mathworks.com/videos/introducing-tables-and-categorical-arrays-79924.html.

Finally, you can learn more about what is meant by an associative array (known as a *map* in MATLAB) by reading the Wikipedia page: https://en.wikipedia.org/wiki/Associative_array.

6.10 Exercises

■ Exercise 6.1

Perform the following tasks in MATLAB: *06.A*

1. Make a cell array called c that contains the following 11 character arrays:

    ```
    'the'
    ' '
    'cow'
    ' '
    'jumped'
    ' '
    'over'
    ' '
    ```

```
'the'
' '
'moon'
```

2. Write code that accesses the *cell* at index 7 in the array c. Make sure that you use the correct type of brackets. Use the function class to confirm that a cell is returned.
3. Write code that accesses the *cells* with odd indices in the array c, i.e., the cells at indices 1, 3, 5, etc. How many cells are returned?
4. Write code that accesses the cells with even indices in the array c.
5. Write code to access the *character array contained* in the cell at index 5 of the array c.
6. Write code that accesses the last character of the character array contained in the cell at index 5 in array c.
7. Make a new cell array d that contains at least three different data types. ■

■ Exercise 6.2

06.A Here are the names of five organs: heart, lung, liver, kidney, pancreas.

1. By padding the names with spaces where necessary, show how you can construct a 2-D char array to contain all of the five names. Assign this to a variable called a.
2. Now show how we can put the five names into a cell array of character arrays. Assign this to a variable called c.
3. Write the code that is needed to replace lung with brain in the 2-D char array a.
4. Write the code that is needed to replace lung with brain in the cell array c. ■

■ Exercise 6.3

06.A In Activity 2.8, you wrote code to determine the *dose escalation pattern* for a phase I clinical trial of a new cytotoxic drug. Recall that the dose escalation pattern was determined by increasing the initial dose by 67%, then by 50%, then by 40%, and, subsequently, by 33% each time.

A common way of applying the dose escalation pattern is the "3+3 rule." With this approach, the trial proceeds using cohorts of three patients in the following way:

- If *none* of the three patients in a cohort experiences significant toxicity from the drug, another cohort of three patients will be tested at the next higher dose.
- If *one* of the three patients in a cohort experiences significant toxicity, another cohort of three patients will be treated at the *same* dose.

- The dose escalation continues until at least *two* patients among a co-hort experience significant toxicity. The recommended dose for phase II of the trial is then defined as the dose just *below* this toxic dose.

The data for phase I of the cytotoxic drug trial are now available, and are contained in the file *phaseI_data.mat*. The file contains a 1×2 cell array. The first cell contains an array of dose levels (determined as described above) for each cohort of three patients. The second cell contains a 2-D array of `logical` values. Each row of this array represents the toxicity results of a single cohort, with `true` indicating that significant toxicity was observed in that patient and `false` indicating no significant toxicity.

Write a MATLAB script *m*-file to determine the recommended dose for phase II of the trial from these data.

Try to make your code robust to unexpected situations, e.g., what happens if the *first* cohort has at least two patients who experience toxicity? You can test your code against this situation using the *phaseI_data2.mat* file provided through the book's web site. You can assume that the phase I trial continued until at least two patients in a cohort experienced significant toxicity. ■

■ Exercise 6.4

In Activity 1.11 and Exercise 2.14, we introduced the concept of an injury severity score (ISS). Recall that ISS quantifies trauma severity and involves each of six body regions (head, face, chest, abdomen, extremities, and external) being assigned an abbreviated injury scale (AIS) score between 0–5, depending on the severity of the trauma in that region. The three most severely injured body regions (i.e., with the highest AIS scores) have their AIS scores squared and added together to produce the ISS.

06.B

In Exercise 2.14, you wrote code to compute the ISS for a single patient from an array of numeric AIS scores. The data file *ais.mat* contains AIS scores for multiple patients in a slightly different format. Write MATLAB code to load in this file; determine the type of the data it contains, and then compute and display the ISS for all patients. ■

■ Exercise 6.5

Chemical elements can be described by their *name* or their *symbol*. Each element also has an *atomic number* (indicating the number of protons in its nucleus).

06.B

1. Show how a structure can be created to represent these three pieces of information for an element. Use the names in italics above for the field names. Give an example for an element of your choice.

2. Repeat the previous part for two more elements, and show how you can collect them into a single array of structures.
3. Show how we can access
   - The name of the second element in the array.
   - The symbols of the first and last elements in the array. ■

■ Exercise 6.6

06.C, 06.D   The table below shows data gathered from a group of subjects who have volunteered for a research study into a new drug.

Study ID	Gender	Age	Diet	Lifestyle	Cholesterol (mmol/L)
13267033	Female	51	Poor	Normal	5.5
18927460	Female	37	Good	Normal	4.2
12666831	Male	66	Poor	Sedentary	6.0
19931878	Female	27	Medium	Active	3.6
11189371	Male	40	Good	Normal	4.8

The researchers would like to exclude from the study any volunteer who has a significant health risk. They have devised the following algorithm for computing a risk score for each volunteer:

- Start from a risk score of 100.
- Subtract 20 for a good diet and 10 for a medium diet.
- Subtract 20 for an active lifestyle and 10 for a normal lifestyle.
- Subtract $(100 - Age)/2$.
- Subtract 10 if the cholesterol level is less than or equal to 5 mmol/L.

Based on the computed risk scores, the researchers would like to exclude any volunteer with a risk score greater than 60.

Write MATLAB code to

1. Enter the data into a MATLAB table.
2. Compute risk scores for all volunteers, and add this as an extra column in the table.
3. Form and display a new table, which contains the same columns as the original table, but only those subjects who have *not* been excluded from the study. You can do this either using iteration or logical indexing. ■

■ Exercise 6.7

06.E   The International classification of diseases (ICD) is a standard system for classifying different types of disease, produced and maintained by the World

Health Organization. The ICD classifies diseases into a number of categories including the following:

- Diseases of the respiratory system
- Diseases of the nervous system
- Diseases of the circulatory system.

The table below shows a list of diseases along with their ICD disease category.

Disease	ICD Category
Acute sinusitis	Respiratory
Coronary atherosclerosis	Circulatory
Epilepsy	Nervous
Pneumonia	Respiratory
Hypertension	Circulatory
Cerebral ischaemia	Nervous
Cerebral palsy	Nervous
Heart valve disease	Circulatory

1. Write MATLAB code to construct a map in which the name of each disease is a key, and its ICD category is the corresponding value. Assign the map to a variable called `diseases`.
2. Write MATLAB code to add *acute bronchitis* to the `diseases` map. Acute bronchitis is a respiratory disease.
3. Write code to iterate over the keys of the map, and for each, display the key/value pair to the command window.
   (Hint: Recall that you can find a list of map keys as follows: `diseases.keys`.) ■

FAMOUS COMPUTER PROGRAMMERS: MARY ALLEN WILKES

Growing up in Maryland, U.S.A. in the 1940s, Mary Allen Wilkes never dreamed that she would one day be a famous computer programmer. In 1950, when her high school geography teacher told her that she should consider becoming a programmer when she grew up, she didn't really know what a computer programmer was.

In her youth, Wilkes dreamed of a legal career, but when she graduated from Wellesley College in 1959, she realized that this ambition was out of her reach. However, her geography teacher's comment stuck in her mind, and so she got her parents to drive her to the Massachusetts Institute of Technology, where she knew they were looking to hire computer programmers. She walked in and asked for a job, and was offered one.

Wilkes quickly became an expert computer programmer, and first worked writing code for the IBM 704, one of the early computers. Later, she worked on developing the LINC, which was intended to be the world's first "personal computer" (in that it would have its own keyboard and screen, and fit in a single office!). Programming existing computers at the time involved a laborious process of creating "punchcards," which were read in using a special machine, but a keyboard and screen would greatly speed this up. In 1964, Wilkes was asked to finish writing the LINC's operating system, but she did not want to go to live in St. Louis, where the project was based. Therefore a LINC was shipped to her parent's house in Baltimore, making her the first home computer user.

Eventually, Wilkes returned to her first love of law. In 1972, she was accepted at Harvard Law School, and spent the next 40 years as a lawyer, but her contributions to the field of computer programming will never be forgotten.

"We had the quaint notion at the time that software should be completely, absolutely free of bugs. Unfortunately it's a notion that never really quite caught on."

Mary Allen Wilkes

File input/output

At the end of this chapter you should be able to:

O7.A Use built-in MATLAB® functions to save and load workspace variables

O7.B Use a range of built-in MATLAB functions to read data from external text or binary files

O7.C Use a range of built-in MATLAB functions to write data to external text or binary files

7.1 Introduction

In Section 1.9, we introduced some MATLAB commands for loading data from external files and for saving data to files. This type of operation is very common, particularly for complex programs that need to process large amounts of data. In this chapter, we will look in more detail at the topic of *file input/output*.

7.2 Recap on basic input/output functions

First, let us recap on what we know already. Section 1.9 introduced the following simple file input/output functions:

- `save`: Save data to a MATLAB *MAT* file.
- `writematrix`: Write data to a delimiter-separated text file.
- `load`: Load data from a MATLAB *MAT* file or delimiter-separated text file.

Therefore we have seen two different types of external files so far: *MAT* files and text files. *MAT*-files are MATLAB-specific and are used to save parts of the MATLAB workspace. When using *MAT* files, the `save` and `load` functions are all that we need. For the rest of Chapter 7, we will focus in more detail on functions that we can use when dealing with other types of file.

175

Table 7.1 Delimiters that can be specified for writing data to text files using `writematrix`.

Argument	Delimiter
`','` `'comma'`	Comma
`' '` `'space'`	Space
`'t'` `'tab'`	Tab
`';'` `'semi'`	Semicolon
`'|'` `'bar'`	Vertical bar

7.3 Simple functions for dealing with text files

We have already seen in Section 1.9 that the `writematrix` command can be used to write numeric data to text files. Text files are files which consist of a sequence of characters, normally stored using the ASCII coding system. We saw that it was possible to change the delimiter used to separate data values in the file, for example, as follows:

```
>> f = [1 2 3 4 9 8 7 6 5];
>> writematrix(f, 'testtab.txt', 'Delimiter', 'tab');
>> writematrix(f, 'testspace.txt', 'Delimiter', ' ');
```

The full list of values that can be supplied along with the `'Delimiter'` argument to `writematrix` is given in Table 7.1.

For reading data from text files, although the `load` function can be used to read numeric data, a more flexible function is `readmatrix`.

■ Example 7.1

07.B

We illustrate the use of `readmatrix` using the example that follows. Physiological data has been exported from an MR scanner to a file called *scanphyslog.txt* in the format shown below.

```
-33 -53 25 -6 0 -1815 0 0 0 0000
-35 -56 24 -9 0 -1815 0 0 0 0000
-37 -59 22 -12 0 -1815 0 0 0 0000
-39 -61 20 -15 0 -1840 0 0 0 0000
-40 -61 18 -18 0 -1840 0 0 0 0000
-40 -60 17 -20 0 -1840 0 0 0 0000
-39 -58 15 -21 0 -1840 0 0 0 0000
-36 -55 15 -21 0 -1840 0 0 0 0000
```

```
-32 -50 15 -20 0 -1871 0 0 0 0000
-26 -44 17 -17 0 -1871 0 0 0 0000
-20 -37 19 -13 0 -1871 0 0 0 0000
-11 -30 23 -8 0 -1871 0 0 0 0000
-1 -22 29 -2 0 -1871 0 0 0 0000
```

The 6th column of this data file represents a signal acquired by a respiratory bellows that is sometimes used for respiratory gating of MR scans. The code example shown below will read the data from the file, and then extract the bellows signal.

```
data = readmatrix('scanphyslog.txt', 'Delimiter', ' ');
bellows = data(:,6);
```

Note that we specify the delimiter to use when reading the data (a space in this case) using the `'Delimiter'` argument followed by the delimiter, similar to `writematrix`. The `readmatrix` function will automatically work with files that use commas as the delimiter, but if we wish it to use an alternative delimiter, we must specify it in this way.

In this example, the `readmatrix` function will return a 13×10 2-D array of the physiological data. The second line of code extracts the 6th column, which represents the bellows data.   ∎

∎ Activity 7.1

The file *patient_data.txt* contains personal data for a group of patients taking part in a clinical trial. There is a row in the file for each patient, and the five columns represent patient ID, age, height, weight, and heart rate. Write a MATLAB script *m*-file that uses the `readmatrix` function to read in these data. Next, write code to input a single integer value n from the keyboard, and display the height and weight data for the n oldest patients.   ∎

07.B

The `readmatrix` and `writematrix` functions can also be used to read/write data using common spreadsheet formats, such as those with *"xlsx"* or *"ods"* file extensions. By simply supplying a file name with one of these extensions, the `readmatrix` and `writematrix` functions will automatically read and write using the specified format.

Finally, there are a number of other data structure-specific built-in file input/output functions in MATLAB. We summarize the most useful ones in Table 7.2. The reader is advised to refer to the MATLAB documentation for further details of these functions.

In general, if we want to read or write numeric data to or from an external file in a standard delimited format, then `readmatrix` and `writematrix` are sufficient. As noted above, these functions also work for common spreadsheet formats, but only if the data are numeric. The `readcell` and `writecell`

Table 7.2 Data structure-specific file input/output functions in MATLAB.	
`c = readcell('filename')`	Read data from a file into a cell array (see Section 6.2).
`writecell(c, 'filename')`	Write a cell array to a file.
`t = readtable('filename')`	Read data from a file into a table (see Section 6.5).
`writetable(t, 'filename')`	Write a table data structure to a file.

functions can be used when we need to store mixed data types in a cell data structure (see Section 6.2). The files produced are simple delimited text files. The `readtable` and `writetable` functions can also handle mixed data types and are intended for storing data in a tabular format, e.g., for a number of rows of data representing individuals and a text heading for each column representing the variables collected (see Section 6.5).

7.4 Reading from files

The `readmatrix` and `writematrix` functions are very useful and will be sufficient for our needs in many situations, but they are limited in their capabilities. One important limitation is that they only work if the data consist of purely numeric values. For more complex data files, these functions will not be flexible enough. In this section, we will introduce some of the more flexible functions provided by MATLAB for file input/output.

Before we proceed, a key concept we should understand is the need to *open* a file before using it, and to *close* it after use. You can think of this as being similar to using a real, physical file: in this case, it is obvious that the file should be opened before looking at or changing its contents. It's the same with computer files: there are specific functions that we can use to open or close external files, and these must be used correctly. (Note that with the `readmatrix` and `writematrix` functions described in the previous section, there was no need to make explicit calls to open or close the files; this was performed automatically for us.)

■ Example 7.2

07.B   Let us illustrate this concept with an example. The piece of code that follows is adapted from the MATLAB documentation for the `while` statement, and it illustrates the concept of opening and closing a file, as well as introducing some useful new functions.

```
% open file
fid = fopen('test.m','r');
if (fid == —1)
    error('File cannot be opened');
end
```

```matlab
% initialize count
count = 0;

% loop until end of file
while ~feof(fid)

    % read line from file
    tline = fgetl(fid);

    % is it blank or a comment?
    if isempty(tline) || strncmp(tline,'%',1)
        continue;
    end

    % count non-blank, non-comment lines
    count = count + 1;
end

% print result
fprintf('%d lines\n',count);

% close file
fclose(fid);
```

This code counts the number of lines in the file *test.m*, except for blank lines and comments. The `fopen` and `fclose` statements open and close a file, respectively. For most MATLAB file input/output functions, it is essential that we open a file before use, and it is good practice to close it afterwards. In addition, it is always a good idea to check that the file was opened successfully. We do this in the above code by checking the value returned by `fopen`: if it is equal to −1, this means that the file could not be opened (most likely because there was a mistake in the file name, or the file was in a different location in the file system). If the file could not be opened, we use the `error` function (see Section 4.3) to report the problem and terminate the program.

The statement `feof` checks to see if the end of file has been reached and will return `true` when this happens; otherwise it returns `false`. Recall that the *tilde* operator (~) means *not* in MATLAB (see Table 2.1).

The `fgetl` function reads (gets) one line from the file. This function returns the line as an array of characters. Therefore the `while` loop will continually read lines from the file, so long as the end of file is not reached. When it is reached, execution passes to after the `while` loop's `end` statement.

Next, two new built-in functions are used together with an `if` statement:

- `isempty(x)`: returns whether or not the array `x` is empty.
- `strncmp(x,y,n)`: returns `true` if the character arrays `x` and `y` are identical up to the first `n` characters.

These two functions are used to check if the current line is either (a) empty, or (b) starts with the MATLAB comment symbol, `%`. If either of these conditions is true (recall that `||` means *OR*; see Table 2.1), then the `continue` statement will cause execution to pass to the beginning of the `while` loop again. If both of these tests fail, we have a line that we want to count, and 1 is added to the `count` variable. Execution then passes back to the beginning of the loop.

The `fprintf` function is used to display the number of lines to the command window. We first introduced `fprintf` in Chapter 4. It can be used as an alternative, and a more flexible version, of the `disp` statement. Here it is used to display some text and a variable (`count`) to the command window (`fprintf` can also be used to write to external files, as we will see later in Section 7.5).

Finally, we make an explicit call to close the file that was opened using the `fclose` command. This tells the operating system that our code does not need to work with the file any more. ■

So, to summarize, we have learned the following new MATLAB functions for file input/output:

- `fid = fopen(filename, permission)`: Open a file called `filename` for reading/writing, giving it the file identifier `fid`. `fid` will equal −1 if the file could not be opened. The `permission` argument is a character array specifying how the file will be used, e.g., 'r' for reading, 'w' for writing, or 'a' for appending.
- `fclose(fid)`: Close a file that we previously opened, with file identifier `fid`.
- `line = fgetl(fid)`: Read a single line from the text file with identifier `fid`, putting the result into the character array variable `line`.
- `feof(fid)`: Return a Boolean value indicating whether the end of the file with identifier `fid` has been reached (`true`) or not reached (`false`).

Now let us consider another situation. Example 7.2 illustrated how entire lines could be read from a text file as arrays of characters. However, it is often the case that a line of data contains a number of individual values which should be stored separately for further processing. In such cases, a more flexible file input function is needed.

■ Example 7.3

07.B

In this example, we would like to read some data about patients' blood pressures from a text file. The file *sys_dia_bp.txt* contains the following data:

```
BP data
51 85
88 141
```

```
67 95
77 111
68 115
99 171
80 121
```

First, there is a header line, which just contains the text "BP data." After this, in each row of the file, the value in the 1st column represents the patient's systolic blood pressure, and the value in the 2nd column represents the diastolic blood pressure of the same patient. The following piece of MATLAB code can be used to read all of the data from the file into a single 2-D array called `data`:

```
% open file
fid = fopen('sys_dia_bp.txt', 'r');
if (fid == -1)
    error('File cannot be opened');
end

% skip header line
line = fgetl(fid);

% read systolic and diastolic blood pressure
data = fscanf(fid, '%d %d', [2 inf])

% close file
fclose(fid);
```

Note that, as before, we open the file before reading any data, and close it after we have finished reading. We use the `fgetl` function that we saw in Example 7.2 to read (and discard) the text header line. Next, in a single line of code, the `fscanf` function is used to read all rows of data from the file. Three arguments are provided to `fscanf`:

- The identifier of the file containing the data: `fid`.
- A *format string* that specifies the format of the data fields to read in: the format specifier `%d` means a signed decimal (base 10) number. `'%d %d'` represents two numbers, so `fscanf` will read numbers from the file in pairs (i.e., a row at a time).
- The dimensions of the output array returned by `fscanf`: `[2 inf]` means that the data will be placed in an array with two rows (representing the pairs of numbers being read in) and an unlimited number of columns, i.e., the function will read pairs of decimal numbers until it reaches the end of the file.

In this example, seven pairs of decimal numbers will be read in, as there are seven lines of numeric data in the input file. Therefore the output variable `data` will be a 2×7 array. ∎

Table 7.3 Format specifiers for use in the format strings of file input/output functions (e.g., `fscanf`).

Specifier	Field Type
%d	Signed decimal integer
%u	Unsigned decimal integer
%f	Floating point number
%c	Character
%s	String (i.e., a sequence of characters)

The type of *format string* that we saw in the above example was first introduced in Section 4.3. As well as reading decimal numbers, format strings enable `fscanf` to be used to read in a range of other data types. A summary of some of the more common *format specifiers* that can be used in the format string is provided in Table 7.3. Note, however, that `fscanf` cannot be used to read *mixed* data types from a single file. For example, it is not possible to read a character array followed by a decimal integer using `fscanf`. The reason for this is that `fscanf` always returns a standard array as its result, i.e., one that cannot contain elements with different data types. It is possible to read a file containing differing *numeric* data types (e.g., signed integer and floating point), but, in this case, `fscanf` will perform an automatic conversion of one of the types to ensure that the resulting array contains only a single type (e.g., it will convert the signed integers to floating point numbers).

It was in fact possible to perform the work of Example 7.3 using the `readmatrix` command, as the following code illustrates:

```
x=readmatrix('sys_dia_bp.txt', 'NumHeaderLines', 1)
```

The final two arguments of `readmatrix` indicate that the file contains one header line, which will be ignored when reading the numeric data. However, `fscanf` is more flexible than `readmatrix` and can be used to read a wider range of data formats, as the example that follows illustrates.

■ Example 7.4

07.B   Suppose now that the file *sys_dia_bp2.txt* contains the following data:

```
BP data
Sys 51 Dia 85
Sys 88 Dia 141
Sys 67 Dia 95
Sys 77 Dia 111
Sys 68 Dia 115
Sys 99 Dia 171
Sys 80 Dia 121
```

These are the same numerical data that we saw in Example 7.3, but with extra text between the numerical values. In biomedical engineering, it is common to have to deal with data exported from a range of devices with different data formats, so this example is quite realistic. It would not be straightforward to use readmatrix to read these data, because each row of data has mixed data types, i.e., character arrays and numbers. However, with fscanf the solution is quite simple.

```
% open file
fid = fopen('sys_dia_bp2.txt');
if (fid == -1)
    error('File cannot be opened');
end

% skip header line
line = fgetl(fid);

% read systolic and diastolic blood pressure
data = fscanf(fid, '%*s %d %*s %d', [2 inf])

% close file
fclose(fid)
```

All we have done here is to modify the format string of the fscanf function: we have added in two extra character array fields before each decimal integer field (these correspond to the Sys and Dia text in the data file). Note the * symbol between the % and s: this causes MATLAB to read the specified field, but to exclude it from the output array. This is necessary, because, although it can read files containing mixed data types, fscanf can only return an array of values of a single type. ∎

■ Activity 7.2

This exercise builds on Activity 7.1, in which you wrote MATLAB code to read in patient data from the text file *patient_data.txt*. On the book's web site, you will also be able to access a second patient data file called *patient_data2.txt*. This also contains patient data (patient ID, age, height, weight, heart rate), but for some different patients in the trial. Modify the script *m*-file you wrote in Activity 7.1 to also read in these data, and combine them with the data from the first file. There were five pieces of data for each of nine patients in the first file and the same data for an additional five patients in this second file, so your final data should be a 14×5 array. Notice that this second file has no line break characters, so you will not be able to use readmatrix, because this uses line breaks to separate the rows of data. ∎

07.B

Sometimes, however, it is desirable to read in data of mixed types and store them all in a single variable. To do this, we need an alternative file input function: textscan. The following example illustrates its use:

■ Example 7.5

07.B   Suppose now that the file *sys_dia_bp3.txt* contains the patients' names and their blood pressure data, and that we wish to read in and store all of the data, i.e., names and blood pressures:

```
Joe Bloggs 51 85
Josephine Bloggs 88 141
Zhang San 67 95
Anders Andersen 77 111
Erika Mustermann 68 115
Ola Nordmann 99 171
Jan Kowalski 80 121
```

It is not possible to read and store all of these data using `fscanf`, because the resulting array could not contain both character array and numeric data. However, `textscan` returns a *cell array*, rather than a normal array (see Section 6.2), and cell arrays *can* contain mixed data types. Therefore it is possible to read these data using `textscan`. Examine the code shown below.

```
% open file
fid = fopen('sys_dia_bp3.txt', 'r');
if (fid == −1)
    error('File cannot be opened');
end

% read systolic and diastolic blood pressure
data = textscan(fid, '%s %s %d %d');

% close file
fclose(fid);
```

Note that `textscan` uses a similar format string to `fscanf`. However, in this case it is permitted to mix up text data fields (`%s`) with decimal number fields (`%d`), and all fields will be included in the returned variable (`data`). The type of the variable `data` returned by `textscan` will now be a 1×4 *cell* array.

The first and second elements of `data` are arrays that contain the first names and second names of the patients, respectively. In fact, since the lengths of the character arrays vary, these first two elements are themselves cell arrays. The third and fourth elements of `data` are normal arrays of integers representing the systolic and diastolic blood pressures of the patients. Try entering this code into a MATLAB script *m*-file, running it, and inspecting the `data` variable. ■

The `textscan` function is quite flexible and can also be used to read in data of mixed types and with multiple delimiters.

■ Example 7.6

For example, let us change the format of the data, in a new file called *sys_dia_bp4.txt*:

07.B

```
Joe Bloggs;51;85
Josephine Bloggs;88;141
Zhang San;67;95
Anders Andersen;77;111
Erika Mustermann;68;115
Ola Nordmann;99;171
Jan Kowalski;80;121
```

If we were to use the code shown in Example 7.5 to read in these data, it would read them incorrectly, because the default delimiter for `textscan` is white space, i.e., spaces, tabs, or newline characters. As a result the 2nd, 3rd, and 4th data fields (i.e., the last name, systolic blood pressure, and diastolic blood pressure) would be read as a single character array. The following code shows how we can change the delimiter(s) for `textscan`:

```
% open file
fid = fopen('sys_dia_bp4.txt', 'r');
if (fid == −1)
    error('File cannot be opened');
end

% read systolic and diastolic blood pressure
data = textscan(fid, '%s %s %d %d', 'delimiter', ' ;');

% close file
fclose(fid);
```

Take care to note that the character array used to specify the delimiter contains multiple characters (a space and a semicolon). This appears after the `'delimiter'` argument. In this case, either of these two characters will be treated as a delimiter, and this allows the data fields to be read in correctly as before.　■

■ Activity 7.3

The file *names.txt* contains the first and last names of a group of patients. Write a MATLAB script *m*-file to read in these data using the `textscan` function, and display the full names of all patients with the last name "Bloggs."

07.B

(Hint: Look at the MATLAB documentation for the `strcmp` *command.)*　■

So far, all our examples have read data from, or written data to, text files. In computer programming, it is common to distinguish between text files and *binary* files. In fact, strictly speaking, all computer files are binary files, since the data are always stored as a series of 1s and 0s (known as binary digits, or *bits*).

However, it is common to refer to files as text files if the sequence of bits can be interpreted as character data using a coding system, such as ASCII. Any other file is referred to as a binary file. Binary files tend to make more efficient use of disk space than text files, but can occasionally be harder to read. Regardless of the advantages and disadvantages of text and binary files, it is a fact that both types of files are in common use, so we need to know how to read from and write to both.

We have already seen how the `load` and `save` commands can be used with .*MAT* files, which are binary files. We will now consider other functions that can be used with binary files and that allow us more control over how the data are interpreted. The main function to know about for reading data from binary files is `fread`. The use of `fread` is illustrated in the following example:

■ Example 7.7

07.B

Suppose that the file *heart_rates.bin* is a binary file containing the following heart rate data stored as a series of 16-bit integers: [65 73 59 101 77 ... 90 68 92]. How can we write code to read these values into a MATLAB array? Consider the program listing shown below.

```
% open file
fid = fopen('heart_rates.bin', 'r');
if (fid == -1)
    error('File cannot be opened');
end

% read data from binary file as 16-bit integers
hr = fread(fid, inf, 'int16')

% close file
fclose(fid);
```

In this example, the `fread` function takes three arguments:

- The identifier of the binary file from which to read the data: `fid`
- The number of data values to read: using `inf` means read until the end of the file
- The data type of the elements to be read: `int16` specifies 16-bit integers (see Section 1.8)

Note that if we didn't specify `int16` as the data type, `fread` would assume that each byte (8 bits) represented a number, and so the data would not be read correctly. Generally, when reading data from binary files, it is important to know and to specify the type of the data we are reading, since different data types use different numbers of bytes on the computer's hard disk. ■

■ Activity 7.4

The file *cholesterol.bin* is a binary file. The first data element in the file is an 8-bit integer specifying how many records follow. Each subsequent record consists of an 8-bit integer representing a patient's age followed by a floating point `double` representing the same patient's total cholesterol level (in mmol/L). Write a MATLAB script *m*-file to read the age/cholesterol data into two array variables.

07.B

■

7.5   Writing to files

Now we will introduce some MATLAB functions that can be used for writing data to binary or text files. For binary files, the main function for writing data is `fwrite`, the use of which is illustrated in the following example:

■ Example 7.8

The code shown below can be used to generate the binary file that was read in using `fread` in Example 7.7.

07.C

```
% define data array
hr = [65 73 59 101 77 90 68 92];

% open file
fid = fopen('heart_rates.bin', 'w');
if (fid == −1)
    error('File cannot be opened');
end

% write data to binary file as 16−bit integers
fwrite(fid, hr, 'int16');

% close file
fclose(fid);
```

In this example, we provide three arguments to `fwrite`:

- A file identifier: `fid`
- An array containing data to be written to the file: `hr`
- The data type to use when writing: `int16`

Note that there is no argument for specifying how much of the array should be written; the entire array will be written. If we only want to write part(s) of the array, we need to modify the value of the second argument ourselves, or use multiple `fwrite` commands. Note also that when using a single `fwrite` command, only values of a single data type can be written. If we want to write mixed data to a binary file, we can use multiple successive `fwrite` commands before closing the file.

■

One of the most flexible functions for writing data to text files in MATLAB is `fprintf`, which we saw earlier in Example 7.2. The `fprintf` function is actually very powerful, and can be used to write data of a range of different data types, as well as specify the formatting of that data. Consider the following example:

■ Example 7.9

07.C

The code shown below can be used to generate the mixed data file that was read in using `textscan` in Example 7.5.

```
% define cell array of names
names = {'Joe Bloggs','Josephine Bloggs', ...
         'Zhang San','Anders Andersen', ...
         'Erika Mustermann','Ola Nordmann', ...
         'Jan Kowalski'};

% define array of blood pressure values
bps = [51 85
       88 141
       67 95
       77 111
       68 115
       99 171
       80 121];

% open file
fid = fopen('sys_dia_bp3.txt', 'w');
if (fid == −1)
    error('File cannot be opened');
end

% write data to file
for x=1:length(names)
    fprintf(fid, '%s %d %d \n', names{x}, bps(x,1), bps(x,2));
end

% close file
fclose(fid);
```

Here we use a cell array called `names` to store the names of the patients. Each element of `names` is a normal 1-D array of characters. Because the names have different lengths, we can't store them using a normal 2-D character array. The blood pressure values are stored using a normal 2-D array of numbers. After opening the file for writing, we use a `for` loop to iterate through all patients (determined by finding the length of the `names` cell array). In each iteration, the `fprintf` function is used to construct and write a line to the file. This is formed from a character array (patient's name), followed by two decimal values (the systolic and diastolic blood pressures).

We supply five arguments to `fprintf` in this example:

- The file identifier: `fid`; if this is omitted the line is displayed in the command window, as we saw in Example 7.2;
- The format string, in this case, containing three fields (a character array and two decimals), and ending with a new-line character (`'\n'`);
- The remaining three arguments contain the data for each field in the format string, so `names{x}` is used for the text field, `bps(x,1)` for the first decimal and `bps(x,2)` for the second decimal.

Finally, after the loop has finished executing, the output file is closed. ■

When specifying a format string for `fprintf`, we can use the same format specifiers that we use with `fscanf` (see Table 7.3). In addition, we can provide extra information to alter the display format of the fields by adding extra characters in between the `%` and the data type identifier. For example, we can specify the width and precision of numeric fields, e.g.,

```
fprintf(fid, '%10.3f \n', x);
```

This code will display the value of the variable `x` as a floating point number using a total field width of 10 characters, which includes 3 digits after the decimal point (i.e., the number will be displayed to 3 decimal places). For example, if the variable `x` contained the square root of two, the line produced by `fprintf` above would consist of the character array: '   1.414' which contains five spaces before the digit 1.

In addition, a number of flags can be added before the field width/decimal place values. Adding the − character will cause the displayed number to be left-justified (the default is right-justified). Adding the + character will cause a plus sign to be displayed for positive values (by default only a minus sign is added for negative values). Adding a space or 0 will cause the number to be left-padded with spaces or zeroes up to the field width. For example, the following statement:

```
fprintf('%010.3f \n', x);
```

is the same as that shown above, except that the number will be padded to the left with zeroes to make the field width 10 characters ('000001.414').

■ Activity 7.5

The file *bp_data.txt* contains systolic blood pressure data. There are four fields for each patient: patient ID, first name, last name, and systolic blood pressure. Write a MATLAB script *m*-file to read in these data, identify any "at risk" patients (those who have a blood pressure greater than 140), and write only the data for the "at risk" patients to a new file, called *bp_data_at_risk.txt*. ■

07.B, 07.C

Table 7.4 Summary of common file input/output functions.

Function	File Type	Operation	Open/Close File?
load	Binary*, text	Read MATLAB variables from a .*MAT* file or a text file of numeric values	N
save	Binary*	Write MATLAB variables to a .*MAT* file	N
readmatrix	Text, spreadsheet	Read delimiter-separated numeric data from a file into a 1-D or 2-D array; can also be used to read common spreadsheet formats	N
writematrix	Text, spreadsheet	Write 1-D/2-D array of numeric data to a delimiter-separated file; can also be used to write common spreadsheet formats	N
readtable	Text	Read a table data structure from a file	N
writetable	Text	Write a table data structure to a file	N
readcell	Text	Read a cell array from a file	N
writecell	Text	Write a cell array to a file	N
fread	Binary	Read data of a single type from a file into a 1-D array	Y
fscanf	Text	Read data of a single type from a file into a 1-D/2-D array	Y
textscan	Text	Read a file containing mixed data types into 1-D/2-D cell array	Y
fgetl	Text	Read a line of data into a 1-D array of characters	Y
feof	Text	Check for the end-of-file condition	Y
fwrite	Binary	Write an array of data of a single type to a file	Y
fprintf	Text	Write array(s) of data of mixed data types to a file	Y

*.MAT *file.*

7.6 Summary

MATLAB provides a range of file input and output functions for interacting with external files. Each function can be useful in different situations, depending on the type(s) of the data, the format of the file, and whether we are dealing with a binary or a text file. Table 7.4 summarizes the key characteristics of the functions that we have discussed in this chapter. Fig. 7.1 provides flow charts to help you to make the right choice about which of the common MATLAB file input/output functions are appropriate for your situation.

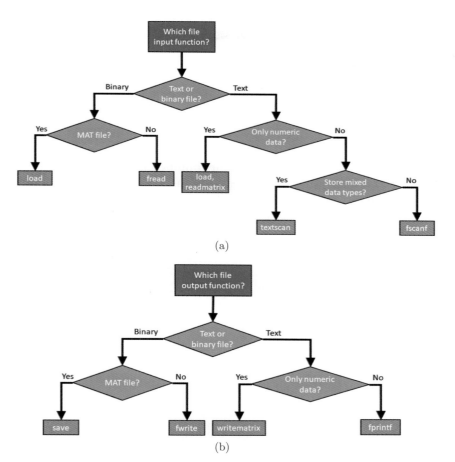

(a)

(b)

FIGURE 7.1
Flow charts to choose which common MATLAB file input/output function is appropriate: (a) input functions; (b) output functions.

7.7 Further resources

- MATLAB documentation on text file functions: http://mathworks.com/help/matlab/text-files.html.
- MATLAB documentation on general file input/output functions: http://mathworks.com/help/matlab/low-level-file-i-o.html.

7.8 Exercises

■ Exercise 7.1

Write a MATLAB script *m*-file to create the following variables in your MATLAB workspace, and initialize them as indicated: *07.A*

- a character variable called x with the value 'a';
- an array variable, called q, containing 4 floating point numbers: 1.23, 2.34, 3.45, and 4.56;
- a Boolean variable, called flag, with the value true.

Now save only the array variable to a *MAT* file, clear the workspace, and load the array in again from the *MAT* file.   ■

■ Exercise 7.2

07.B, 07.C   Write the same 1-D array of floating point numbers that you defined in Exercise 7.1 to a text file using the writematrix function. Use semicolons as the delimiters in the file. Then clear the workspace and read the array in again using readmatrix.   ■

■ Exercise 7.3

07.B, 07.C   Again, working with the array from Exercise 7.1, this time, write it to an "xlsx" spreadsheet file using the writematrix function. Then clear the workspace, and read the data in again using readmatrix.   ■

■ Exercise 7.4

07.B   The equation for computing a person's body mass index (BMI) is

$$\text{BMI} = \frac{\text{mass}}{\text{height}^2},$$

where the mass is in kilograms, and the height is in meters. A BMI of less than 18.5 is classified as underweight, between 18.5 and 25 is normal, more than 25 and less than 30 is overweight, and 30 or over is obese. You are provided with two text files: *bmi_data1.txt* and *bmi_data2.txt*. These both contain data about patients' height and weight (preceded by an integer patient ID), but in different formats. Write a MATLAB script *m*-file to read in both sets of data, combine them, and then report the patient ID and BMI classification of all patients who do not have a normal BMI.   ■

■ Exercise 7.5

07.B   When MR scans are performed a log file is typically produced containing supplementary information on how the scan was acquired. An example of such a file (*logcurrent.log*) is available to you. One piece of information that this file contains is the date and time at which the scan started. This information is contained in the 2nd and 3rd words of a line in the file that contains the text "Scan starts." Write a MATLAB script *m*-file to display the start date and time for the scan that produced the *logcurrent.log* file.

(Hint: Look at the MATLAB documentation for the `strfind` *and* `strsplit` *functions.)* ■

■ Exercise 7.6

Whenever a patient is scanned in an MR scanner, as well as the log file mentioned in Exercise 7.5, the image data are saved in files that can then be exported for subsequent processing. One format for these image data, which is used on Philips MR scanners, is the PAR/REC format. With PAR/REC, the image data themselves are stored in the REC file, whereas the PAR file is a text file that contains scan details to indicate how the REC file data should be interpreted. An example of a real PAR file (*patient_scan.par*) is available to you through the book's web site. Write a MATLAB script *m*-file to extract from the PAR file, and display the following information: the version of the scanner software (which is included at the end of the 8th line), the patient name, and the scan resolution. ■

07.B

■ Exercise 7.7

Measuring blood sugar levels is an important part of diabetes diagnosis and management. Ten patients had their post-prandial blood sugar level measured, and the results were 5.86, 8.71, 4.83, 7.05, 8.25, 7.87, 7.14, 6.83, 6.38, and 5.77 mmol/L. Write a MATLAB script *m*-file to write these data to a binary file using an appropriate data type. Your program should then clear the MATLAB workspace and read in the data again. ■

07.B, 07.C

■ Exercise 7.8

Write a MATLAB script *m*-file to read in a text file, and write out a new text file that is identical to the input file, except that all upper case letters have been converted to lower case. You can test your code using the file *text.txt*. ■

07.B, 07.C

■ Exercise 7.9

Write a MATLAB script *m*-file to read in another *m*-file, ignoring all lines that are blank or contain comments and removing any leading white space on each line. It should write the remaining statements to a new *m*-file. Note that a comment is any line in which the first non-white space character is a %.

07.B, 07.C

(Hint: The MATLAB command `strtrim` *can be used to remove leading and trailing white space from a character array.)* ■

FAMOUS COMPUTER PROGRAMMERS: GRACE HOPPER

Grace Hopper was a pioneering American computer scientist, who, due to the breadth of her achievements, is sometimes referred to as "Amazing Grace." She was born in 1906 in New York City and studied mathematics and physics at undergraduate level before gaining a PhD in mathematics in 1934.

During World War II, she volunteered to serve in the United States Navy. However, she was considered to be too old and not heavy enough for regular duty, and so was assigned to work on the Navy's Computation Project at Harvard University, which was working to develop an early electro-mechanical computing machine. It was on this project that Grace became one of the world's first computer programmers. She is also commonly credited with popularizing the term "debugging" for fixing programming problems (inspired by an actual moth that was removed from the computer).

After the war finished, she requested to transfer to the regular Navy, but was turned down due to her age. She continued to work at Harvard on the Computation Project until 1949. Grace then joined the Eckert–Mauchly Computer Corporation as a senior mathematician, and joined the team developing the UNIVAC I (the second commercial computer produced in the United States). Whilst working on the UNIVAC I project, she developed the world's first compiler (a program for converting a higher level language into machine readable instructions). These days almost all programming is done in such higher level languages, but, at the time, it was a major advance and Grace had trouble convincing people of its importance. In her own words, "I had a running compiler and nobody would touch it. They told me computers could only do arithmetic." Eventually others caught on to Grace's idea, and in 1959 she was a technical consultant to the committee that defined the new language COBOL.

She always remained a Navy reservist, and by the time of her retirement, at the age of 79, she held the rank of rear admiral. She died in 1992, aged 85, and was interred with full military honors in Arlington National Cemetery.

> "To me programming is more than an important practical art. It is also a gigantic undertaking in the foundations of knowledge."

Grace Hopper

Program design

8.1 Introduction

So far, we have learned how to make use of a range of MATLAB programming features to develop increasingly sophisticated programs. In Chapter 4, we looked at how to define our own functions for performing specific tasks. However, as we start to address more and more complex problems, it is unlikely that they will be easily solved by writing one big function. Rather, an efficient and elegant implementation is likely to consist of a number of functions that interact by making calls to each other and passing data as arguments. *Program design* refers to the process of deciding which functions to write and how they should interact.

When tackling a larger programming problem, it is generally not a good idea to start coding straight away: time invested in producing and documenting a high quality structured design will produce significant benefits in the long run, primarily by reducing time spent writing code and maintaining it. This chapter will introduce some tools that will be useful when designing programs to solve larger and more complex problems.

We also return to the important topic of program testing. Testing, or verifying, that a program meets its requirements is an essential part of the development process. In Chapter 5, we saw how an incremental approach to writing programs, with testing of the incomplete program after each stage, can actually speed up and improve the development process. In this chapter, we will discuss how this philosophy of incremental development interacts with a structured design process.

In computer science, it is common to identify two distinct approaches to program design: *top-down* design and *bottom-up* design. In this chapter, we will discuss the meanings of these two concepts. However, in procedural programming (see Section 1.2), which is the subject of this book, top-down design has been the traditional approach, so we will focus mainly on this, illustrating it with a simple case study.

8.2   Top-down design

With a top-down design approach, the emphasis is on planning and having a complete understanding of the system before writing code. Top-down design is also often referred to as *stepwise refinement*. The basic idea is to take a hierarchical approach to identifying program "modules." By a "module" we simply mean a part of the overall program that performs a specific task. For example, in MATLAB, we can consider a function to be a "module."

In top-down design, we start off by breaking down a complex problem into a number of simpler sub-problems. Each sub-problem will be addressed by a separate program module, which will have its requirements separately specified, including how it should interact with the *main* part of the program (i.e., the top level of the hierarchy). The interaction is typically specified in terms of what data are passed to a given module from the main module, and what data are passed back from the module to the main module.

Next, each identified module can be broken down further into sub-modules, which will also have their behavior specified, including what data are passed between the new sub-modules and their parent modules. This process continues until the sub-modules are simple enough to be easy to implement (i.e., to write as code).

The result of this *stepwise refinement* is a hierarchy of modules and sub-modules that represents the structure of the program to be implemented. This, together with specifications for the behavior of each module, represents the program design. Normally, no coding will begin until a sufficient level of detail has been reached in the design of at least some part of the system. Often, the complete structured design will be produced before any coding takes place.

We will illustrate this process with a case study.

■ Example 8.1

The file *patient_data.txt* contains data for a group of patients registered at a general practitioner's (GP)[1] medical surgery (patient ID, name, date of birth). A second file, *patient_examinations.txt*, contains data about physical examinations that patients at the surgery have undergone (patient ID, age, height, weight, systolic/diastolic blood pressure, total cholesterol level). There may be multiple examinations for each patient, or none.

08.A, 08.B, 08.C, 08.D

The GP has decided to implement a program in which "high risk" patients are invited for further tests. The initial definition for "high risk" patients is those who have an age that was greater than or equal to 75 at their most recent examination.

We are required to design and implement a program to read in the data from these files and display the patient IDs and names of all high risk patients. If there are no high-risk patients the program should terminate with an appropriate message.   ■

Because this is a reasonably complex problem to address, and it is not immediately obvious how we should go about solving it, we will start by breaking the problem down into simpler sub-problems. In software design, this process is known as *factoring*, and the first time we apply it to the overall problem it is known as *first-level factoring*.

When performing top-down design, it is also important to document our design in some way. This documentation can help to guide the code development process and ensure that different members of a software team are working toward a common goal. There are several graphical and textual techniques that can be used to document designs. In this chapter, we will introduce a common approach based on a combination of one graphical and one textual technique: namely *structure charts* and *pseudocode*. We note, however, that the principles that we aim to get across in this chapter are not specific to these techniques, and that other approaches could be used in their place.

Step 1: first-level factoring

Structure charts are a graphical technique for representing top-down designs. In a structure chart, boxes are used to represent program modules, and connections between modules are shown using arrows between the boxes. Structure charts also allow *data flows* between modules to be indicated by circles, with arrows indicating the direction of flow. In MATLAB, such data flows can correspond to arguments being passed into and back from functions.

---

[1]The term "GP" is used in the UK, and some other countries, to describe a doctor working as a primary care physician.

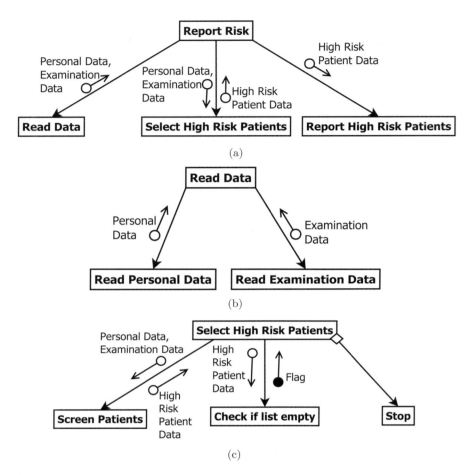

FIGURE 8.1
Structure charts representing the design of the risk screening program: (a) **Step 1:** first-level factoring; (b) **Step 2:** further factoring of *Read Data*; (c) **Step 3:** further factoring of *Select High Risk Patients*. Empty circles represent data, filled circles represent control. The diamond in (c) represents conditional execution.

Fig. 8.1a shows the structure chart after first-level factoring of the problem in Example 8.1 has been performed. The main program (which we have called *Report Risk*) is at the top level of the hierarchy. Below this, we have broken down our initial problem into three sub-problems. We are saying that the requirements of our program can be met by executing three modules one after the other: first, we read all personal and examination data from the files (*Read Data*), then we perform the risk screening (*Select High Risk Patients*), and finally we display the details of the selected patients (*Report High Risk Patients*).

Notice that the arrows that link the three sub-modules to the parent (main) module are annotated with the data that need to be passed between them. We have used empty circles with arrows to indicate data flowing in the direction of the arrow. For example, the *Read Data* module should pass the personal and examination data that was read back to the *Report Risk* program. These data are then passed to the *Select High Risk Patients* module for screening, which returns a list of the selected high-risk patients. These data are passed to the *Report High Risk Patients* module.

Currently, none of these three new modules are simple enough to be implemented easily, so we proceed by further breaking down one of them (*Read Data*). This is known as *further factoring*.

Step 2: further factoring of *Read Data*

Fig. 8.1b shows the additional structure chart for the further factoring of *Read Data*. Now, we are saying that the problem of reading in all data can be solved by first reading in the personal data, and then reading in the examination data. Note the data flowing between the parent module (*Read Data*) and its two sub-modules.

Step 3: further factoring of *Select High Risk Patients*

Now we proceed with further factoring of the *Select High Risk Patients* module identified in **Step 1**. Fig. 8.1c shows the structure chart that represents the operation of this module. This specifies how the problem of selecting the high-risk patients can be solved: first, the screening operation is performed, then we check if the list of patients is empty or not, and finally the program stops if the list was empty. This time there is some new notation in the structure chart. In addition to the empty circles with arrows representing data, we also have a filled circle. A filled circle with an arrow in a structure chart indicates *control* information. Control information is normally a binary value that has some effect on future program execution. For example, in Fig. 8.1c, the control information, called `Flag`, is passed back from the *Check if list empty* sub-module: this will be true if there are no high-risk patients and false otherwise. This value is then used to decide whether or not to execute the next sub-module, *Stop*. The *conditional execution* of the *Stop* sub-module is indicated by the diamond symbol, where its line joins the *Select High Risk Patients* module.

As an aside, as well as conditional execution (which corresponds to conditional statements in procedural languages, such as MATLAB, such as `if` statements; see Chapter 2), structure charts can also represent *iteration*. For instance, suppose that we wanted our program to screen data from multiple GP surgeries, i.e., the *Read Data*, *Select High Risk Patients*, and *Report High Risk Patients* modules together, should be iteratively executed. Fig. 8.2 shows how the structure chart shown in Fig. 8.1a should be modified to indicate this iteration. The

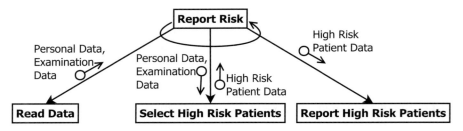

FIGURE 8.2
Structure chart showing iteration of sub-modules.

Table 8.1 Summary of structure chart notation.	
Symbol	**Meaning**
Box: $\boxed{x}$	Module performing task x; see Fig. 8.1.
Arrow joining boxes: ⟶	One module calls another; see Fig. 8.1.
Annotated empty circle: x ○⟶	Data flow, i.e., data x flows between modules in direction of arrow; see Fig. 8.1.
Annotated filled circle: x ●⟶	Control flow, i.e., control information x flows between modules in direction of arrow; see Fig. 8.1c.
Diamond: ◇⟶	Conditional execution, i.e., execution of called module depends on control information; see Fig. 8.1c.
Curved arrow: ⤺	Iteration, i.e., called modules covered by arrow are executed multiple times; see Fig. 8.2.

curved arrow covering the three modules indicates that they should be executed multiple times.

Table 8.1 provides a summary of the notation we have introduced for structure charts.

Step 4: write pseudocode

To return to our case study, we now proceed by examining which modules might need further elaboration. Looking at Fig. 8.1b (i.e., the *Read Data* module), we might conclude that this structure chart has enough detail already to proceed to implementation. However, there are two other modules that do require further analysis: *Report High Risk Patients* from the first-level factoring, and the new sub-module from **Step 3** (*Screen Patients*). Although it would be possible to continue to use structure charts to further break down these problems, we will now introduce a new way of representing program structure: *pseudocode*. Pseudocode is simply an informal description of the steps involved in executing a computer program, often written in something similar to plain English. Pseudocode is commonly used when we get down to the lower-level details of how modules work internally, rather than the interactions of higher-level operations commonly represented by structure charts.

For example, the pseudocode below shows more detail on how the *Screen Patients* module works (which was identified in the structure chart shown in Fig. 8.1c).

PSEUDOCODE - Screen Patients:

```
For each PatientID in Personal Data
  Look up all records matching PatientID in Examination Data
  HighRisk = false
  For each record
    If Age >= 75
      HighRisk = true
  If HighRisk = true
    Add PatientID, Name to High Risk Patients List
Return High Risk Patients List
```

This pseudocode indicates the steps involved in screening for high-risk patients. We first iterate over all patients. For each patient, we look up all corresponding examination records (i.e., those with the same Patient ID). Next, we initialize a HighRisk variable to false. Then, for each of the corresponding examination records, we check the age of the patient at the time of the examination, and if it is greater than or equal to 75, we set HighRisk to true. This way, if the age at *any* examination was greater than or equal to 75, then HighRisk will be true, otherwise it will be false. If HighRisk is true, we add the patient details to the high-risk patients list. Finally, we return the list as the output of the module.

Note that this pseudocode is starting to look more like computer code, with variables being defined and simple operations to act on them. However, we haven't yet written any code, so our design could be implemented in any language of our choice, including MATLAB.

Pseudocode for the remaining module of our design (*Report High Risk Patients*) is shown below.

PSEUDOCODE - Report High Risk Patients:

```
For each record in High Risk Patients List
  Display PatientID, Name
```

In summary, a typical top-down design process would involve first-level factoring (e.g., Fig. 8.1a), followed by further factoring (e.g., Figs. 8.1b and c) as needed, until a certain level of detail is reached. Then some of the bottom level modules in the structure charts would be expanded by writing pseudocode. The choice of when to switch from structure charts to pseudocode is subjective and a matter of personal preference: you may prefer to use entirely structure charts, or entirely pseudocode, but it is common to use a mixture of the two, as we have in this example. Either way, once the design is complete and documented, coding can begin.

8.2.1   Incremental development and test stubs

We now have a design that can be used as the starting point for producing an implementation. Regardless of the programming language used, we need a general approach for producing our implementation. In Section 5.2, we introduced the concept of *incremental development*, in which software is developed in small pieces, with frequent testing of the developing program. In Example 5.1, a program for cardiac dyssynchrony index calculation was developed incrementally, and the entire program was tested after each new piece of code was added. However, this was for a relatively simple example, in which each part of the overall program consisted of just a few lines of code. Now that we are tackling larger, more complex problems, it is more likely that our operations will be functions that perform quite sophisticated tasks. So this begs the question: how can we perform incremental development and testing when implementing a top-down design of a larger program?

The answer to this question lies in the concept of a *test stub*. A test stub is a simple function or line of code that acts as a temporary replacement for more complex, yet-to-be-implemented code. For instance, in our case study, we might use test stubs for the *Select High Risk Patients* and *Report High Risk Patients* modules whilst we are developing and testing the code for the *Read Data* module. The test stub for *Select High Risk Patients* could, for example, always return a list containing a single patient, which could be "hard-coded," i.e., just directly assigned to the appropriate variable. The test stub for *Report High Risk Patients* could always display a simple message. The code listing shown below illustrates such a program under development.

ReportRisk.m:

```
% GP risk screening program

% Read Data (under development in another file)
[personalData, examData] = ReadData();

% Select High Risk Patients (test stub)
highRisk = '123 Joe Bloggs';

% Report High Risk Patients (test stub)
fprintf('High Risk:\n');
```

Once the *Read Data* module has been developed and tested, the test stub for *Select High Risk Patients* could be replaced with an implementation and tested. Finally, the test stub for *Report High Risk Patients* could be replaced with its implementation. Final testing would then verify the operation of the complete program.

8.3   Bottom-up design

Although top-down design is the most common approach when developing procedural programs, an alternative approach is *bottom-up design*. The philosophy of bottom-up design can be best understood by drawing an analogy with children's building blocks. When building a structure from building blocks, we would probably have a general idea of what we wanted to make to begin with (e.g., a house), but it is common to start off by building some small components of the overall structure (e.g., the walls). Once enough of the components are complete, they can be fitted together.

The same principle applies to bottom-up software design: we begin with a general idea of the system to be developed (not specified in detail), and then start off by developing some small self-contained modules. When enough of the modules are complete, they can be combined to form larger subsystems. After each combination step, the developing program can be tested. Like the top-down approach discussed in Section 8.2.1, this can be thought of as a form of incremental development (see Section 5.2): the program is developed module-by-module, with some increments consisting of combining existing modules to form larger modules. This process is repeated until the complete system has been produced.

A key feature of development using bottom-up design is early coding and testing. Whereas, with top-down design, no coding is performed until a certain level of detail is reached in the design, with bottom-up design, coding starts quite early. Therefore there is a risk that modules may be coded without having a clear idea of how they link to other parts of the system, and that such linking may not be as easy as first thought. On the other hand, a bottom-up approach is well-suited to benefiting from *code reuse*. Code reuse refers to taking advantage of previously developed and tested, modular pieces of code. It is generally considered to be a good thing in software development, since it reduces the amount of duplicated effort in writing the same or similar code modules, and it also reduces the chances of faults being introduced into programs (since the reused modules have been extensively tested already).

To further illustrate the concept of bottom-up design, let's return to our GP surgery risk screening program case study. To develop this program using bottom-up design, we would first consider what types of function/module we might need to write to develop a program for risk screening. An initial obvious choice might be a module to find examinations for a given patient. So we would start off by specifying the requirements for this module, designing it using structure charts and/or pseudocode, and then writing the code and testing it. Next, we might choose to expand this module to check if any of the ages at the examinations were greater than or equal to 75, and to build up a list of high-risk patients. This module would be specified/designed/coded/tested. Next, we might write a module to check if the list is empty, and terminate the program if it is. This new module would be combined with the existing module to make a new module for selecting high-risk patients. This process of expanding and combining modules would continue until all modules were completed and tested, forming the final program.

8.4   A combined approach

Although top-down and bottom-up design seem to represent opposing philosophies of software design, most modern software design techniques combine both approaches. Though a thorough understanding of the complete system is usually considered necessary to produce a high quality design, leading theoretically to a more top-down approach, most software projects attempt to maximize code reuse. This reuse of existing code modules introduces more of a bottom-up flavor. You may find it useful to adopt such an approach when designing your own software, i.e., take a predominantly top-down approach to the software design process (e.g., using structure charts and pseudocode), but always keep in mind the possibilities for code reuse. Where possible, try to include modules in your design that can be reused from other programs you have written in the past or from freely available code on the internet.

8.5   Alternative design approaches

In this chapter, we have introduced the basics of software design using structure charts and pseudocode. However, in reality software design is a large and complex field, and many authors have written entire books on the subject. Alternative design notations and approaches do exist, and we will not attempt to cover them all here. Our aim in introducing this topic was not to argue for a particular notation or approach, but rather to give a flavor for the field and to illustrate the key point that investing time in designing a program before writing code can be a good idea.

■ Activity 8.1

08.A, 08.B, 08.C, 08.D

Write a MATLAB implementation of the design we produced for Example 8.1 (i.e., the GP surgery risk screening program). You will need to consider whether to implement each module using a sequence of statements in the main script *m*-file or as a separate MATLAB function. Use an incremental development approach, and test stubs when developing your solution.   ■

■ Activity 8.2

08.A, 08.B, 08.C, 08.D

The GP now wishes to expand her definition of a "high-risk" patient by also including those patients who have a body mass index (BMI) of less than 18.5 or greater than 25 at their most recent examination. Recall that a patient's

BMI can be calculated using the formula

$$BMI = mass/height^2,$$

where the height is specified in meters and the weight in kilograms.

Modify your program design and implementation from Activity 8.1 to meet this new requirement. ∎

∎ Activity 8.3

Now modify your program design and implementation from Activity 8.2 so that the program also reports the reason for a patient being high-risk, i.e., either because of their age or their BMI, or both. ∎

08.A, 08.B, 08.C, 08.D

∎ Activity 8.4

Adjust your program design and implementation from Activity 8.3 so that the program also reports a patient as high-risk if their total cholesterol level is greater than 5 mmol/L. Again, the reason for being reported as high-risk should be displayed. ∎

08.A, 08.B, 08.C, 08.D

∎ Activity 8.5

Finally, modify your program design and implementation from Activity 8.4 so that the program will also consider the patients' systolic and diastolic blood pressures. The rules should be the following:

08.A, 08.B, 08.C, 08.D

- If the systolic blood pressure is ≥120 mmHg, but <140 mmHg, or the diastolic blood pressure is ≥80 mmHg, but <90 mmHg, the patient is classified as having *pre-high blood pressure*. These patients should only be high-risk if another one of the risk criteria is met. The pre-high blood pressure should also be reported as a reason for being at risk.
- If the systolic blood pressure is ≥140 mmHg, or the diastolic blood pressure is ≥90 mmHg, the patient is classified as having *high blood pressure*. These patients should be reported as high-risk, even if no other criteria are met. ∎

8.6   Summary

There are two opposing philosophies to approaching the software design process: *top-down* design and *bottom-up* design.

Top-down design is also known as *stepwise refinement* and involves successively breaking down problems into simpler sub-problems, stopping when the sub-problems are simple enough to tackle on their own. The process of identifying sub-problems and the interactions between them is referred to as *factoring*. Top-down designs can be documented using a graphical technique, such as *structure charts*, and/or text-based techniques, such as *pseudocode*. Top-down designs can be incrementally developed by making use of *test stubs*, which are simple functions or lines of code that act as a temporary replacement for the full, yet to be implemented code.

Bottom-up design involves starting off with a general idea of the system to be developed, but not documenting it in detail. Rather, small, self-contained modules that are likely to be useful are identified, their requirements specified, a design produced, and code written and tested. These low-level modules are successively combined to form higher-level modules. This process continues until the final program has been formed and tested. With bottom-up designs, there is a risk that modules may be coded without having a clear idea of how they might link to other parts of the system. On the other hand, bottom-up approaches are well-suited to the concept of *code reuse*, (making use of previously written and tested code modules).

Modern software design approaches generally feature a combination of top-down and bottom-up design: a predominantly top-down approach is taken, but care is taken to identify modules that can be reused from previous implementations. As a general rule, for larger programs, it's a good idea not to begin coding straight away. It's almost always a good idea to plan our design on paper first using tools such as structure charts and pseudocode.

8.7   Further resources

- Top-down vs. bottom-up design: http://en.wikipedia.org/wiki/Top-down_and_bottom-up_design.
- Test stubs: http://en.wikipedia.org/wiki/Test_stub.
- Structure charts: http://en.wikipedia.org/wiki/Structure_chart.

8.8   Exercises

■ Exercise 8.1

08.A, 08.B, 08.C, 08.D

In Activity 3.3, we introduced the concept of an *ejection fraction* (EF) in cardiology. Recall that the *end diastolic volume* (EDV) and *end systolic volume* (ESV) of the heart's left ventricle refer to the volume (in mL) of blood just prior to the heart beat and at the end of the heart beat, respectively. The EF is the fraction of blood pumped out from the heart in each beat, and can be computed from the EDV and ESV as follows:

$$EF = \frac{EDV - ESV}{EDV}.$$

The normal range of EF values in healthy adults is 50%–65%.

The text file *heart_data.txt* contains measured EDV and ESV values for a number of patients. The first few lines of the file are shown below:

```
1111 120 61
2222 136 49
3333 89 64
4444 111 42
5555 160 119
```

The first column represents a patient ID, the second column the EDV, and the third column the ESV.

A program should read in these data from the file, compute the EF for each patient, and select only those patients who have an EF outside the normal range. The patient ID, EDV, ESV, and EF for only these abnormal patients should then be written to a new text file, called *at_risk.txt*. The format of this new file should be as follows:

```
1111 120 61 0.49167
3333 89 64 0.2809
5555 160 119 0.25625
```

Design and implement a program to meet these requirements. Use structure charts and pseudocode in your design, and incremental development/test stubs when writing your implementation. ∎

■ Exercise 8.2

The *arterial resistivity index* (ARI) is a measure of pulsatile blood flow that indicates the resistance to blood flow through an artery. The ARI is a common way of characterizing the arterial waveform and is used in applications such as detection and monitoring of kidney disease, detection of pre-eclampsia during pregnancy, and evaluation of the success of liver transplants. The equation for computing ARI is

08.A, 08.B, 08.C, 08.D

$$ARI = \frac{(v_{systole} - v_{diastole})}{v_{systole}},$$

where $v_{systole}$ and $v_{diastole}$ are the velocities of the blood flow during cardiac *systole* (i.e., contraction) and *diastole* (i.e., relaxation), respectively.

Arterial velocity is commonly measured using *Doppler ultrasound*, which works by measuring the change in frequency of reflected sound waves due to the Doppler effect.

The file *doppler.mat* contains a series of velocity measurements made from the carotid artery of a human volunteer. The file contains two variables: `doppler` contains the Doppler velocity measurements in cm/s, and `t` contains the times in milliseconds of these measurements. The peaks in the velocity measurements represent the blood flow at systole, whereas the troughs represent the flow at diastole.

A program is required to read in these data and display them as a plot similar to that shown in the figure below.

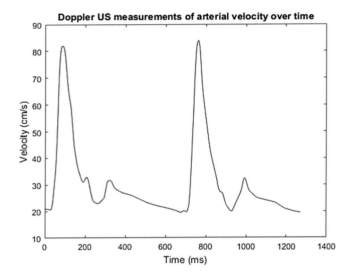

The program should then use the data to compute the ARI. It can be assumed that the systolic velocity is the maximum velocity in the sequence, and the diastolic velocity is the minimum.

Design and implement a program to meet these requirements. Use structure charts and pseudocode in your design, and incremental development/test stubs when writing your implementation.   ■

FAMOUS COMPUTER PROGRAMMERS: EDSGER DIJKSTRA

Edsger Dijkstra was a Dutch computer scientist, who was born in 1930 in Rotterdam. Originally, he wanted to study law, and had the ambition to represent the Netherlands at the United Nations. However, his family persuaded him to turn his attentions to science. He then planned to become a theoretical physicist, and went to study mathematics and physics at the University of Leiden.

In 1951, a chance event changed Edsger Dijkstra's life. He saw an advertisement for a 3-week course on computer programming at the University of Cambridge in the UK, and he decided to attend. Due to the scarcity of such skills at the time, this resulted in Dijkstra getting a part-time job as a programmer in the Computation Department of the Mathematical Centre in Amsterdam. Subsequently, he reflected that "in '55 after three years of programming, while I was still a student, I concluded that the intellectual challenge of programming was greater than the intellectual challenge of theoretical physics, and as a result I chose programming." From that moment on he dedicated his life to computer science.

Dijkstra married in 1957, and was required to state his profession on the marriage certificate. He wrote that he was a Computer Programmer, but this was not accepted, because the authorities claimed that there was no such profession. He had to change it to Theoretical Physicist.

In 1962 Dijkstra became a Professor of Mathematics at Eindhoven University, because it had no computer science department at the time. Whilst there, he built a team of computer scientists, and performed some pioneering work on computers and computer programming. He is considered to be one of the very earliest computer programmers and has been one of the most influential figures in computer science since its inception.

Despite having been responsible for much of the technology underlying computer programming, Dijkstra resisted the personal use of computers for many decades. He would produce most of his articles by hand, and when teaching he would use chalk and a blackboard, rather than overhead slides.

Edsger Dijkstra died in 2002 at the age of 72.

"How do we convince people that in programming simplicity and clarity - in short: what mathematicians call 'elegance' - are not a dispensable luxury, but a crucial matter that decides between success and failure?"

Edsger Dijkstra

CHAPTER 9

Visualization

At the end of this chapter you should be able to:

O9.A Use MATLAB® to produce pie chart visualizations
O9.B Use MATLAB to visualize multiple plots
O9.C Use MATLAB to produce 3-D plots
O9.D Use MATLAB to read/write/display/manipulate imaging data

9.1 Introduction

MATLAB is a powerful application for numerical processing and visualization of complex datasets. So far in this book, we have seen how to create data using variables (Chapter 1) or to read data from external files (Chapter 7), as well as how to process data using programming constructs (Chapters 2 and 4). In Chapter 1, we also introduced the basics of data visualization in MATLAB. In this chapter, we will build on this knowledge and introduce some built-in MATLAB functions for performing more sophisticated visualizations of complex datasets, including images.

9.2 Visualization

In Chapter 1, we saw how to use the MATLAB `plot` function to produce line plots. We also introduced a number of MATLAB commands to annotate plots produced in this way (e.g., `title`, `xlabel`, `ylabel`, `legend`, etc.). Most of these annotation commands are applicable to other types of plot, as we will see later in this chapter. However, `plot` is only suitable for relatively simple data visualizations, such as visualizing the relationship between two variables or plotting a one-dimensional (1-D) signal. In this chapter, we will introduce further built-in MATLAB functions that provide more flexibility when producing data visualizations.

211

MATLAB® Programming for Biomedical Engineers and Scientists. https://doi.org/10.1016/B978-0-32-385773-4.00018-6
Copyright © 2023 Elsevier Ltd. All rights reserved.

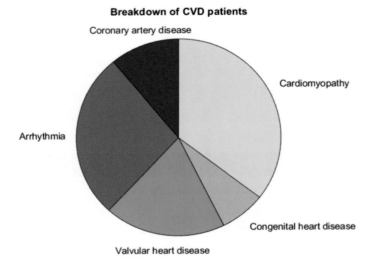

FIGURE 9.1
Generating a pie chart in MATLAB using the `pie` command.

9.2.1 Pie charts

The first type of visualization we will introduce is the pie chart. Pie charts can be used to quickly visualize the relationships between different quantities, showing them as fractional parts of the total quantity.

■ Example 9.1

09.A Consider the code below, in which numbers of cardiovascular disease (CVD) patients of different disease types, who are registered at a hospital, are stored in the numerical array `CVD_numbers`, and the corresponding disease descriptions are stored in the cell array `CVD_labels`. The MATLAB `pie` function takes these two variables as arguments and produces the visualization shown in Fig. 9.1.

```
CVD_numbers = [179, 440, 312, 118, 575];
CVD_labels = {'Coronary artery disease', ...
              'Arrhythmia', ...
              'Valvular heart disease', ...
              'Congenital heart disease', ...
              'Cardiomyopathy'};
pie(CVD_numbers, CVD_labels);
title('Breakdown of CVD patients');
```

■ Activity 9.1

09.A The table below shows the breakdown of a cohort of hospital patients by blood pressure status (see Exercise 2.5).

Status	Number of Patients
Low	36
Ideal	79
Pre-high	64
High	12

Use MATLAB to produce a pie chart to visualize these data. ∎

9.2.2 Visualizing multiple datasets

It is often useful to be able to view multiple visualizations at once, for example, to enable us to compare the variations of different variables over time. There are a number of ways in which we can do this using MATLAB.

∎ Example 9.2

The easiest way to visualize different variables is to carry out multiple plots, either within the same figure or in separate figures. For example, in Section 1.10, we saw how to display multiple plots on the same figure using the plot function:

09.B

```
x = 0:0.1:2*pi;
y = sin(x);
y2 = cos(x);
plot(x,y,'-b', x,y2,'--r');
title('Sine and cosine curves');
legend('Sine','Cosine');
```

The plot function can take arguments in sets of three (i.e., *x* data, *y* data, line/marker style). Each set will represent a different line plot in the figure. ∎

∎ Example 9.3

Alternatively, if we have a large number of plots to display, perhaps within a loop in our program, it may be more convenient to make separate calls to the plot command to produce our figure. We can do this using the MATLAB hold command, as the following example illustrates:

09.B

```
x = 0:0.1:2*pi;
y = sin(x);
y2 = cos(x);
plot(x,y,'-b');
hold on;
plot(x,y2,'--r');
title('Sine and cosine curves');
legend('Sine', 'Cosine');
hold off;
```

The command `hold on` tells MATLAB to display all subsequent plots on the current figure, without overwriting its existing contents (the default behavior is to clear the current figure before displaying the new plot). This behavior is turned off using the command `hold off`. ∎

■ Example 9.4

09.B   Sometimes, displaying too many plots on the same figure can make visualization more difficult. Alternatively, the plots may represent completely different variables on the x and/or y axes, which cannot easily be plotted together in a single figure. In such cases, we can use multiple figures. We can achieve this in MATLAB through the use of the `figure` command, e.g.,

```
x = 0:0.1:2*pi;
y = sin(x);
y2 = cos(x);
plot(x,y);
title('Sine curve');
figure;
plot(x,y2);
title('Cosine curve');
```

Here, the `figure` command tells MATLAB to create a new figure window (keeping any existing figure windows intact) and to make it the current figure. Any subsequent plotting commands will be displayed in the new figure. ∎

■ Example 9.5

09.B   Creating multiple figures as in the previous examples can be useful, but we might prefer not to generate too many figure windows. An alternative way of generating multiple separate plots is to use subplots within a single figure. The MATLAB `subplot` command allows us to do this, as this example illustrates.

```
x = 0:0.1:2*pi;
y = sin(x);
y2 = cos(x);
subplot(2,1,1);
plot(x,y);
title('Sine curve');
subplot(2,1,2);
plot(x,y2);
title('Cosine curve');
```

This code produces the output shown in Fig. 9.2. From the code shown above, we can see that `subplot` takes three arguments. The first two are the numbers of rows and columns in the rectangular grid of subplots that will

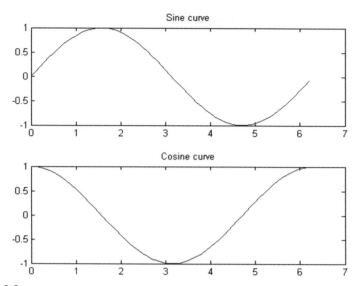

FIGURE 9.2
Example of the use of the `subplot` command to generate multiple plots in a single figure.

be created (in this case, 2 rows and 1 column). The third argument is the number of the current subplot within this grid, i.e., where subsequent plots will be displayed. MATLAB numbers its subplots using the "row major" convention: the first subplot is at row 1, column 1; the second subplot is at row 1, column 2, and so on. ■

Notice how, in Fig. 9.2, the two subplots had a common variable on their *x*-axes, and that the limits of the axes were the same (i.e., 0 to 7). In cases like this, it is a good idea to position the subplots vertically (i.e., one above the other), as shown in Fig. 9.2, rather than horizontally, to enable the plots to be more easily compared visually.

■ Example 9.6

Finally, there is occasionally a requirement to display multiple plots on the same figure with different scales and/or units on their *y*-axes. The example that follows illustrates this. This is a program that visualizes two exponentially decaying oscillating functions with very different scales. The output of the program is shown in Fig. 9.3.

09.B

```
% generate data
x = 0:0.01:20;
y1 = 200*exp(-0.05*x).*sin(x);
y2 = 0.8*exp(-0.5*x).*sin(10*x);

% produce plots
```

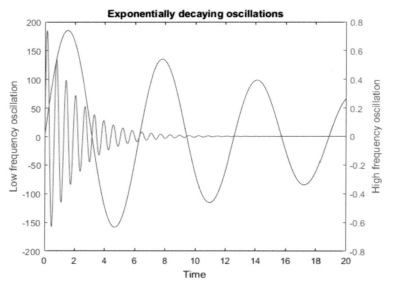

FIGURE 9.3
Example of the use of the `yyaxis` command to generate multiple plots in a single figure with different y-axes.

```
figure;
yyaxis left;
plot(x,y1);
yyaxis right;
plot(x,y2);

% annotate plots
title('Exponentially decaying oscillations');
xlabel('Time');
yyaxis left;
ylabel('Low frequency oscillation');
yyaxis right;
ylabel('High frequency oscillation');
```

The built-in MATLAB function `yyaxis` is used to display a single plot with two different y-axes: the *left* and *right* axes. The command `yyaxis left` means that all subsequent plots and annotations will relate to the left-hand y-axis, whereas `yyaxis right` activates the right-hand y-axis. Note that the scales of the y-axes are determined automatically by the ranges of the two datasets (`y1` and `y2`). The two datasets share a common x-axis. ∎

■ Activity 9.2

09.B   The file *patient_data.txt* contains four pieces of data for multiple patients: whether they are a smoker or not ('Y'/'N'), their age, their resting heart rate, and their systolic blood pressure. Write a program to read in these data, and produce separate arrays of age and heart rate data for smokers and non-smokers.

Produce plots of age against heart rate for smokers and for non-smokers,

1. On the same figure with a single use of the `plot` command.
2. On multiple figures using the `figure` command.
3. On the same figure using the `hold` command.
4. On subplots using the `subplot` command.   ∎

∎ Activity 9.3

Electromyography (EMG) involves measurement of muscle activity using elec-   09.B
trodes, which are normally attached to the skin surface. One application of EMG is in *gait analysis* (see Activity 1.6).

The file *emg.mat* contains EMG data of the left tibialis anterior muscle ac-quired from a patient with a neurological disorder that affects movement. The file contains the following variables:

- `times`: The times of the EMG measurements (in seconds).
- `LTA`: The EMG measurements from the left tibialis anterior muscle.

In addition, the *emg.mat* file contains two additional variables representing information about specific gait events that have been identified:

- `foot_off`: An array containing the timings (in seconds) of "foot off" events (i.e., when the toe leaves the ground).
- `foot_strike`: An array containing the timings (in seconds) of "foot strike" events (i.e., when the heel hits the ground).

Produce a plot of time (on the *x*-axis) against EMG measurement (on the *y*-axis), and annotate the plot appropriately. In addition, add lines indicating the timings of the gait events provided. Add a legend to indicate the mean-ings of the different lines in your graph. The final figure should look similar to that shown in Fig. 9.4.   ∎

9.2.3   3-D plotting

Often when working in biomedical applications, it is useful to be able to visual-ize and analyze data that have several different values per "individual" datum. For example, we might want to visualize points in 3-D space: in this case, the individual data are the points, and each point has three associated values (its *x*, *y*, and *z* coordinates). Alternatively, we might have a range of measurements related to hospital patients' health (e.g., blood pressure, cholesterol level, body mass index (BMI), etc.): in this case, the individuals are the patients, and there

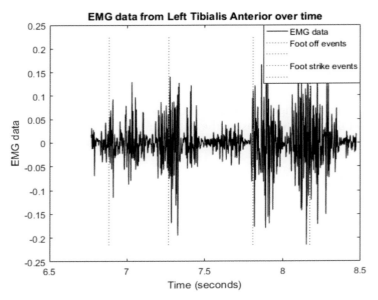

FIGURE 9.4
A plot of EMG measurements showing specific gait events.

will be multiple measurement values for each patient. This type of data is called *multivariate* data.

■ Example 9.7

09.C   The first command we will discuss for visualizing multivariate data is the `plot3` function. The example that follows (adapted from the MATLAB documentation) illustrates the use of `plot3`. This program will display a helix, and its output is shown in Fig. 9.5.

```
% the z-coordinate of the helix
t = 0:0.05:10*pi; % 5 times around a circle

% the x-coordinate of the helix
st = sin(t);

% the y-coordinate of the helix
ct = cos(t);

% 3-D plot
figure;
plot3(st,ct,t, '.b')
title('Helix');
xlabel('X');
ylabel('Y');
zlabel('Z');
```

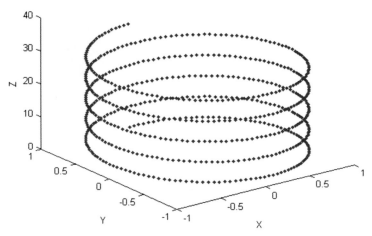

FIGURE 9.5
Output of the helix program shown in Example 9.7.

The first three (non-comment) program statements set up the array variables for storing the x, y, and z coordinates of the helix. These three arrays will all have the same number of elements, which corresponds to the number of 3-D points. Each 3-D point will take its coordinates from the corresponding elements in the arrays. The `plot3` function takes four arguments: the first three are the arrays of coordinates, and the fourth (optional) argument specifies the line/marker appearance, i.e., the same as for the `plot` function (see Section 1.10). Note that the same annotation commands (`title`, `xlabel`, etc.) can be used with figures produced by `plot3`, but that now we have an extra axis to annotate using `zlabel`.   ■

■ **Activity 9.4**

Using the same *patient_data.txt* file you used in Activity 9.2, write a MATLAB program to produce a 3-D plot of the patient data. The three coordinates of the plot should be the age, heart rate, and blood pressure of the patients, and the smokers and non-smokers should be displayed using different symbols on the same plot. Annotate your figure appropriately by, for example, inserting suitable labels on the axes.   ■

09.C

MATLAB also provides two built-in functions for visualizing multivariate data as *surfaces*: `mesh` and `surf`. Examples of such visualizations are shown in Figs. 9.6a and b, which are the outputs of the program shown in Example 9.8

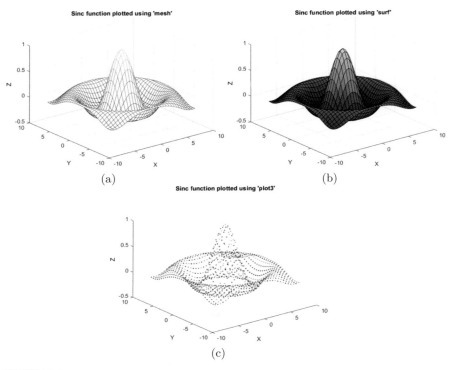

FIGURE 9.6

Output of the sinc function program shown in Example 9.8: (a) using the `mesh` command; (b) using the `surf` command. Compare these to (c), which uses the `plot3` command. It is easier to visualize the surface when using `mesh` or `surf`.

below. This type of plot is commonly used for visualizing data in which there is a single z value for each point on an x-y plane. (Note that this isn't the case for the helix data in Example 9.7, as a given pair of x,y values can have multiple z values.) For comparison, Fig. 9.6c shows how the same visualization would have looked if we had just used the `plot3` command. We can see that the surface is more easily visualized using `mesh` or `surf`.

■ Example 9.8

09.C   In this example (adapted from the MATLAB documentation), a mesh plot is produced to illustrate the *sinc* function (i.e., *sin(x)/x*).

```
% create arrays representing a grid of x/y values
[X,Y] = meshgrid(-8:.5:8, -8:.5:8);

% define sinc function
R = sqrt(X.^2 + Y.^2) + eps;
Z = sin(R)./R;
```

```
% mesh plot
mesh(X,Y,Z)
%surf(X,Y,Z)
title('Sinc function');
xlabel('X');
ylabel('Y');
zlabel('Z');
```

The built-in MATLAB `meshgrid` command creates two rectangular arrays of values to define the *x-y* grid, over which the mesh is plotted. One array indicates the *x*-coordinate at each grid point, and the corresponding element in the other array indicates the grid point's *y*-coordinate. See Section 9.2.4 for further details of the `meshgrid` command. Arrays of this form, together with a corresponding array of *z* values, are the arguments required by the `mesh` command.

The values for the *z*-coordinate array are computed by applying the *sinc* function to the distances from the origin of each grid point. Note the use of the `.^` operator to square the *x* and *y* coordinates: this is necessary to ensure that the `^` operator is applied element-wise to the array (rather than performing a *matrix power* operation, see Section 1.6). The `eps` constant in MATLAB returns a very small floating point number and is necessary to avoid a division-by-zero error at the origin of the *x-y* plane.

The `surf` command has the same form as `mesh`, as we can see from the commented-out command in the code above. The result of using `surf` is to produce a filled surface, rather than a wireframe mesh (see Fig. 9.6b). ∎

9.2.4   The `meshgrid` command

The `meshgrid` command introduced in the previous example is useful for 3-D plotting of gridded data. This section describes in more detail how it works. As we saw, `meshgrid` generates arrays containing the coordinate values of all points on a regular grid. The syntax for making a 2-D grid is

```
[X,Y] = meshgrid(x, y)
```

The input arguments are two 1-D vectors containing the coordinate values required along each dimension: one for *x* and one for *y*. In this case, the output arguments, X and Y, are 2-D arrays with the same shape as the grid, containing *all* the grid points' *x* and *y* coordinates. Each one is generated by replicating the input 1-D vectors, horizontally or vertically. An example call to make a small grid is as follows:

```
>> [X, Y] = meshgrid(1:4, 2:3)
X =
      1      2      3      4
      1      2      3      4
```

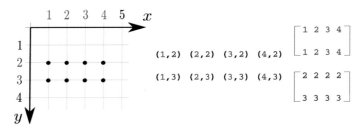

FIGURE 9.7
An illustration of the use of the `meshgrid` command. The grid required is shown by the black dots on the left-hand axes (with y pointing down). The grid ranges from $x = 1$ to $x = 4$ and $y = 2$ to $y = 3$. The coordinates for all required grid points are shown in the middle of the figure, with x-coordinates shown in red and y-coordinates shown in blue. The outputs from $[X,Y]$ = `meshgrid(1:4, 2:3)` are two separate arrays shown on the right of the figure, one containing the grid x coordinates and one containing the grid y coordinates.

```
Y =
      2       2       2       2
      3       3       3       3
```

Note that the y coordinates increase as we read *down* the array, because MATLAB arrays are indexed starting from the top left. This is equivalent to viewing the x-y axes, as shown in Fig. 9.7, which illustrates how the matrices X and Y in the above call are generated.

■ Activity 9.5

09.B, 09.C

A (1-D) Gaussian (or *normal*) distribution is defined by the following equation:

$$\frac{1}{\sigma\sqrt{2\pi}}e^{\frac{-(x-\mu)^2}{2\sigma^2}}, \tag{9.1}$$

where x is the distance from the mean μ of the distribution. The parameter σ is the standard deviation of the distribution, which affects how "spread out" the distribution will be. An example of such a distribution is shown in Fig. 9.8 (left).

1. Write a MATLAB function to display a figure of a 1-D Gaussian distribution. The function should take the following arguments: the standard deviation and the mean, and the minimum/maximum x axis values for the plot. You can experiment with different standard deviations, but to begin with, try a value of 3 with a mean of 0, and plot the distribution between x coordinates of -10 and $+10$.
2. Write another MATLAB function to display a 2-D Gaussian distribution. To calculate the function for a 2-D point, you should replace the

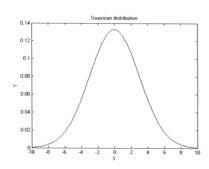
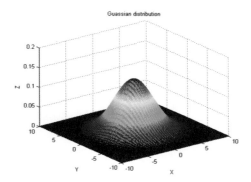

FIGURE 9.8
1-D and 2-D Gaussian distributions.

$(x - \mu)$ term in Eq. (9.1) with the distance of the point from the center of the distribution (for example, the origin, $(0, 0)$). See Fig. 9.8 (right) for an example. This function should take the following arguments: the standard deviation of the distribution, the x/y coordinates of the center of the distribution, and the minimum/maximum axis values. You can experiment with different standard deviations, but, to begin with, try a value of 3, and plot the distribution between the x/y coordinates ranging from -10 to $+10$, and use a center of $(0, 0)$. ∎

9.2.5 Modifying axis limits

We saw in Section 1.10 how it is possible to modify the limits of a plot using the axis command. The example we saw then modified the x and y axis limits in a 2-D plot, but the same function can be used to modify limits in a 3-D plot. Continuing with Example 9.8, we can modify the x, y, and z limits of the mesh plot as follows:

```
>> axis([0 10 −5 5 0 1]).
```

This results in the plot shown in Fig. 9.9, which has the x-axis limited to the range 0 to 10, the y-axis limited to the range -5 to 5, and the z-axis limited to the range 0 to 1.

Alternatively, individual axis limits can be defined separately using the xlim, ylim, and zlim commands, as shown below.

```
>> xlim([0 10])
>> ylim([−5 5])
>> zlim([0 1])
```

These commands have the same effect as the single axis command.

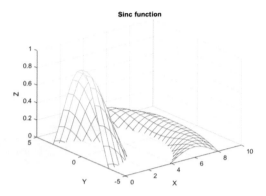

FIGURE 9.9
Modifying 3-D axis limits using `xlim`, `ylim`, `zlim`, or `axis`.

9.2.6 Imaging data

The use of images in biomedical engineering is increasingly common. Medical images can be acquired using modalities, such as magnetic resonance (MR), X-ray computed tomography (CT), and ultrasound (US). In this section, we will introduce how to read, write, and display imaging data using MATLAB. Chapter 11 will deal with images in MATLAB in more detail.

■ Example 9.9

09.D   The following example illustrates the use of several built-in image-related MATLAB commands:

```
% read image
im = imread('fluoro.tif');

% display image
imshow(im);

% modify image
im(100:120,100:120) = 0;

% save image (saves 'im' array)
imwrite(im, 'fluoro_imwrite.tif');

% save image (saves figure as displayed)
saveas(gcf, 'fluoro_saveas.tif');
```

The `imread` command reads imaging data from a file with a name specified by the single argument. The value returned is a 2-D array of pixel intensities. The *fluoro.tif* file is a *fluoroscopy* image (i.e., a low dose real-time X-ray) and is available from the book's web site.

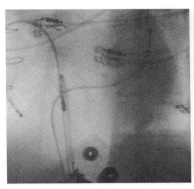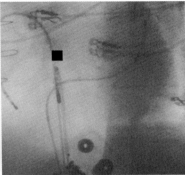

FIGURE 9.10
A fluoroscopy image described in Example 9.9 shown before (left) and after (right) some of its pixels are modified.

Images are treated as normal 2-D arrays in MATLAB. The intensities can be of different types, such as `int8`, `single`, or `double`, but within a single image, all must have the same type, as with all MATLAB arrays. Imaging data can be manipulated just like any other MATLAB array, as we can see from the third non-comment line in the above example, in which a rectangular block of pixels has its intensities set to 0.

The `imshow` command can be used to display a 2-D image in a figure window. The single argument is the 2-D array of pixel intensities.

This example shows two ways in which the image can be written to an external file. The `imwrite` command takes a 2-D array and a file name as its only arguments, so the image data written to the file depends only on the array and not on any figure currently being displayed. The `saveas` command is used in this example to save the data in the current figure (`gcf` means *get current figure*). Therefore any changes made to the displayed figure will also be saved, but any changes made to the original array since it was displayed (as in the example above) will not be saved. In this example, `saveas` saves the figure as an image, but it can also be used to save in the native MATLAB figure format, which has a ".fig" file extension. Illustrations of the image before and after modifying its pixel data are shown in Fig. 9.10. ∎

Both the `imread` and `imwrite` commands can handle a range of different imaging formats, such as `bmp`, `png`, and `tif`. For a full list of supported formats, type `imformats` at the MATLAB command window. Normally, just adding the extension to the file name is enough to tell MATLAB which format we want to use.

In medical imaging, it is common for images to be 3-D datasets, rather than 2-D (e.g., many MR and CT images and some US images). Unfortunately, MAT-

LAB doesn't have any built-in functions for visualizing 3-D images. However, several have been written by the MATLAB user community and been made available for free download from the MathWorks File Exchange web site. For example, the `imshow3Dfull` function allows simple interactive slice-by-slice visualization of 3-D image datasets. See Section 9.4 at the end of this chapter for details.

■ **Activity 9.6**

09.D   *Digital subtraction angiography* is an imaging technique, in which two fluoroscopy images are acquired: one before injection of a contrast agent and one shortly after. The pre-contrast image (the *mask* image) is subtracted from the contrast enhanced image (the *live* image) to provide enhanced visualization of blood vessels.

Write a MATLAB script to

- Ask a user to choose mask and live image files using a file selection dialog box*.
- Load in the mask and live image files.
- Subtract the mask from the live image.
- Display the resulting digital subtraction angiography image.

Sample mask and live images (*dsa_mask.tif* and *dsa_live.tif*) are available for download from the book's web site.

(*Hint: Look at the MATLAB documentation for the `uigetfile` function.)   ■

9.3   Summary

MATLAB provides a number of built-in functions for performing more sophisticated data visualizations. The `pie` command can be used to generate pie charts to visualize the relationships between different quantities. The `plot`, `hold`, `figure`, `subplot`, and `yyaxis` commands can be used to visualize multiple datasets at the same time. 3-D plots can be produced using the `plot3`, `mesh`, and `surf` commands. 2-D grids of coordinate values suitable for use with `mesh` and `surf` can be generated using the `meshgrid` command.

Imaging data can be read and assigned to 2-D array variables using `imread`, and displayed using `imshow`. An array can be saved to an image file using `imwrite`, and a figure that is currently being displayed can be saved using `saveas`. A number of 3-D image visualization tools are available for free download from the MathWorks File Exchange web site.

We will return to the topic of data visualization, in the context of statistical analysis, in Chapter 13.

9.4 Further resources

- MATLAB documentation on pie charts: https://mathworks.com/help/matlab/ref/pie.html.
- MATLAB documentation on 2-D and 3-D plots: https://mathworks.com/help/matlab/2-and-3d-plots.html.
- MATLAB documentation on image file operations: https://mathworks.com/help/matlab/images_btfntr_-1.html.
- MathWorks File Exchange: https://mathworks.com/matlabcentral/fileexchange.
- `imshow3Dfull` function: https://mathworks.com/matlabcentral/fileexchange/47463-imshow3dfull--3d-imshow-in-3-views-.

9.5 Exercises

■ Exercise 9.1

The table below shows the numbers of patients diagnosed with three different types of cancer at a hospital over a 3 year period.

09.A

Type of cancer	2018	2019	2020
Lung	1016	1028	988
Prostate	370	411	500
Breast	539	490	507

1. Produce a pie chart to visualize the 2018 figures.
2. Produce a single figure containing three pie charts, one for each year. *(Hint: see the example in the MATLAB documentation at* https://www.mathworks.com/help/matlab/ref/pie.html#buipz21*).* ■

■ Exercise 9.2

In Exercise 5.4, we introduced data measured using an oscillatory blood pressure monitoring device. The data (contained in the file *blood_pressure.mat*) comprise an array (p) of pressure measurements in mmHg and an array (peaks) containing zeros and ones, with a value of one, where peaks in the pressure oscillations were identified (zero otherwise). Produce a plot of both datasets in the same figure, but with two different *y* axes for the two arrays. Annotate your figure appropriately. ■

09.B

■ Exercise 9.3

The process of aligning 3-D medical images is known as *image registration*. When assessing how well two images have been registered (aligned), it is

09.B

possible to use a variety of measures. A research team has evaluated the registrations of a number of image pairs using two different measures. The first is the value of the *normalized cross correlation* (a measure of image similarity) between each image pair. The second is the *landmark error*, which is determined by observers manually placing corresponding anatomical landmarks in each image pair and computing the distance between them after registration. A good registration should give a lower landmark error. The values of these measures for 10 separate experiments are provided in the files *ncc.txt* and *points.txt*. Both files contain 10 rows arranged in two columns that represent the experiment number and the respective evaluation measure.

Write a MATLAB program to read in these data, and produce a single plot showing the values of both measures (on the *y*-axis) against the experiment number (on the *x*-axis). Annotate the *x* and *y* axes of your plot, and add a suitable title. Your final figure should look like the one shown below.

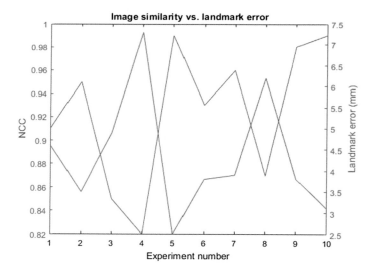

■ Exercise 9.4

09.B
This exercise continues the case study that we have already seen in Examples 1.3 and 5.1. Recall that we have data representing radial displacements of the left ventricular myocardium of the heart. We have these data separately for the 17 American Heart Association (AHA) segments. Now we wish to visualize the displacements for all segments in a single plot.

The input to your program will be a file containing radial displacement data broken down by AHA segment. The basic format of the file is:

```
tres
s1t1 s1t2 ... s1tR
s2t1 s2t2 ... s2tR
...
s17t1 s17t2 ... s17tR,
```

where `tres` represents the *temporal resolution* of the radial displacement data, i.e., the amount of time (in milliseconds) between subsequent displacement measurements. Each other entry in the file represents a radial displacement measurement for a particular segment (`s`) at a particular time (`t`) in the cardiac cycle. `R` is the number of radial displacements recorded in a particular cardiac cycle. For the purposes of this exercise, you can assume that `R` will always be equal to 20, and there will always be 17 segments.

The following sample input files are provided:

- *radial_motion_aha_pat_01.txt*
- *radial_motion_aha_pat_02.txt*
- *radial_motion_aha_pat_03.txt*

Your task is to write a MATLAB script that reads in the data from a specified file and displays a single figure that plots the radial displacement for each AHA segment against time in milliseconds. The output of your script for the first sample input file is shown in the figure below.

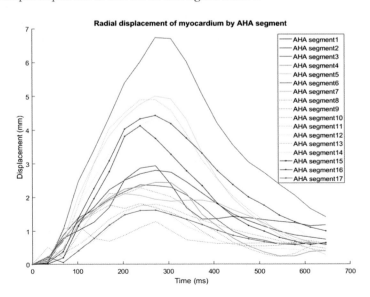

Note the following features of this figure that your code should reproduce:

- The lines for each segment should be displayed using a different style/color.
- A legend should be included indicating which line corresponds to which segment (in the input file, the data for segment 1 are first, segment 2 second, and so on).

Try to use an iteration statement in your solution.   ∎

■ Exercise 9.5

09.B   In Activity 1.6, we introduced knee flexion data that were used in gait analysis. The file *joint_flexion.mat* contains the same knee data, and data from the hip and ankle. Specifically, the file contains the following variables:

- `times`: The timings of the flexion values (in seconds).
- `angles`: A matrix in which the first column represents flexion (i.e., angle) of the left hip, the second column is flexion of the left knee, and the third column is flexion of the left ankle.
- `foot_off`: An array containing the timings (in seconds) of "foot off" events (i.e., when the toe leaves the ground).
- `foot_strike`: An array containing the timings (in seconds) of "foot strike" events (i.e., when the heel hits the ground).

Produce a plot of joint flexion against time for all three joints on the same figure, and annotate the plot appropriately. As in Activity 1.6, add lines indicating the timings of the "foot off" and "foot strike" events, and a legend indicating the meanings of the different lines.   ∎

■ Exercise 9.6

09.C   Optical tracking systems have a range of applications in biomedical engineering, from gait analysis to image-guided surgery. This technology typically uses a pair of cameras, which localize markers attached to the object being tracked. The 3-D location of a marker can be determined from a pair of 2-D image pixel locations using triangulation.

A company is developing a new optical tracking system and wish to evaluate its tracking accuracy. They have tested the system on "phantom" data (i.e., data acquired from a specially designed test object), and have collected 3-D localization error data. The data consist of errors along the x, y, and z axes for 50 separate tests, where the x/y axes are parallel to the image plane of one of the cameras, and the z axis is the direction of the same camera.

These error data are available in the file *optical_tracking.mat*. Write a MATLAB script to produce a 3-D plot of the x, y, and z errors, to enable visualization of the directional dependency of the errors. What do you observe? Can you offer an explanation?   ∎

■ Exercise 9.7

In Exercise 9.3, we introduced the concept of *image registration*. When evaluating the behavior of image registration algorithms, it can be useful to visualize the "similarity landscape" of a pair of images. This means we evaluate a similarity measure at a number of different registration parameters, and then visualize the variation of the similarity measure as a function of the registration parameters.

09.C

Suppose that we have done this for a pair of images. x and y translations ranging from -10 mm to $+10$ mm have been applied to one of the images, and the resulting normalized cross correlation (NCC) similarity measure has been computed. These data are available to you in the file *similarity_measure.mat*. The file contains a 21×21 array, representing the NCC values for x and y translations from -10 mm to $+10$ mm, in steps of 1 mm.

First, write a MATLAB script to visualize the NCC data as a function of x and y translation. Annotate your plot appropriately. Next, write MATLAB code to find the peak NCC value and the corresponding x and y translations.

(Hint: Look at the MATLAB documentation for the `ind2sub` *command.)*   ∎

■ Exercise 9.8

Write a MATLAB program that will read in an image from a file and display it, and then allow the user to *annotate* the image (i.e., add text to it at specified locations). To perform the annotation, the user should be prompted to enter text, then click their mouse in the image to indicate where they want it to appear. The user should be allowed to add multiple annotations in this way, and the program should terminate and save the annotated image with a new file name when empty text is entered.

09.D

A sample image file, called *brain_mr.tif*, which you can use to test your program, is available through the book's web site.

(Hint: Look at the MATLAB documentation for the `gtext` *command.)*   ∎

■ Exercise 9.9

Try visualizing 3-D imaging data using the `imshow3Dfull` function. The MATLAB code for `imshow3Dfull` can be downloaded from the MathWorks File Exchange (see Section 9.4). Sample 3-D imaging data can be downloaded from the book's web site as the *MAT* file *MR_heart.mat*.   ∎

09.D

FAMOUS COMPUTER PROGRAMMERS: DENNIS RITCHIE

Dennis Ritchie was an American computer scientist who was born in 1941 in Bronxville, New York. While a student at Harvard University, he attended a talk on Harvard's computer system and became fascinated with computers and, in particular, how they could be programmed. As a result, he worked in his spare time to learn more, and even got a part-time job working at Massachusetts Institute of Technology in their computer labs. In 1968, he received a PhD in Computer Science from Harvard.

Probably Dennis Ritchie's main contribution to the field of computing and computer programming was his early work on operating systems. Operating systems are the software that enable a computer's resources to be accessed by users. In the early days, most computers took up entire rooms, and their operating systems were very complex and difficult to use. Smaller, portable computers were starting to be developed, but Ritchie saw that there was a need for a simpler operating system to bring their power to a wider audience. Therefore in 1969, he worked with Ken Thompson to develop UNIX, which was much cheaper, simpler, and easier to use than any previous operating system. It was hugely influential and is still widely used today.

But this wasn't Ritchie's only major contribution. UNIX was written in machine code, which is specific to the machine it is written for. Ritchie realized that for UNIX to be "ported" to other machines, it must be written in a higher-level programming language, which could be easily compiled on different computer platforms. This led to him making his second major contribution to computing: the invention of the C programming language in 1973. C has proved to be one of the most widely used languages in the history of computer programming, and was also the precursor to C++, another of the most popular programming languages.

Dennis Ritchie died in 2011 at the age of 70 in New Jersey. He died less than a week after the Apple co-founder, Steve Jobs. Whereas Jobs' death was (rightly) met with widespread media coverage and tributes, Ritchie's passing went almost unnoticed in the mainstream media. However, to those in the know, Ritchie's contribution to computing was arguably the greater.

"Ritchie was under the radar. His name was not a household name at all, but ... if you had a microscope and could look in a computer, you'd see his work everywhere inside."

Paul Ceruzzi

Code efficiency

10.1 Introduction

When code becomes reasonably long or complex, it may become necessary to consider how efficient it is. There are two senses in which we can talk about efficiency of code: in terms of the time it takes to run and in terms of the amount of computer memory it needs when running. In some cases, a program needs to be very time-efficient and should run and produce results quickly, for example, in software that is used by cars to manage the running of the engine or software used for instrument control in robotic surgery. In other cases, a program may need to be very economical with the amount of memory it uses, for example, in programs running on devices where memory is very limited, such as an embedded processor in a domestic appliance or in a wearable ECG monitor. This chapter will consider some aspects of efficiency, focusing on MATLAB®-specific examples, although some of the issues raised will apply to most languages.

10.2 Time and memory efficiency

As a rule of thumb, to assess the memory efficiency of a piece of code, we need to focus on the variables it has, in particular, on the largest arrays that it uses. Recall that MATLAB defaults to double precision when assigning numeric

233

values (see Section 3.3), which means we need eight bytes per element in a numeric array (unless we override the default behavior).

To assess how time-efficient a piece of code is, we need to break it down into *fundamental instructions*. We make the simplifying assumption that these all roughly take the same amount of time each. Fundamental operations or instructions can include:

- Assigning a value to a variable, e.g., `a = 7;`
- Looking up a value of a particular element in an array, e.g., `myArray(3)`
- Comparing a pair of values, e.g., `if (a > 10)...`
- Arithmetic operations, such as adding and multiplying, e.g., `a * 4`

■ Example 10.1

O10.A

We consider a very simple example to illustrate some aspects of code efficiency. The following is a trivial function that takes one argument (`N`), which is treated as an integer, and adds up all the integers from 1 up to `N` and reports the result:

```
function sumOfIntegers(N)
% Usage: sumOfIntegers(N)
% Add up all the integers from 1 up to N and report
% the result.

a = 1:N;
total = 0;

for j = 1:N
  total = total + a(j);
end

fprintf('Sum of first %u integers = %u \n', N, total);
```

To keep things simple, we consider the first two assignments (to the variables `a` and `total`) as one assignment each.[1] Then, there is a `for` loop, which has `N` iterations. Within each iteration there are the following:

- An array look-up: `a(j)`,
- A basic arithmetic operation: `total + a(j)`,
- An assignment: `total = total + a(j);`

Therefore we have three operations per iteration, which gives a total of $3N$ operations for the entire loop. Counting in this way, the estimate is that the function will need $3N + 2$ fundamental operations. As we can see, the number of operations that the function will carry out depends upon the number passed in as the argument `N`. For large values of `N`, more operations are carried out. ■

---

[1] Really, the assignment to variable `a` is an array assignment, so it requires allocating more memory than for a single numeric scalar value. This can add a time overhead, which we ignore here.

10.2.1   Timing commands in MATLAB

Recall that the built-in commands `tic` and `toc` can be used to measure execution time (see Section 2.7). They can be used as follows:

```
tic
% some
% commands
% here
toc
```

If we just have a single command, we can use commas or semi-colons to separate them, for example,

```
>> tic; a = sin( exp(3) ); toc
```

This could give output similar to

```
Elapsed time is 0.000387 seconds
```

The commands can also be separated with commas:

```
>> tic, a = sin( exp(3) ), toc
```

The output for the second command could be as follows:

```
a =
    0.9445

Elapsed time is 0.000355 seconds.
```

MATLAB reports that the above example took 355 microseconds. The same example, repeated on a different machine, is likely to run in a different amount of time, due to differences in the hardware and the operating system. Even repeating the same operation on the same machine can lead to different times being reported, because the load on the operating system can change between runs.

■ **Example 10.2**

Returning to the `sumOfIntegers` function from Example 10.1, we can measure how long the function takes to execute for various values of N. Let us try values of one thousand and one million for N. We'll focus on the main loop in the function. We can add a `tic/toc` pair around the loop, so it becomes

010.A

```
% ...
tic
for j = 1:N
  total = total + a(j);
end
toc
% ...
```

■ Activity 10.1

O10.A   Download the code for the modified function described in Example 10.2. It is called *sumOfIntegers_2.m*. Call the function from the command window using different values of *N*. In particular, call it for $N = 1000$ and for $N = 1,000,000$, and make a note of the time taken; you should confirm that the second run takes approximately 1000 times as long as the first.   ■

The ratio in the times given will not be exactly 1 to 1000 for a number of reasons, perhaps due to variations in the demands on the operating system between runs, or because MATLAB might optimize the code in some way.

10.2.2   Assessing memory efficiency

Recall that the `whos` function identifies which variables are available in the workspace and how much memory they are using (see Section 1.8). This can be particularly useful when entering commands that define variables in the command window.

When writing a function of our own, however, we need to adopt a different strategy. We can inspect the code directly to find which variables will require the most memory. In the case of the function `sumOfIntegers` in Example 10.1, the variable that requires the most memory is the variable a, declared in the command `a = 1:N;`. This statement declares and assigns an array of length N when the function is run. Therefore the amount of memory used by the function depends upon the argument given to it. Larger values of N will mean that more memory is needed.

Another way to inspect the amount of memory used by a function is more direct. We can run the function and use the debugger to stop the execution and look directly at how much memory is being used. For example, we can place a break point on the last line in `sumOfIntegers`, and call it with a value of one thousand:

```
>> sumOfIntegers(1000);
```

The code stops on the last line, and a call to `whos` gives the following output:

```
K>> whos
  Name        Size            Bytes  Class     Attributes

  N           1x1                 8  double
  a           1x1000           8000  double
  j           1x1                 8  double
  total       1x1                 8  double
```

This lists the four variables (N, a, j, and total) and how many bytes each one uses. Note that the variables listed are only those within the scope of the function as it is currently being run, and any workspace variables, for example, will not be listed (see Section 4.10).

■ Example 10.3

Returning once again to the trivial function `sumOfIntegers`, an obvious way to avoid a memory overhead is to simply avoid the allocation of the array. For the example, this is quite simple: the code can be re-written as

010.B

```
function sumOfIntegers(N)
% ... usage
% ..
total = 0;

for j = 1:N
  total = total + j;
end
% ...
```
■

■ Activity 10.2

Download the original code from Example 10.1 and the code for the modified function described in Example 10.3. They are called *sumOfIntegers.m* and *sumOfIntegers_3.m*, respectively. Call each one from the command window using the same value of *N*. For each one, place a breakpoint on the last line so that the running halts there in the debugger. When the code stops at this breakpoint, for each version, type `whos` in the command window. Use this to confirm that the new version is more memory-efficient than the first. ■

010.B

The above example illustrates an easy win; the code did not need an array at all. But not all examples are as straightforward as this.

It is possible to make excessive memory demands unintentionally. The built-in `ones` or `zeros` functions are commonly used to quickly initialize an array containing only 1's or 0's. By default, when only one argument is given to these functions, it is assumed that the user requires a square matrix (i.e., a two-dimensional [2-D] array).

■ Example 10.4 (A very, very memory-inefficient piece of code)

If we want a long *vector* of zeros, i.e., a 1-D array, say a vector that has a length of 100,000, it is very easy to make a mistake by calling the `zeros` function as follows:

010.B

```
myVector =  zeros(100000);   % Bad! do not do this...
```

MATLAB interprets this as a request for a square *matrix* of size 100,000 × 100,000. Therefore it tries to allocate enough memory for 100,000 ×

$100,000 = 10^{10}$ `double` values. Each `double` value requires eight bytes, because that is the default numeric precision in MATLAB. This means that the user is asking for 80,000,000,000 bytes, which is a little over 74 GB, because

$$
\begin{array}{cccc}
 & \text{bytes/KB} & \text{KB/MB} & \text{MB/GB} \\
74 \times & 1024 & \times \quad 1024 & \times \quad 1024 & = 79{,}456{,}894{,}976 \text{ bytes.}
\end{array}
$$

Most ordinary computers do not have 74 GB of Random Access Memory (RAM) available, so typing such a command into MATLAB is very inadvisable and likely to cause a crash. ∎

■ Activity 10.3

010.B

Assign to an array using the functions `ones` or `zeros`, using reasonable values for the size of the array, e.g.,

```
>> a = zeros(100);
```

and confirm that the variables in the workspace have the size you expect. Carry out the following commands:

```
>> b = ones(80);
```

```
>> c = ones(80, 80);
```

```
>> d = ones(80, 1);
```

and decide which variables take up the same amount of space in memory. ∎

10.3   Tips for improving time-efficiency

A couple of tips for improving time-efficiency are given below. These are to pre-allocate arrays before using them and to avoid loops by using vector style operations.

10.3.1   Pre-allocating arrays

MATLAB is fairly relaxed about the length of an array and will adjust it automatically if the user decides to assign a value to an element that is beyond its current limits. For example, the command

```
x = randi(3, 1, 6);
```

will generate a length 1×6 row vector of random integers drawn from the set $\{1, 2, 3\}$, for example, $[3, 1, 3, 1, 2, 2]$. If we make an assignment beyond the end of the array, e.g., `x(10) = 5`, then it is automatically resized to a size of 1×10. The "gap" between the end of the original array and the new array will be filled with zeros. This means the resulting vector will become $[3, 1, 3, 1, 2, 2, 0, 0, 0, 5]$.

As discussed in Section 2.7, this can be convenient and enables the user to avoid worrying about whether the element being written to is within the array (yet) or not. This is, however, a costly operation in terms of time. There is an overhead involved in resizing an array, and this is made worse if it is done within a loop.

The code snippet that follow illustrates this. It calculates the first 10 cubic numbers:

```
% Start off with an empty array
cubes = [];
for j = 1:10
  cubes(j) = j^3;
end
```

The code starts with an empty array. Then, in the first iteration, it assigns to the element at index 1 (which doesn't exist yet). So the array is resized to accommodate this element. The second iteration assigns to the element at index 2, and again the array needs to be resized, etc. In each iteration, we are assigning a value to an element in the array `cubes`, whose index is one more than its current size. Every iteration leads to a resizing of the array, and each resize has a cost associated with it.

As we saw in Section 2.7, we can avoid this cost by *pre-allocating* the array to achieve much better time efficiency:

```
cubes = zeros(1, 10);
for j = 1:10
  cubes(j) = j^3;
end
```

10.3.2 Avoiding loops

Among the most time-consuming parts of a program are loops such as `for` or `while` loops. This is especially true if the loops are *nested*, for example,

```
for j = 1:N
  for k = 1:M
    % Some code carried out here
  end
end
```

In this case, the code within the central loop is executed $M \times N$ times (unless a `break` command is contained in the central section of code and can be reached). In other words, nesting loops have a multiplicative effect on the number of operations needed. This is why nesting loops should only be done when it is unavoidable, especially if three or more loops are nested.

One way to avoid loops is by using built-in functions. In the previous version of the `sumOfIntegers` function (Example 10.3), we can replace the entire loop

```
for j = 1:N
...
end
```

with a simple call to the built-in function sum:

```
total = sum(1:N);
```

Here, the array initialization code 1:N is used directly as an input argument to the sum function (without actually assigning it to an array). This single line will run faster than the original loop. In fact, many built-in functions can accept array arguments, e.g., sin, exp, etc., and we can use this fact to avoid unnecessary loops.

■ Example 10.5

O10.A, O10.B

The code below constructs 50 points on the graph $y = \sin(x)$ between $x = 0$ and $x = 2\pi$.

```
x = linspace(0, 2*pi, 50);
y = zeros(size(x));
for j = 1:50
  y(j) = sin(x(j));
end
```

It tries to improve efficiency by pre-allocating the array y to be the same size as x. However, since the built-in sin function in MATLAB accepts an array argument, we can simply replace the loop with a one-line command to generate the array variable y directly. This is done by applying the sin function element-wise to array x:

```
x = linspace(0, 2*pi, 50);
y = sin(x);
```

■

Operating directly on arrays in this way, to avoid loops, is often called *vectorization* or *vectorized calculation*. It can also be applied when we have multiple arrays to process, rather than just one as in the previous example.

■ Example 10.6

O10.A

Suppose we have collected the widths, lengths, and heights of four cuboids, and we want to calculate their volumes. Say the dimensions are collected into three arrays:

```
w = [1, 2, 2, 8];
l = [2, 3, 5, 3];
h = [1, 3, 2, 4];
```

We could use a loop to find the volumes:

```
v = zeros(1,4);
for j = 1:4
  v(j) = w(j) * l(j) * h(j);
end
disp(v)
```

which gives the output:

```
    2    18    20    96
```

However, this code can be speeded up by *vectorization*. We use the dot operator to perform element-wise multiplication of our dimension data:

```
v = w .* l .* h;
```

achieving the same result with improved time-efficiency.   ■

Remember, the above is a "toy" example (there are only four data points in it for a start), so the speed-up will be slight. But if the data arrays were much larger, then the time saved would be more significant.

■ Activity 10.4

Re-write the following code so that it avoids using a loop: *O10.A, O10.B*

```
N = 10;

a = zeros(1, N);
b = zeros(1, N);

for i = 1:N
  a(i) = i;
  b(i) = a(i) * a(i);
end
```
■

■ Example 10.7

Recall that the surface plot of the sinc function that we saw in Example 9.8   *O10.A*
relied on the generation of two arrays X and Y using the meshgrid function, and then calculating the height of the function in an array Z. The code used vectorization and is repeated here:

```
% create arrays representing a grid of x/y values
[X,Y] = meshgrid(-8:.5:8, -8:.5:8);

% define sinc function in a good way
R = sqrt(X.^2 + Y.^2) + eps;
Z = sin(R)./R;
```

The input arrays are both 33×33 arrays, and the resulting Z array has the same size, because the element-wise operations ensure this.   ■

■ **Activity 10.5**

O10.A

Re-write the code from Example 10.7 so that it uses nested loops instead. Which of the versions, with or without loops, is more readable? Which version is most efficient? ■

10.3.3 Logical indexing

Another way to speed up code, and avoid loops, is to use *logical indexing*. We first introduced the concept of logical indexing in Section 3.4.1. To recap, logical indexing allows us to index into any array using another array of logical values, which should have the same size as the first array. The values returned will be those from the first array, where there is a corresponding true value in the logical array. Logical indexing can be very useful when we want to modify some of the elements in an array, depending on whether they meet a particular condition. We review this concept below in a pair of examples.

Before looking at the examples, we show one way to generate data that we can use. The built-in `magic` function returns a square array of numbers with the "magic" property that the rows, columns, and diagonals all add up to the same value. The command

```
>> A = magic(6)
ans =
    35     1     6    26    19    24
     3    32     7    21    23    25
    31     9     2    22    27    20
     8    28    33    17    10    15
    30     5    34    12    14    16
     4    36    29    13    18    11
```

produces an array A, representing a magic square, which contains all the numbers from 1 to 36 (6 × 6) inclusive. You can check whether the column, row, and diagonal sums are equal.

The next example illustrates code that does *not* use logical indexing. This is then followed by an example that does.

■ **Example 10.8 (No logical indexing)**

O10.A

Here, we create a new array B with the same size as A. B has a value of one everywhere, unless the corresponding element in A is divisible by 3, in which case, the value is set to zero. We can do this by looping over all the elements in A, testing each one in turn and assigning the value to B:

```
N = 6;
% NxN magic square
A = magic(N);
% Initialise B to be the same size as A with ones
B = ones(N);
```

```
% Set B to one where A is divisible by 3
% Loop over all elements
for row = 1:N
  for col = 1:N
    % Get current element's value.
    value = A(row,col);
    % Is it a multiple of 3?
    if ( mod(value, 3) == 0 )
      B(row, col) = 0;
    end
  end
end

disp(B)
```

∎

■ Activity 10.6

Download the code for Example 10.8 from the website, and run it. Confirm *010.A*
that this code produces the array B, as shown below. Note the pattern in
the values of B, and confirm that it matches what you expect, based on the
values in the array A.

```
1    1    0    1    1    0
0    1    1    0    1    1
1    0    1    1    0    1
1    1    0    1    1    0
0    1    1    0    1    1
1    0    1    1    0    1
```

Note how the code loops over all the elements, testing each one separately
to see if it is divisible by 3. This is done with the test

```
if ( mod(value, 3) == 0 ) ...
```

The mod function returns the remainder when value is divided by 3. ∎

We can obtain a speed-up in the above code by using logical indexing, a vectorized operation that allows loops to be discarded.

■ Example 10.9 (Using logical indexing)

Example 10.8 can be carried out with logical indexing as follows: *010.A*

```
N = 6;
A = magic(N);
B = ones(N);

indices = mod(A, 3) == 0;
B(indices) = 0;

disp(B)
```

The original pair of nested loops have been replaced by two lines. There is quite a lot happening in these lines, so let's look at them in more detail. The call mod(A,3) applies the function mod *element-wise* to A to calculate the remainder for every element after dividing by 3.

Next, an equality test (mod(A,3)==0) compares every element in the array of remainders with zero. This is again an element-wise step, and it results in an array of *logical* values: true values indicate where the remainder was zero, otherwise the values are false.

The array of logical values (with true indicating the elements of interest) is assigned to a variable indices. This is a logical indexing variable, and it is used in the next line. It is applied directly to the array B with the round (indexing) brackets. Logical indexing causes the assignment to 0 to be carried out for all elements in the array B, for which the indices variable is true.

These two lines of code can even be replaced with a single line by avoiding the use of the variable indices:

```
B(mod(A, 3) == 0) = 0;
```
■

■ Activity 10.7

O10.A　Modify the code in Example 10.9 so that it starts with a magic square of size 5×5 and produces a second array that has a value of zero, where the magic square is odd, and a value of 10 where the magic square is even. ■

Example 10.9 uses much more dense and compact code and is much faster than the original longer code, which used of a pair of nested loops. The difference between them illustrates an important point:

Code that has been optimized for efficiency tends to be harder to read.

Generally, it is better to start with code that might be slow and inefficient, but where it is clear what it is meant to do. Optimizing the code should only be done (a) if it is considered important and (b) once we are sure that the code is running correctly. This illustrates a second key point:

It is more important that code runs correctly than that it runs efficiently.

These points may seem obvious, but it is fairly common for coders to forget them during routine programming.

10.3.4　A few more tips for efficient code

These tips are adapted from the MATLAB help sections:

- Keep *m*-files small. If a file gets large, then break it up, and call one file from another.

- Convert script files into functions whenever possible. Functions generally run a little faster than scripts. See Sections 4.2 and 4.4 for more information on function and script files.
- Keep functions themselves as small and as simple as possible. If a function gets very long, look for parts of it that can be put into a new function, and call it from the first one. (Recall that the second function can still be in the same file as the first.)
- Simple utility functions that are called from many other functions can be put into their own *m*-files so that they can be accessed easily (their location will need to be in the MATLAB path (see Section 4.8).
- *To repeat*: Avoid loops where possible. Try to use vectorization and element-wise operations.
- *To repeat*: Pre-allocate arrays when it is known how large they will be. Try to avoid growing a loop with each iteration of an array if possible, especially if there are likely to be many iterations. If the size of an array is not known in advance, it is sometimes possible to pre-allocate it to a size that is safely larger than what will be needed.

■ Activity 10.8

In Exercise 4.10, you wrote code to perform respiratory gating using a bellows signal. Make a copy of your solution to Exercise 4.10 (or the one from the book's web site), and adapt it so that the gating operation is implemented in two different ways: using a loop and using logical indexing. Add commands to measure the time taken to run each implementation.

O10.A, O10.C

Note that the code to compute the low and high bounds of the gating window should remain unchanged; you only need to modify the code for the gating operation itself (i.e., computing the indices of all elements in the full bellows array that are within the gating window).

Of the two implementations, which is the most time-efficient and which do you think is the most easy to understand? ■

10.4   Recursive and dynamic programming

The way some problems are defined can lead to time-consuming or memory-demanding solutions when a program is written to solve them. A recursively defined problem is one such type of problem. When writing a program to solve a recursive problem, it can be easy to write a solution that naively ends up having a high computational cost. Dynamic programming is one way to reduce this cost. Before discussing this in more detail, we will give a recap of recursive programs, focusing particularly on the number of function calls that are made.

The example that follows is a continuation of Activity 4.11 that looked at the factorial function. Here, we will concentrate on the number of times that the function is called and with what arguments.

■ Example 10.10

010.D   The code for evaluating the factorial $n!$ of a number n is repeated here, but we have changed the name slightly so that, for brevity, it is $f(n)$.

```
function result = f(n)
% f(n) : return the factorial of n

if (n<=0)
   % Stopping condition.
   result = 1;
else
   result = n * f(n-1);
end

end
```

If we want to evaluate f for $n = 5$, then the call we make will be $f(5)$, and this will in turn make a call to evaluate $f(4)$, which in turn makes a call to evaluate $f(3)$, and so on. The sequence of calls can be written

$$f(5) \rightarrow f(4) \rightarrow f(3) \rightarrow f(2) \rightarrow f(1) \rightarrow f(0).$$

The final call to $f(0)$ is dealt with by the *stopping condition* (see Section 4.11) in the `if` clause, and no further recursive calls are made. In other words, after the original call to $f(5)$, a further 5 calls to the function are made recursively before stopping. In the general case, to evaluate $n!$, we will carry out n recursive calls of the function f.   ■

This type of analysis of recursive function calls can be performed for many recursive implementations. In the case of the factorial function above, the number of recursive calls was not excessive, and so the implementation shown was not particularly inefficient. Now let's consider another example, which *does* turn out to have an inefficient solution if implemented naively.

Full binary trees: We define a *full* binary tree as one with a set of nodes, of which one is a *root* node, and every node in the tree is either a *leaf* node or is the *parent* of exactly two *child* nodes. A leaf node is one with no children. In the example shown in Fig. 10.1a, there are seven nodes shown as gray circles. Each arrow connects a parent node above a child node. The root node is at the top of the tree, and there are four leaf nodes and two intermediate nodes (neither a root nor a leaf).

(a)   (b)

FIGURE 10.1

Examples of full binary trees with seven nodes. (a) With 3 leaf nodes in the left sub-tree and 1 in right sub-tree. (b) With 2 leaf nodes in each of the left and right sub-trees.

A natural question to ask is *how many different full binary trees exist for a given number of leaves*? The above example is one with four leaf nodes. How can we calculate how many such trees are possible with four leaf nodes?

One way to answer this question is to look at *sub-trees* below the root node. In the above example, the sub-tree below the root node on the left contains three leaf nodes, and the sub-tree on the right contains the remaining single leaf node (i.e., the right-hand sub-tree consists of one node only).

In another tree with four leaf nodes, we could, for example, have two leaves in the left sub-tree and two leaves in the right sub-tree. An example is shown in Fig. 10.1b.

Considering the sub-trees on either side of the root node provides us with a way to find the total number of trees for a given number of leaf nodes. For the four leaf node example, we can have the following pairs of numbers of leaves assigned to each sub-tree beneath the root node:

$(1, 3)$ $(2, 2)$ $(3, 1)$.

Let T_k be the number of trees with k leaf nodes. For each pairing above, we can work out the number of trees with that pairing by multiplying the number of possible sub-trees on either side of the root. For example, if we have a $(3, 1)$ split for the leaf nodes, then the total number of trees with this split will equal

\# trees with three leaves $\times$ \# trees with one leaf $= T_3 \, T_1$.

We can therefore obtain the total number of full binary trees with four leaves by adding up the expressions over all possible splits:

$$T_4 = T_1 T_3 + T_2 T_2 + T_3 T_1.$$

By writing the formula for T_4 in this way, we can start to see the recursive nature of a function that would work for *any* number of nodes. The value of T_4 above depends on the previous values of T_k for $k = 1, 2, 3$. We can write a similar expression for five leaves, six leaves, etc.:

$$T_5 = T_1 T_4 + T_2 T_3 + T_3 T_2 + T_4 T_1$$
$$T_6 = T_1 T_5 + T_2 T_4 + T_3 T_3 + T_4 T_2 + T_5 T_1$$
$$\ldots$$

A general formula for n leaves can be written as

$$T_N = \sum_{k=1}^{N-1} T_k \, T_{N-k}.$$

We haven't actually calculated any values of T_k yet, but note that if we have only a single node, i.e., $k = 1$, the root and the leaf are the same node. So there is only one possible tree, which means that $T_1 = 1$. This can be used for the stopping condition (ground case) of our recursive definition.

■ **Example 10.11 (A naive and inefficient recursive program to determine the number of trees)**

010.D   We can see from the definition of T_N above that it is recursive, and this leads to an obvious possible recursive program to evaluate it:

```
function nTrees = T(N)
% Stopping condition:
if N == 1
  nTrees = 1;
  return
end

% N > 1 here.
% Initialise value to calculate.
nTrees = 0;

% The main loop.
for k = 1:N-1
  nTrees = nTrees + T(k) * T(N-k);
end

end
```

The code has a stopping condition of $N = 1$, in which case, there is a single leaf node. (Recall the return keyword that we introduced in Section 4.5.) This would correspond to a single node in the whole tree, and so there is only one possible tree. For larger values of N, recursive calls are made for smaller values (k and $N - k$), and the results are multiplied together and accumulated.

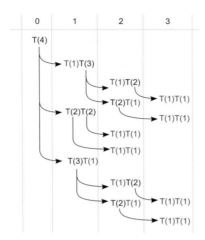

FIGURE 10.2
A visualization of the calls made when counting the number of full binary trees with four nodes.

> N.B. This code has been written compactly and without usage or many comments to save space. This is not an example of perfect programming. ■

In the above example, consider the case $N = 4$. We can write out in full the calls that are made:

$$T(4) = T(1)T(3) + T(2)T(2) + T(3)T(1).$$

We see that T is called twice with the stopping condition ($N = 1$), twice with $N = 2$, and twice with $N = 3$. Note that these calls will lead to further calls:

- Each call with $N = 3$ leads to calls $T(1)T(2) + T(2)T(1)$, i.e., two further calls with $N = 1$ and $N = 2$ each.
- Each call with $N = 2$ leads to calls $T(1)T(1)$.
- Each call with $N = 1$ returns the stopping condition value and ends the recursion.

A visualization of the calls made when evaluating $T(4)$ with this function is given in Fig. 10.2. The numbers along the top indicate the "depth" of the recursion for each call. Recursion depth 0 corresponds to the original (first) call, which, in turn, makes calls to evaluate three pairs of terms at recursion depth 1 to find their products and add the results. These make calls at recursion depth 2, and so on. The calls when $N = 1$ are marked in red, as they match the stopping condition, at which each path of recursive calls is ended.

Adding up all these calls, including those within recursive ones, we can find the total number of calls. There is 1 call with $N = 4$, 2 calls with $N = 3$, 6 calls with $N = 2$, and 18 calls with $N = 1$. In other words, there are many repeated calls for smaller values of N that are really unnecessary, because they are repeatedly calculating the number of trees for the same value of N. This shows that the naive recursive implementation can lead to a very inefficient solution, in which a lot of work is duplicated.

10.4.1   A note on the depth of recursive function calls

We saw in the previous example that the call to evaluate $T(4)$ led to recursive calls to a depth of three. In general, for this example, a call to evaluate $T(N)$ will lead to a depth of $N - 1$ in the final level of recursion. There are practical limits when running recursive functions, and MATLAB has a built-in limit that stops the levels of recursion exceeding a certain number (typically 500). This is sensible, because each call uses up a certain amount of memory, and if too many calls are made, then this would use up all of the computer's memory. Also, it is possible for a programmer to write a recursive function that, perhaps due to a bug, fails to hit its stopping condition. In this case, we should guard against the function recursively calling itself an infinite number of times. In general, we should take care when using recursive functions, and avoid using them when the depth of recursion is likely to be high.

■ Activity 10.9

010.D   *Pharmacokinetics* is a branch of pharmacology that studies the fate of chemicals (such as drugs) administered to the body. When doctors administer a drug to a patient, the kidneys act to filter it out of the blood stream, so the level of the drug in the body will continually drop unless extra doses are given. When administering regular doses of a drug, doctors are interested in modeling what the level of the drug will be in the body over time. This activity addresses this issue.

We first make a simplifying assumption: each day, the kidneys will filter out a fixed proportion of the drug that is still in the body. This assumption leads to an intuitive recursive definition of the problem of modeling drug levels in the body:

$$\text{level}_n = \text{decay} \times \text{level}_{n-1} + \text{dose}.$$

That is, the level of drug at day n is equal to the level at day $n - 1$ multiplied by a factor by which the level decays each day (due to filtering by the kidneys), plus the new dose administered each day. We assume that the level at the first day is zero (i.e., this is the stopping condition).

Write a recursive MATLAB function m-file to compute the level of drug in the body at day n, given a decay factor and a daily dose as additional arguments. Write a script m-file that reads in the dose, decay factor, and a maximum number of days from the command window, and then uses the function to determine the drug level at all days up to and including the maximum day. The script should then produce a plot of drug level against day.

Analyze your solution to determine the maximum depth of recursive call and the number of recursive calls that will be made for a given value of the day n.

Also write an iterative version of the function, and compare the time efficiency of the two implementations. Which is the most time efficient and which is the easiest to understand?   ■

10.5 Dynamic programming to improve performance

The previous example that counted full binary trees is one that can benefit from *dynamic programming*. This is because it is computed by evaluating the same function on sub-problems of the original problem (in our case, on sub-trees of the original tree).

The basic idea of dynamic programming is to store results that might need to be repeatedly calculated. This is known as *memoization*. We keep a "memo" of the results that have already been calculated, and that may be useful later. Values of the function are only directly evaluated if they have not previously been stored (memoized).

■ Example 10.12 (Using dynamic programming for the tree counting example)

To reduce the number of evaluations on the above tree counting example, we will write the code differently using a dynamic programming approach. We begin by writing the top-level function. This initializes an array for the memoized values and calls a helper function, called `count`:

O10.D, O10.E

```
function nTrees = countTreesDynamic(N)

memoValues = -1 * ones(1,N);
nTrees = count(N, memoValues);

end
```

The helper function accepts two arguments: the value of N for which the tree count is required and the array of stored values. The stored value array (`memoVals`) is initialized so that every entry is -1. We use a value of -1 to indicate that the result has not yet been calculated. At the start, this is the case for all of the possible values of N. Now we can consider the helper function. The start of this function looks like this:

```
function [nTrees, memoVals] = count(N, memoVals)
% Helper function for counting the trees. Keeps
```

```
% a memoised list of values to prevent too many
% calculations.

% N > 1. Check if we have already
% calculated for this value of N.
if memoVals(N) > −1
  nTrees = memoVals(N);
  return
end
% ... continues below ...
```

Note that the function accepts the number of leaves N and the array for the stored values and returns two items: the number of trees and the array of stored values. The array of stored values also appears as a return value, because it may be updated during a call. The function begins by seeing if the required value has already been calculated. This is done by checking if the value stored at index N in the memoVals array is greater than -1 (the initialization value of each entry). If it is greater than -1, then the value was calculated earlier, and we can assign it directly to the output argument nTrees and return. Job done. There is no need to update the memoVals array.

Now let's look at the next part of the function:

```
% ...
% stopping condition:
if N == 1
  nTrees = 1;
  memoVals(N) = nTrees;
  return
end
% ...
```

This part deals with the stopping condition when there is only one leaf in the tree. In this case, there is only one possible tree, and this is the count returned. Importantly, this count is stored in the memoVals array (at index 1) so that we can use it in any later repeated calls with $N = 1$.

The remaining parts of the function are only reached if $N > 1$ and the number of trees for N has not yet been calculated and memoized.

```
% ...
% No recorded value. Need to calculate from scratch.
% Initialise:
nTrees = 0;

for k=1:N−1
  [nLeft , memoVals] = count(k, memoVals);
  [nRight, memoVals] = count(N−k, memoVals);
  nTrees = nTrees + nLeft * nRight;
end
```

```
% Store value we have just calculated so we don't
% need to do this again.
memoVals(N) = nTrees;

end
```

Here, we make two recursive calls to the same function. Each time, we collect both arguments, the count of the trees (nLeft/nRight) and also the (possibly updated) memoVals array. During the loop, the number of trees is calculated, and after it is complete, we carry out the important step of storing the result in the array at the index N. ∎

In the exercises, we will compare the speed of this memoizing version of the tree counting function with the speed of the naive recursive implementation. It is sufficient to say here that the memoized version is much faster, because it avoids a lot of unnecessary duplication of calculations of values, which have previously been calculated and stored.

We note that the use of a memoized array of previously calculated and cached results has led to an increase in memory use. In the above case, this has been a small price to pay compared to the gain in speed, but, in general, we should bear in mind the memory used when seeking to adopt this approach to ensure that the memory usage of the program remains reasonable.

10.6 Summary

This chapter has covered some aspects of code efficiency and general tips to help improve it. These form useful advice, but it is important to maintain good coding practice (see Chapter 5). The general advice when coding is to proceed in two broad stages:

- Write code that is correct, clear, readable, and tested.
- Where possible, only consider performance and efficiency when the above step is complete. In particular:
  - Decide whether or not improvements in efficiency are necessary.
  - Only implement them if they are, by rewriting appropriate parts of the code.

The reason for this is that code that has been optimized to be as efficient as possible tends to be less readable and harder to debug when errors arise. Code that takes up more space and uses more operations might be less efficient, but is more likely to be clearer and easier to follow.

To stress this point: it is more important to have code that is *correct* than code that is fast or memory-efficient. Once we are confident that the code is correct, we can consider changes for optimizing efficiency.

In other words, it is almost always okay to start off with code that works and is slow. We can then incrementally change it so that it becomes quicker, only changing small parts of the code at a time (in line with the incremental development model outlined in Section 5.2).

In the second half of the chapter, we looked specifically at recursive functions and how, if implemented naively, they can be inefficient by carrying out a lot of unnecessary calculation. We took an idea from dynamic programming, in which we store or cache the results of calculations, to avoid such duplication. Generally, this will lead to significant improvements in the calculation of recursive functions, but we should remain aware that it can cause an increase in memory usage.

10.7 Further resources

- The MATLAB built-in help has a number of tips on measuring and improving the performance of programs from a time or memory point of view. A good place to look is under *MATLAB → Software Development Tools → Performance and Memory*.
- Dynamic programming applies to a number of other areas:
  - For a good general description of how it applies to a number of interesting problems, follow the links from http://people.cs.clemson.edu/~bcdean/dp_practice/.
  - A more general tutorial can be found here: http://www.codechef.com/wiki/tutorial-dynamic-programming.
- The number of trees with a given number of leaves is an example of an application of Catalan numbers, which are interesting in their own right and have a diverse number of other applications: http://en.wikipedia.org/wiki/Catalan_number.

10.8 Exercises

■ Exercise 10.1

O10.B

a. Calculate how much memory will be required by the following assignment:

```
x = ones(1000).
```

b. Explain why a call to `z = 3 * zeros(1000000)` is likely to cause problems when it is run. ■

■ Exercise 10.2

O10.A, O10.B

a. Give two reasons why the following code is not efficient:

```
myArr = 0;
```

```
N = 100000;

for n = 2:N
    myArr(n) = 2 + myArr(n−1);
end
```

b. Re-write the above code so that it runs more efficiently.
c. Use the built-in timer functions to estimate the speed-up given when the above code is re-written. ∎

■ Exercise 10.3

Re-write the following code so that it becomes more time efficient: *010.A*

```
N = 20;

a = [];
for j = 1:N
    a(j) = j * pi / N;
end

for k = 1:numel(a)
    b(k) = sin(a(k));
end

for m = 1:numel(a)
    c(m) = b(m) / a(m);
end

plot(a, b, 'r')
hold on
plot(a, c, 'k')
legend('sin', 'sinc')
```

∎

■ Exercise 10.4

010.A

a. Use the built-in random function `rand` to generate a 5×6 array of uniformly distributed random numbers, and assign it to a variable called A.
b. Use loop(s) to check the elements in A. If an element is less than 0.6, replace it with a value of 0. Do this inefficiently using the same approach as in Example 10.8.
c. Use logical indexing to achieve the same result as in the previous part, i.e., by avoiding loops.
d. Use logical indexing to set the value of any elements in A that are greater than 0.7 to a new value of −1.
e. Use logical indexing to set any values in A that are greater than 0.7 OR less than 0.3 to −1.
   (*Hint: Use the bitwise logical or operator: |*). ∎

■ Exercise 10.5

O10.A, O10.B, O10.C

a. Write a MATLAB function that takes a single integer input argument N and computes a 2-D array representing the multiplication table for integers from 1 up to N. For example, using the `disp` command to display the table for $N = 5$ should show the following array:

```
1  2  3  4  5
2  4  6  8 10
3  6  9 12 15
4  8 12 16 20
5 10 15 20 25
```

Use loops in your implementation, and focus on writing clear, understandable code that works.

b. Now write another implementation, this time optimized for speed. Run and time both implementations for values of N, from 1 to 200, and show the time results for both methods.

(Hints: You can use matrix multiplication or the `repmat` command to obtain faster implementations. The commands `tic` and `toc` can be used for timing.)

c. Write code to run both functions for all values of N from 1 to 200, calculate the total time taken for each one. Write code to display the time results for both methods.

d. Finally, write a function that just *displays* a multiplication table for a given size, and that uses very little memory (i.e., one that is memory-efficient). N.B. It does not need to store the table.   ■

■ Exercise 10.6

O10.A, O10.C

In Exercise 6.3, you wrote a MATLAB script *m*-file to determine the recommended drug dose based on the results of a phase I clinical trial. This involved searching through a series of toxicity results for cohorts of three patients to find the first instance in which two or more of the three patients experienced significant toxicity.

Make a copy of your solution to Exercise 6.3 (or the one from the book's web site), and adapt it so that this searching is performed in two different ways. One should use a loop, and the other should use logical indexing. It should be possible to write the version that uses logical indexing with a single line of code. Make sure that both implementations give exactly the same results when applied to the data in the file *phase1_data.mat*. Add commands to your script *m*-file to measure the time taken to run each of the two implementations.

Leave the remainder of the code unchanged, i.e., the code that finds the recommended dose, once the first cohort with two or more significant toxicity results has been found.

Which implementation is the most time-efficient? Which is the easiest to understand? ∎

■ Exercise 10.7

In Exercise 4.7, you wrote a function to determine the indices of all peaks and troughs in a given array, and demonstrated its use for gait analysis. Make a copy of your solution to Exercise 4.7 (or the one from the book's web site), and adapt it so that there are two different functions that each identify the peaks and troughs, but in different ways. One should use a loop, and the other should use logical indexing. Make sure that both give exactly the same results when applied to analyzing the z-coordinates of the ankle marker as in Exercise 4.7 (i.e., the file *LAnkle_tracking.mat*). Add commands to your main script *m*-file to call and measure the time taken to run each of the two functions.

O10.A, O10.C

Which of the two implementations is the most time-efficient? Which is the easiest to understand? ∎

■ Exercise 10.8

Go to the book's web site, and download both versions of the code that counts the number of full binary trees with N leaves:

O10.A, O10.D, O10.E

- The naive recursive version (Example 10.11).
- The version that uses memoization (Example 10.12).

a. Write a MATLAB script to measure the execution times of the two implementations for a single value of N. Be careful when choosing the value of N. When using the naive version, a value of N greater than around 14 could lead to a long wait!

b. Extend your script to compute the execution times for the two implementations for $N = 1 \ldots 10$, and plot both sets of values on the same graph against N. What do you notice about the times taken by each method as N increases? ∎

■ Exercise 10.9

The Fibonacci sequence is the series of integers $1, 1, 2, 3, 5, 8, \ldots$, and it is defined mathematically and recursively by the formula

O10.A, O10.D, O10.E

$$f_1 = 1, \qquad f_2 = 1, \qquad f_{n+1} = f_n + f_{n-1}, \text{ for } n > 2.$$

Recall that Exercise 4.11 involved writing a recursive implementation of a function to compute Fibonacci numbers. The Fibonacci sequence has applications in a wide range of fields, including biology [5].

a. Write a function that directly implements the recursive Fibonacci formula. Name the function *fibonacciRecursive*, and save it to a function *m*-file.

b. Each call to the function will make further calls recursively with smaller values of n. If the function is called originally (recursion depth 0) with an input argument of $n = 5$, work out how many calls there are for the function with arguments of $n = 4, 3, 2,$ and 1, and how many calls in total.

c. Extend this to work out the number of all such calls to the function when we start with higher values of n, i.e., $n = 6$, $n = 7$, etc.

d. One way to avoid the overhead of multiple calls which evaluate the same value of the function in Exercise 10.9 is to avoid recursive calls altogether. One way to do this is to calculate the values in a "bottom up" way, starting with the values for the lowest values of n and "building up." This way, we can visit each value of n once only.

By defining values for the current and previous terms in the sequence (or otherwise), write a function that loops upwards to the required value of n. Within each iteration, the following will need to be done:

- Calculate the value of the next term in the sequence.
- Increment a counter that keeps track of how far we have moved along the sequence.
- Update variables holding the previous and current terms based on the incremented counter.
  (Hint: Use the value calculated in the first step).
- Check if the loop has reached the required value of n, stopping if this is the case, and returning the required value.

N.B. Such *bottom-up evaluation* is another technique used in dynamic programming (as well as memoization).

e. Estimate the speed up of the dynamic programming bottom-up evaluation of the Fibonacci code compared with the previous recursive version. How many times faster is the new code? It may be advisable not to call the previous version with a value much higher than 30 or so. ∎

FAMOUS COMPUTER PROGRAMMERS: TIM BERNERS-LEE

Tim Berners-Lee is a British computer scientist who is best known as the inventor of the World Wide Web (WWW). He was born in 1955 in London. His parents worked on the first commercially built computer, the Ferranti Mark 1. Tim was a keen trainspotter as a child, and he learned about electronics from playing with his model railway. Later, he went on to study physics at Queen's College, Oxford, graduating in 1976.

From 1980, he worked as a computer programmer at the European Nuclear Research Lab, CERN, in Geneva, Switzerland. While there, he wrote a computer program called ENQUIRE, which was a piece of software for storing information together with associations between them; in writing this program, Berners-Lee had actually invented the concept of "hypertext" that is so common on the WWW today. After working in the UK for a few years, in 1984, he returned to CERN. In 1989, he saw an opportunity to utilize the concept of hypertext to make the internet more accessible. The result was the WWW. In his own words, "creating the web was really an act of desperation, because the situation without it was very difficult." He wrote software for the first WWW server, and developed specifications for HTML and HTTP, which are the basis for almost all web sites in the world today. Throughout, he has made his work available freely without patent, and has earned no royalties from the success of the WWW.

In 1994, he founded the World Wide Web Consortium (W3C), the not-for-profit organization that coordinates the development of the WWW. He received a knighthood from the UK government in 2004, and in 2012, he was a part of the London Olympics opening ceremony, which celebrated his immense and selfless contribution to the development of the internet as we know it today, all of which was made possible by computer programming.

He is currently a Professor in the Computer Science Department at the University of Oxford in the UK and works on a variety of projects related to the WWW and net neutrality.

"When somebody has learned how to program a computer ... You're joining a group of people who can do incredible things. They can make the computer do anything they can imagine."

Tim Berners-Lee

CHAPTER 11

Signal and image processing

LEARNING OBJECTIVES

At the end of this chapter you should be able to:

O11.A Load, save, and visualize signal and image data with MATLAB® commands or, in the case of image data, with the available interactive tools

O11.B Carry out some of the fundamental one-dimensional (1-D) signal processing steps with MATLAB

O11.C Carry out some of the fundamental image processing steps with MATLAB

11.1 Introduction

Signals and images form an important part of general data analysis and have a particularly important role to play in many biomedical applications. We can carry out a number of tasks involving them: collecting the raw data, storing and retrieving data from files, visualization, processing, and so on.

It is very common to collect biomedical signals that vary in time. These can be generated, for example, by monitoring physiological processes in the human body. We can measure the heart rate, blood glucose levels, electrical activity in the brain, and so on. Each of these can be represented by a time-varying signal, which we can process and analyze to derive further information and devise biomarkers relating to the health of patients. Fig. 11.1 shows some examples of such time-varying signals.

An image can also be viewed as a type of general signal. For example, an image acquired with an everyday digital camera represents a set of measurements that are arranged in *space* on a 2-D grid of pixel locations. This bears much similarity with signals acquired on a 1-D regular "grid" of *time* points. Images are an important source of information in biomedical applications, as they can assist clinicians in making diagnoses or monitoring treatment. Biomedical examples of 2-D images include microscopy images of cells, photographs of skin for detecting melanomas, and real-time ultrasound images, which can be used for many purposes, including diagnosing heart disease or looking for fetal

MATLAB® Programming for Biomedical Engineers and Scientists. https://doi.org/10.1016/B978-0-32-385773-4.00020-4

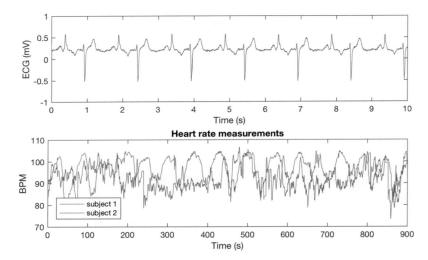

FIGURE 11.1

Top: ECG measurements for a 10-second interval. Bottom: Heart rate measurements over a 15-minute interval for two different subjects. BPM = beats per minute.

anomalies. 2-D images are either projections (e.g., microscopy or photographic images) or represent "slices" through the 3-D anatomy (e.g., ultrasound). As well as 2-D images, it is also possible to acquire 3-D imaging data. Magnetic resonance (MR) imaging can be used to acquire such 3-D images, as can positron emission tomography (PET), which shows metabolic function by imaging the concentration of an injected radiotracer. Modern ultrasound scanners can also acquire 3-D images, and can even dynamically acquire multiple 3-D images of an organ over time. This is an example of 4-D medical image data (i.e., three spatial dimensions and one time dimension). Fig. 11.2 shows some examples of different types of medical images.

Processing images is an important part of developing biomedical applications and is typically carried out using computer programming. Specialist software as well as general programs, such as MATLAB can carry out a variety of tasks that manipulate and alter images in ways that can provide useful information. This chapter will focus on some of the basic steps of signal and image manipulation and processing. A few methods will be described in general terms, and we will also show how they can be carried out in MATLAB.

11.2 Storing and reading 1-D signals

1-D signals, such as time series data, have a very simple structure and can be easily stored in text files with, say, one measurement on each line of the file. Typically, we also need "meta-data" to indicate what units are used for the measurements. We also need to know the sampling frequency, i.e., the time

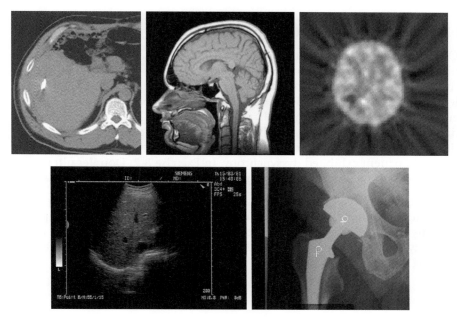

FIGURE 11.2
Examples of medical images: from top-left, slices taken from a computed tomography (CT) image of the abdomen, an MR image of the head, a single PET volume of the head, a 2-D ultrasound image of the abdomen, and a 2-D X-ray of a hip.

step between adjacent measurements. If the measurements are not obtained at regularly spaced time points, then we can store the times at which each measurement was taken, as well as the value of the measurement. This remains a relatively simple structure and can, for example, be stored in a "comma-separated values" (*.csv*) file, with two columns, one for the times and one for the measurements. Of course, it is also possible to store data for a 1-D signal in a binary file format. The different methods for saving and reading both text and binary format data files were discussed in Chapter 7.

11.3 Processing 1-D signals

We have, in fact, already seen a number of different examples in the book, in which we analyzed and extracted information from 1-D signals.

For example, in Activity 1.6, Exercise 2.9, and Exercise 4.7, we looked at data to represent the gait of a person when walking. In particular, we treated the z-coordinate of the ankle (its height) as a time-varying 1-D signal and searched for peaks and troughs in the data. In Activity 6.2, we also visualized these height data. In Exercise 4.9, we looked at heart rate data to find times at which there were abnormal changes. We looked at time series data from an oscillatory monitoring device in Exercise 5.4, and examined how we could use it to obtain blood pressure and pulse rate information.

Other examples include Exercise 4.10, which looked at a 1-D signal from a respiratory bellows. This was further analyzed in Exercise 5.5, by smoothing the time-varying signal to reduce noise prior to finding peaks and estimating the length of the respiratory cycle. We considered data on blood sugar levels in Exercise 5.6 to identify time intervals, where the average level indicates possible diabetes. In Example 5.1, we analyzed measurements of radial heart contraction (a time series for each of a number of different heart regions) to identify if a given patient exhibited dyssynchrony. In Exercise 8.2, Doppler ultrasound data, representing blood flow velocity, were used to estimate arterial resistivity.

These examples serve to illustrate that time series data in biomedical applications are very common, and the nature of the data and the processing applied can vary quite widely. In the next section, we consider in more detail a particular type of processing that is widely used to process both time series and imaging data.

11.4   Convolution

Convolution is an important and widely used tool in signal processing. It is a general method that can re-express many of the operations applied to signals, such as smoothing, finding peaks, etc. Convolution takes an array of signal values and a second array containing the values of a *kernel*. The kernel and the signal are *convolved* to produce a modified signal.

By way of an example, recall the signal from a respiratory bellows in Exercise 5.5. After reading in the raw data, the aim was to find peaks in the signal so that different breathing cycles could be distinguished. We assumed that a peak in the signal occurs when a signal value is greater than the values before it and after it (see Exercise 4.7). The problem is that some peaks detected in this way may be due to random noise and fluctuations in the data and may not be located at the boundaries of breathing cycles. This was why we *smoothed* the data, so that the effect of noise was reduced and true peaks could be found more easily. Fig. 11.3 shows a plot of the raw bellows signal (red) and the smoothed data (blue). At $t = 4.6\ s$, we can see an example of a "false" peak in the original signal.

■ Example 11.1

011.A, 011.B   The bellows plot used a signal sampled at 5 Hz, i.e., with a time resolution of 0.2 s. The code to produce the smoothed data is shown below.

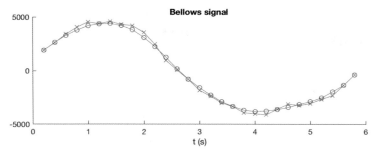

FIGURE 11.3
Data from a respiratory bellows measured over time: (red) original signal, (blue) smoothed signal.

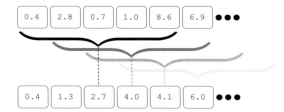

FIGURE 11.4
A sliding window is used to smooth a sequence of signal values. This illustrates a window of width five. Each successive set of five values are averaged to give the value in the output signal. The first few values of the input and output signals are shown. N.B.: For this window, the third value in the output is the first one where the window "covers" a full set of input values. The first two values need to be calculated differently (see text).

```
% Load data
load('bellows_raw.mat');

% Smooth signal (See 'doc smooth')
bellows_smooth = smooth(bellows_raw);
```
∎

The smooth function is part of the curve fitting toolbox in MATLAB, and it works by calculating a moving average of values in a "sliding window." The window has a specific width and moves over the original signal. At each position, the average of the values it covers is calculated and assigned to the current value of the output signal.

By default, the smooth function uses a window width covering five values; this is illustrated in Fig. 11.4. Note in the figure how the mean value of a set of successive values is assigned at the central location, for example, $(0.4 + 2.8 + 0.7 + 1.0 + 8.6)/5$ gives 2.7, $(2.8 + 0.7 + 1.0 + 8.6 + 6.9)/5$ gives 4.0, and so on. For a width 5 window, this means that the third value in the output sequence is the first one that has a full set of input values covered by the window. The first and second values are handled differently, and how we handle them is

FIGURE 11.5
Smoothing a 1-D signal using a shifting row of weights. For each position of the weights, the original signal values are multiplied by the corresponding weights and the results summed.

a matter of choice. In the case of the `smooth` function, the first value in the output sequence is equal to the first of the input sequence, and the second is the mean of the first three input values (rather than five). The values at the end of the sequence are treated similarly.

In general, when processing 1-D signal data in this way, it is common for values at (or near) the start and end, i.e., the boundaries, to be handled in a slightly different way from other non-boundary values.

To introduce the idea of convolution, note that finding the mean of a set of five adjacent values is equivalent to multiplying each by 0.2 and adding the results. We can picture a row of five weights, each equal to 0.2, being placed against the original sequence at successive positions. For each position, the values in the original sequence are multiplied by the corresponding weights and added together. After this step, the row of weights shifts forward one place, and the process is repeated. This is illustrated in Fig. 11.5.

This is the fundamental idea of convolution. A set of weights in a *kernel* are multiplied by sets of adjacent values in the signal, and the result is summed to give an output value. In simple terms, the *convolution of the signal with the kernel* is this "multiply and add" process, applied after repeated shifts of the kernel.

■ Activity 11.1

011.B   Now we can look at the process of smoothing using the MATLAB `conv` function. Type the following into a script and run it:

```
% The original 1-D signal.
u = [0.4, 2.8, 0.7, 1.0, 8.6, 6.9, 3.3, 10.2];
% The kernel.
kernel = [0.2, 0.2, 0.2, 0.2, 0.2];
% Do the convolution.
w = conv(u, kernel);
% What does the 'smooth' function give?
v = smooth(u);
```

Now we can compare the outputs from the `conv` and `smooth` functions. After executing the code above, the variable `w` should contain the following values:

```
0.08 0.64 0.78 0.98 2.7   4.0   4.1   6.0   5.8   4.08   2.7   2.04
```

The variable v contains

```
0.4   1.3   2.7   4.0   4.1   6.0   6.8   10.2
```

We can see immediately that there are differences between the two resulting outputs. They even have different lengths. This is partly due to differences in the way the boundaries are handled, as we will see below. Before reading on, can you see which parts of the two outputs are the same? ∎

To understand the output from conv, recall that the kernel used has a length of five. When we slide the kernel across the array containing the original signal, we start with little overlap between the kernel and the signal. At the outset, there is an overlap of a single value, as shown below, where the top line shows the signal, and the second line shows the kernel:

```
                0.4  2.8  0.7  1.0  8.6  6.9  3.3 10.2
0.2  0.2  0.2  0.2  0.2
```

In this position, applying the kernel leads to the first element in the output array w, which is computed as the product of the first signal value and the rightmost kernel element: $0.08 = 0.4 \times 0.2$. The other kernel elements do not overlap the signal and are effectively multiplied by zero.

Applying successive shifts, the first four values of the output array w will still be obtained from a partial overlap of the kernel and the signal:

```
                                              Result
                0.4  2.8  0.7  1.0  8.6  6.9  3.3 10.2
0.2  0.2  0.2  0.2  0.2                        -> 0.08

                0.4  2.8  0.7  1.0  8.6  6.9  3.3 10.2
     0.2  0.2  0.2  0.2  0.2                   -> 0.64

                0.4  2.8  0.7  1.0  8.6  6.9  3.3 10.2
          0.2  0.2  0.2  0.2  0.2              -> 0.78

                0.4  2.8  0.7  1.0  8.6  6.9  3.3 10.2
               0.2  0.2  0.2  0.2  0.2         -> 0.98
```

■ **Activity 11.2**

Show the calculation that needs to be made to give the third value in the output (0.78).   *011.B* ∎

After computing the first four values, we can finally carry out convolution with the kernel fully overlapping the signal. This is shown in the following positions, which give the next four values in `w`:

```
                                           Result
0.4  2.8  0.7  1.0  8.6  6.9  3.3 10.2
0.2  0.2  0.2  0.2  0.2                  -> 2.7

0.4  2.8  0.7  1.0  8.6  6.9  3.3 10.2
     0.2  0.2  0.2  0.2  0.2             -> 4.0

0.4  2.8  0.7  1.0  8.6  6.9  3.3 10.2
          0.2  0.2  0.2  0.2  0.2        -> 4.1

0.4  2.8  0.7  1.0  8.6  6.9  3.3 10.2
               0.2  0.2  0.2  0.2  0.2   -> 6.0
```

Note that, for this set of values, where the kernel fully overlaps the signal, the output from `conv` (variable `w`) matches the output from `smooth` (variable `v`).

Beyond this position, the remaining values in the convolution output will also result from partial overlaps of the kernel and will be calculated in the same way as the first four values above.

In summary, the outputs from `conv` and `smooth` have different lengths, and the values at the boundaries are handled differently. It is reasonable to now ask: *How can we reconcile the outputs from the two different functions* `smooth` *and* `conv`?

■ Example 11.2

011.B

To make the outputs comparable, we can use an optional argument to the MATLAB `conv` function that specifies which part of the convolution output to calculate. We can achieve this with the following:

```
w2 = conv(u, kernel, 'same');
```

This option restricts the convolution to be calculated over a portion of the output that matches the length of the input signal (i.e., the variable `u`).

The resulting variable `w2` now looks like this

```
0.78   0.98   2.7   4.0   4.1   6.0   5.8   4.08
```

which we can compare to the output from the `smooth` function (variable `v` from the earlier call), which we repeat here:

```
0.4    1.3    2.7   4.0   4.1   6.0   6.8   10.2
```

■

Here, we can see that the two outputs now match on the central portion (from the 3rd to the 6th elements, i.e., where the kernel has full overlap). The values at the ends still differ, because they are handled differently by each function, but this is not so significant in general, especially as the signals we typically work with will be much longer than the one given in this example, and the majority of values in the array will not be near the boundary.

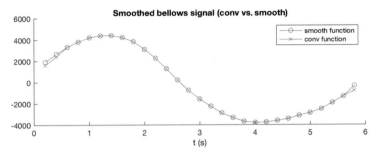

FIGURE 11.6
Smoothing the bellows signal using the MATLAB `smooth` and `conv` functions.

■ Example 11.3

We can repeat the smoothing of the bellows signal data in Example 11.1   *011.B*
using the `conv` function and compare the two results.

```
% ... previous code as before
% Smooth signal (See 'doc smooth')
bellows_smooth = smooth(bellows_raw);

% Smooth using 'conv' function
kernel = [0.2, 0.2, 0.2, 0.2, 0.2];
bellows_smooth_2 = conv(bellows_raw, kernel, 'same');
```
■

If we plot the two resulting outputs, we can see how they are the same at all values, except for the sets of four values near each boundary (see Fig. 11.6).

11.4.1   Convolution: more detail

So far, we have only considered a very simple kernel to illustrate the concept of convolution. It simply consisted of five equal values:

```
kernel = [0.2, 0.2, 0.2, 0.2, 0.2];
```

This has allowed us to ignore a further detail of how convolution is actually carried out. Specifically, when a kernel is convolved with an array, it is actually *reversed* before "sliding" over the input signal values. As our kernel only contained a single repeated value, this reversal had no effect.

■ Activity 11.3

We consider a very simple example that illustrates this reversal of the kernel   *011.B*
by using some simple integer arrays for our signal and kernel.

```
u = [1, 2, 1, 4, 3];
kernel = [1, 2, 3];
result = conv(u, kernel, 'same');
```

Displaying the variable `result` that contains the output of the convolution should now give

```
4     8     12     14     18
```

■

The calculations involved are illustrated below for each of the output values. In this case, the first and last values in the result are obtained with a partial overlap of the kernel. The important thing to note is that the kernel that is passed in [1,2,3] is reversed before being shifted across the input array:

```
        1  2  1  4  3
     3  2  1                  -> 4
     --------------------------
        1  2  1  4  3
        3  2  1               -> 8
     --------------------------
        1  2  1  4  3
           3  2  1            -> 12
     --------------------------
        1  2  1  4  3
              3  2  1         -> 14
     --------------------------
        1  2  1  4  3
                 3  2  1  -> 18
     --------------------------
```

For example, the third value is obtained by calculating $2 \times 3 + 1 \times 2 + 4 \times 1 = 12$.

The gradient of a signal: Given a set of measurements in a 1-D signal and the time step between the measurements, we can estimate the *gradient* at a point by using the *central difference*. If the nth value in a sequence is y_n, and the time step between values is Δ_t, we can estimate the gradient g_n at that point using the values y_{n-1} and y_{n+1} and the formula

$$g_n \approx \frac{y_{n+1} - y_{n-1}}{2\Delta_t}.$$

■ **Example 11.4**

011.A, 011.B

The file *pos-z.csv* is available on the book's web site and contains the *z*-coordinates of a surgical robot arm with measurements taken every 400 *ms* (i.e., $\Delta_t = 0.4$ s). We will estimate the gradient of these *z*-values, i.e., their rate of change with respect to time.

```
z_data = readmatrix('pos-z.csv');
```

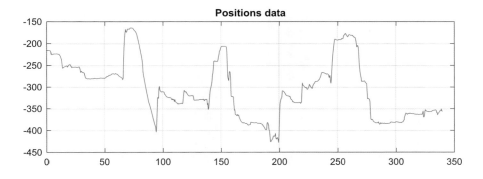

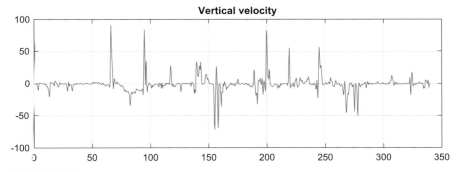

FIGURE 11.7
Surgical robot arm data: (top) vertical position, (bottom) vertical velocity.

Let the time-step variable be set by `delta_t = 0.4;`. We can now estimate the gradient of the signal at all points using convolution:

```
% A kernel to use with conv.
kernel = [−1, 0, 1] / (2 * delta_t);

% Flip the kernel before passing to conv
kernel = fliplr(kernel);

deriv_est = conv(z_data, kernel, 'same');
```

Note that the kernel was flipped left to right before passing to `conv`. At the end points, not enough data are available to estimate the gradient (the kernel does not fully overlap the data). We can therefore arbitrarily set the values at the end points to zero.

```
deriv_est(1) = 0, deriv_est(end) = 0;
```                                                                ∎

The full code for the above example is provided on the book's web site. In this case, the data represent z-positions, so the gradient represents vertical velocity. If we plot the z-positions and the vertical velocity for the data file specified, we obtain the plot shown in Fig. 11.7.

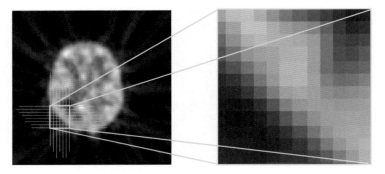

FIGURE 11.8
A PET image showing the individual pixels. These are the discrete locations at which we have a measurement.

11.5 Storing and reading image data

Image data have a slightly more complicated structure than 1-D signal data, such as time series signals. This means that we have to take a little more consideration of what the image data represent. For example, images can contain a *lot* of data, and this means we may need to be careful about what data type is used when storing an image. In general, we need to consider the following points:

- Discrete locations:
  - Image data are stored for a finite number of discrete locations.
  - These locations are nearly always arranged on a regular grid, and are often called *pixels*. The term pixel comes from "picture element." In the case of 3-D medical images, the term *voxel* ("volume element") is often used.
  - The image has a value (or intensity) associated with each pixel, and we can visualize each pixel as a block of uniform color or a block of uniform shading (see Fig. 11.8).
- Discrete intensities:
  - As well as having discrete locations, the intensity value at each pixel is also usually discrete, i.e., it can take on one of a limited number of values.
- Data type:
  - The most common way to store the intensity of each pixel is using an integer data type, often between 0 and 255.
  - This is because a number between 0 and 255 can be represented by an unsigned integer type in a single byte (8 bits), which makes it easier to store intensities for many pixels in large images.
  - It is also possible to use other data types, such as floating point numbers or similar.

11.6 Accessing images in MATLAB

11.6.1 Color versus gray scale images

Whereas everyday images tend to be color[1] images, medical images tend to be gray scale[2] images. A color image typically uses three numbers to represent the red, green, and blue intensities associated with each pixel. Gray scale images only need to store a single value to represent a shade of gray between 0 (black) and a maximum value (white).

Storing pixel data for a gray scale image is relatively easy. The simplest way is to store an unsigned one-byte integer with 0 representing black and 255 (the maximum value) representing white. An integer between 0 and 255 will be shown as an intermediate shade of gray.

It is also possible for an image to have pixels that can be only black (minimum value) or white (maximum value), i.e., with no intermediate shades of gray. These are known as *binary* images.

Storing pixel data for a color image can be done in more than one way. Perhaps the simplest is to repeat the same approach used for gray scale images, but with a separate number to represent the intensity of each of the red, green, and blue (R, G, and B) channels. In other words, we can use three bytes for each pixel, with one byte each for the red, green, and blue intensities. For example, (0, 0, 255) would represent pure blue; (0, 255, 0) would represent pure green. Mixtures such as (255, 0, 255), which is magenta, are also possible. A shade of magenta with a lower intensity can be represented by (127, 0, 127), etc. Using a numeric value for each of the R, G, and B intensities is sometimes called a *true color* representation.

Another way to store pixel data for a color image is to use a *color map*. A color map is a list of triples representing the RGB colors that appear in the image. In MATLAB, we can store a color map as an array with three columns. The number of rows will need to be sufficient to capture all the colors in the image. To represent an individual pixel, we only need to store a single integer, which indexes the row in the color map that contains the pixel's color. This method of using a color map and integers to look up rows in the map is called *indexed color* representation, and it is useful for images with a reasonably limited range of colors.

11.6.2 Getting information about an image

We can use the built-in MATLAB function `imfinfo` to get information relating to an image file. First, we specify the name of the file, and then a structure

---

[1] *colour* in the UK.
[2] *grey scale* in the UK.

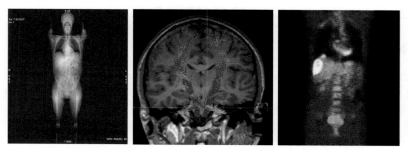

FIGURE 11.9
Medical images: (left) a CT scout image, (middle) a slice through a MR brain image (image generated using MRtrix: http://www.mrtrix.org/), (right) a PET scan (from Prof. Dr. Hojjat Ahmadzadehfar).

variable (see Section 6.3) is returned, which contains information on different aspects of the image, for example, its size. There are three example images provided via the book's web site (see Fig. 11.9):

- CT_ScoutView.jpg: A CT scout image (i.e., an X-ray used to plan a CT scan) of a body.
- mr-fibre-tracks.png: A slice taken from a 3-D MR image of a brain, showing tracks of white matter in color.
- pet-image-liver.png: A PET image of the thorax and abdomen, showing metabolic activity in the liver.

■ Example 11.5

O11.A Make sure that the images are in the current MATLAB path (see Section 4.8). We can use imfinfo to find information about an image as shown below.

```
>> imfinfo('mr-fibre-tracks.png')

ans =
struct with fields:
        Filename: '/PATH/TO/FILE/mr-fibre-tracks.png'
     FileModDate: '08-Sep-2014 10:48:42'
        FileSize: 192215
          Format: 'png'
   FormatVersion:
           Width: 465
          Height: 464
        BitDepth: 24
       ColorType: 'truecolor'
  and some further information follows ...
```

We can see that the size of the pixel grid is 465×464, and that the image is a true color image. The struct has a BitDepth field, which says that 24 bits (i.e., three bytes) are used for each pixel.   ■

■ **Activity 11.4**

Use the `imfinfo` function to list the information for the other two provided images: `CT_ScoutView.jpg` and `pet-image-liver.png`.   ■

O11.A

11.6.3   Viewing an image

The simplest way to view an image from the command window is to use the `imshow` function,[3] giving the name of the image file as an argument.

■ **Example 11.6**

O11.A

```
>> imshow('mr-fibre-tracks.png')
```

The above call should result in the display shown below. Here, we see that `imshow` has loaded the image into a standard plotting window. The same tools that we can use for graph plots are also available in this window.

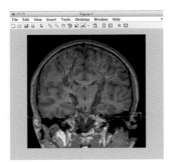

■

Another way to view an image is to use the *interactive* image viewer provided by the image processing toolbox. This provides more ways to control how an image is viewed and also includes some tools for manipulating and editing the image. We can start it by calling `imtool` with the name of the image file:

```
>> imtool('mr-fibre-tracks.png')
```

This command loads the image into the interactive viewer with a *Tools* menu that offers some basic ways of interacting with the image. For example, under *Tools → Image Information*, we can access the same information as we obtained using `imfinfo` above.

---

[3]The `imshow` function belongs to the image processing toolbox. This toolbox is not part of the default MATLAB installation and needs a separate license. If a call to `imshow` fails, it may be because the toolbox is not licensed for your machine.

It is also possible to start the viewer without loading an image by simply typing `imtool` at the command window. This will open the viewer with a blank screen, and an image can be loaded by going to *File → Open*.

■ Activity 11.5

011.A

The images *CT_ScoutView.jpg*, *mr-fibre-tracks.png*, and *pet-image-liver.png* are available on the book's web site. Try the different methods above for loading and viewing these images. ■

11.6.4   Accessing the pixel data for an image

For simplicity, we will assume from now on that all images being described are gray scale, and that we can simply represent each pixel's intensity by one number. The example image *CT_ScoutView.jpg* is a single channel gray scale image with an unsigned integer stored for each pixel. This means that we can store the intensities for all pixels in a single array. We obtain the pixel data using a call to the `imread` function[4]:

```
>> x = imread('CT_ScoutView.jpg');
>> whos
  Name          Size              Bytes  Class    Attributes

  x             1024x880          901120  uint8
```

Here, we load the pixel intensities into the array x. A call to `whos` tells us that the size of the array is 1024 × 880 pixels, and its data type is `uint8`, i.e., unsigned 8-bit (1-byte) integers.

As we saw in Section 9.2.6, we can use the `imshow` function to display the pixel data in array x directly (i.e., without using the image file name). In other words, the command

```
>> imshow(x);
```

will produce the same result as `imshow('CT_ScoutView.jpg')`.

11.6.5   Viewing and saving a sub-region of an image

We do not have to view the whole image, we can focus on a sub-region or "window" inside the image. In the interactive viewer (`imtool`), we can use the zoom tool to adjust our view, zooming in and out as necessary.

We can also focus on a specific region using explicit commands. Picking out a sub-region is one of the simplest tasks we can carry out in image processing.

---

[4] `imread` is a built-in MATLAB function, not part of the image processing toolbox.

■ Example 11.7

As the image data are stored in an array, we can set ranges of row and column   *011.A, 011.C*
indices to specify the sub-region that we are interested in. For example,

```
>> x = imread('CT_ScoutView.jpg');
>> imshow(x(25:350,350:500))
```

This code displays a region of the CT scout image around the head and
shoulders. Note that the rows specified in the example above (from 25 to
350) are counted starting from the top of the image. The columns (from
350 to 500) are counted from the left.   ■

We can save our *cropped* version of the image data into a new image with the
following command:

```
>> imwrite(x(25:350,350:500), 'ct—scout—crop.png')
```

This uses the built-in `imwrite` function that we saw in Section 9.2.6. Recall that
this function can take two arguments: the array of image data to save (in this
case, a sub-region of x) and a character array containing the file name to save it
to.

■ Activity 11.6

Experiment with different crop regions and save or display the results. Use   *011.A, 011.C*
the command window and the tool built into the interactive `imtool` viewer.
Can you choose a region which identifies the chest area in the CT scout
image?   ■

11.7   Image processing

Now that we have seen the ways in which we can load gray scale image data,
we will look at some simple operations that can be applied to modify the data.

11.7.1   Binarizing a gray scale image and saving the result

Converting a gray scale image to a binary image produces an image with one
of two values at each pixel (e.g., "on"/"off," "true"/"false," 0/1). We can use
logical tests to do this in MATLAB. For example, using the same CT scout image
described earlier:

```
x = imread('CT_ScoutView.jpg');
y = x > 50 & x < 250;
```

This code first reads the image data into array x. The next line applies a pair
of logical tests (element-wise) to x and applies the logical *AND* (&) operator
to the results. This statement will produce an array (which is assigned to y) in

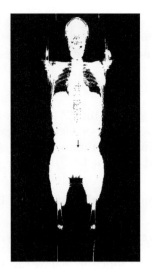

FIGURE 11.10
A binarized version of the CT scout image.

which a value of `true` or 1 is present for those pixels in the original image with an intensity between 50 and 250 (exclusive).

A call to `whos` shows that `x` contains unsigned 8-bit integers and `y` is a logical array:

```
>> whos
   Name        Size            Bytes  Class      Attributes

   x        1024x880           901120  uint8
   y        1024x880           901120  logical
```

The resulting binarized image, displayed using `imshow(y)`, is shown in Fig. 11.10.

■ Activity 11.7

011.A, 011.C   Generate and display your own binary image from the CT scout image using your own ranges of values to decide which pixels are set to 0 and which are set to 1.   ■

It is also possible to save the resulting array to a new image using the `imwrite` function.

```
% Save our binary (logical) data
>> imwrite(y, 'ct-binary.png')
```

We can now use `imfinfo` to obtain general image information on the file that was saved. The salient parts of its output are given below:

```
>> imfinfo('ct-binary.png')

            ans =
            struct with fields:
                    Filename: '/PATH/TO/FILE/ct-binary.png'
                    ...
                      Format: 'png'
                       Width: 880
                    ...
                      Height: 1024
                    BitDepth: 1
                   ColorType: 'grayscale'
                    ...
```

Note that MATLAB has saved the file to a gray scale image that uses one bit for each pixel. This is because one bit is sufficient to indicate whether a pixel in a binary image is off (0) or on (1).

11.7.2   Threshold-based operations

Threshold-based operations take an image and update each individual pixel, depending on whether its value is above (or below) a user chosen *threshold value*.

As a simple example, we can define a threshold value $tVal$ and set any pixel's intensity to zero in the new image, if the intensity at the same location in the old image is less than $tVal$, otherwise we leave the value the same. At row i and column j, we denote the intensities of pixels in the old and new images by $old(i, j)$ and $new(i, j)$, respectively. The thresholding operation is then defined as

$$new(i, j) = \begin{cases} 0 & \text{if } old(i, j) < tVal \\ old(i, j) & \text{otherwise.} \end{cases}$$

We say the above operation "thresholds to zero" (i.e., sets to zero any pixel below the threshold). In MATLAB, we can easily do this with array operations.

■ Example 11.8

Here, we set values to zero in the CT scout image if they are below 100.   *O11.A, O11.C*

```
x = imread('CT_ScoutView.jpg');
y = x;
tVal = 100;
y(x < tVal) = 0;
% View a sub-window of the
% input and output images.
imshow(x(25:350,350:500))
figure, imshow(y(25:350,350:500))
```

The last two lines display a sub-window of each of the input and output (old and new) images, so that we can see the effect of applying the threshold. The resulting images are shown below.

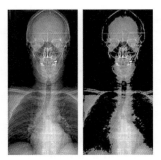

We can invert the way the threshold operation works, i.e., we can set to zero any pixel with a value *above* the threshold (rather than below). In other words,

$$new(i, j) = \begin{cases} 0 & \text{if } old(i, j) > tVal \\ old(i, j) & \text{otherwise.} \end{cases}$$

We can alternatively use the threshold to limit or truncate the values in an image, for example, by setting a pixel intensity to the threshold value if it was previously greater than the threshold:

$$new(i, j) = \begin{cases} tVal & \text{if } old(i, j) > tVal \\ old(i, j) & \text{otherwise.} \end{cases}$$

The value assigned to a pixel above (or below) a threshold can be another choice for the user. In other words, as well as choosing a threshold value $tVal$, the user can set a substitute value $newVal$ to apply to pixels above (or below) the threshold. For example,

$$new(i, j) = \begin{cases} newVal & \text{if } old(i, j) > tVal \\ old(i, j) & \text{otherwise.} \end{cases}$$

■ Activity 11.8

011.A, 011.C Make your own threshold-based filter based on one of the types shown above. Apply it to the CT scout image, and save or view the result. ■

11.7.3 Chaining operations

We have only considered single operations so far, but we can easily build up more complex ones by chaining them together. For example, we can apply more than one threshold operation to an image by making the output of one operation the input to the next. Again, we start with the gray scale CT scout image:

```
x = imread('CT_ScoutView.jpg');
y = x;
y(y < 50) = 0;
y(y > 200) = 0;
y(y > 0) = 255;
```

The first two lines read the image and copy it into an array variable y. The next three lines apply a chain of threshold operations to the array y, updating y on each line and modifying it again on the next.

■ Activity 11.9

Run the code above, and compare it with the output of the binarizing code in Section 11.7.1. What do you notice? ■

011.A, 011.C

This was a simple chain of operations, but such chains can be made as complicated as we like. A chain of operations is often called a *pipeline*.

11.7.4   Image data type, value range, and display

We need to be aware of some of the assumptions that MATLAB makes about the data types for images when reading and displaying them. For example, in Example 11.7, we read data for a gray scale image into an array x, and then used imshow to display (a part of) the image. If we inspect the contents of the variable x in this case, we can see that the data type was uint8.

When such an array, i.e., one with an integer data type, is passed to imshow, an assumption is made that the full range of the image is from 0 to 255. This means that, when displayed, a value of 0 will correspond to a black pixel, a value of 255 will correspond to a white pixel, and intermediate values will correspond to intermediate shades of gray. Any value above 255 is *clamped* and shown as a white pixel. Similarly, any value below 0 is clamped and shown as a black pixel.

Sometimes, we may need to apply a processing operation to the image data that leads to an array with a floating point data type (e.g., double). If we then pass a floating point array to imshow, it will still make the assumption that black pixels correspond to a 0 value. However, it will assume that white pixels correspond to a value of 1 (or any higher value). This can lead to unexpected results, so care needs to be taken, as shown in the example below.

■ Example 11.9

Hypothetically, assume that we need to take the square root of some data for an image. In what follows, we load the CT scout image into an array x,

011.A, 011.C

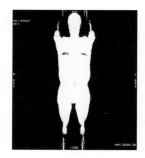 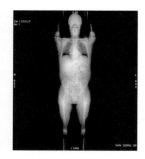

FIGURE 11.11

Displaying floating point versions of a processed CT scout images: (left) without scaling to the range 0–1, (right) with scaling to 0–1.

as before. Recall that x contains 8-bit unsigned integers (uint8). To take the square root of the array values, we first need to convert them to a double data type before passing them to the sqrt function. This is done in the second line. The result is stored in a variable y, and then displayed in the last line.

```
x = imread('CT_ScoutView.jpg');
y = sqrt(double(x));
imshow(y)
```

In this case, most of the non-zero pixels in the original image will now be shown as fully white pixels. This is because the original data ranged up to 255, and most non-zero pixel values, even after taking their square root, will still be greater than one. All such pixels will be shown as white by imshow.

To properly display such a double type array, we need to scale it so that it ranges from zero to one *before* passing it to imshow. This can be done in one line, as follows:

```
imshow(y / max(y(:)))
```

∎

The outputs from the two calls to imshow in the example above are shown in Fig. 11.11. On the left is the result of the first call (without scaling the floating point square root data), and on the right is the result of the second call (after scaling to a range of 0 to 1).

11.8 Image filtering

All of the operations applied to images that we have considered have been *point processing operations*. For such operations, the intensity of a pixel after the operation depends only on the same pixel's intensity before the operation. In

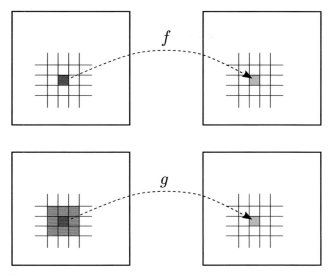

FIGURE 11.12
Top: a point processing operation denoted by f in which the new pixel intensity depends only on the old pixel intensity. Bottom: a filter operation denoted by g gives a new pixel intensity that depends on the old intensities in a neighborhood.

general, for a point processing operation, we can write

$$new(i, j) = f(old(i, j))$$

for some function f. See Fig. 11.12 (top).

Filter operations are different, because the new pixel intensity depends on all of the intensities in the original image that are in some *neighborhood* of the pixel. One simple neighborhood is a 3×3 window centered on the pixel. As illustrated in Fig. 11.12 (bottom), the pixel intensity in the output depends on all intensities in the neighborhood. Denoting the intensities in the 3×3 neighborhood as $\{v_1, v_2, \ldots, v_9\}$, we can write

$$new(i, j) = g(v_1, v_2, \ldots, v_9)$$

for some function g that operates, in this case, on nine input intensities.

11.8.1 The mean filtering operation

A very simple filtering operation is when we use the *mean* of the neighborhood values to obtain the new pixel value. That is, in the case of the 3×3 neighborhood, the function g above finds the mean of $\{v_1, v_2, \ldots, v_9\}$. In other words, for a 3×3 window,

$$new(i, j) = g(v_1, v_2, \ldots, v_9) = \frac{1}{9}(v_1 + \ldots + v_9).$$

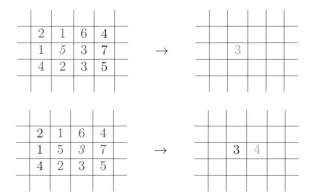

FIGURE 11.13
Computing new intensities using a mean filter.

For example, the grid in Fig. 11.13 (top) shows a small part of an image. We assume that the image extends further around this part, but we are ignoring it for now. The pixel with an intensity of 5 (in red italic) in the original image is assigned a new intensity of 3, because the mean of its neighborhood is 3:

$$\frac{1}{9}(2 + 1 + 6 + 1 + 5 + 3 + 4 + 2 + 3) = 3.$$

To work out the new intensity for the next pixel, which has an original intensity of 3, we need to "slide" the window by one pixel and recalculate the mean, as shown in Fig. 11.13 (bottom).

$$\frac{1}{9}(1 + 6 + 4 + 5 + 3 + 7 + 2 + 3 + 5) = 4$$

To calculate the new intensity in the next pixel, the window needs to slide again to cover a new neighborhood, and the process is repeated.

■ **Activity 11.10**

011.C   The pixels at the border of the image do not have a 3 × 3 window of pixels that completely surround them. Decide on a strategy for dealing with these border pixels when applying a mean filter. More than one strategy is possible.   ■

11.8.2   The actual filter used

If we expand the calculation for the pixel shown in Fig. 11.13 (bottom), it can be written as

$$\frac{1}{9} \times 1 + \frac{1}{9} \times 6 + \frac{1}{9} \times 4 + \frac{1}{9} \times 5 + \frac{1}{9} \times 3 + \frac{1}{9} \times 7 + \frac{1}{9} \times 2 + \frac{1}{9} \times 3 + \frac{1}{9} \times 5,$$

FIGURE 11.14
Applying a mean filter in 2-D.

i.e., the mean filter produces a new pixel intensity by taking a weighted sum of the intensities in the original image.

The weights can be arranged in a 3×3 array, which we call the *filter*,

$$\begin{pmatrix} \frac{1}{9} & \frac{1}{9} & \frac{1}{9} \\ \frac{1}{9} & \frac{1}{9} & \frac{1}{9} \\ \frac{1}{9} & \frac{1}{9} & \frac{1}{9} \end{pmatrix}.$$

In exactly the same way that we did for the convolutions in 1-D in Section 11.4, we can view this filter as sliding over the image and being used to weight and sum the corresponding pixel intensities (see Fig. 11.14). The only difference is that the filter slides in *two* directions (horizontally and vertically). For instance, we can slide the filter horizontally across a row, from left to right, until we reach the end of the row. We can then restart the process on the next row down, and so on.

In this case, the operation is a little trivial, because all of the weights are equal (to 1/9). But we can vary the weights to obtain different effects. For example, we can use a filter that gives a higher weight to the central pixel in the neighborhood and lower weights to the others:

$$\begin{pmatrix} \frac{1}{12} & \frac{1}{12} & \frac{1}{12} \\ \frac{1}{12} & \frac{1}{3} & \frac{1}{12} \\ \frac{1}{12} & \frac{1}{12} & \frac{1}{12} \end{pmatrix}.$$

11.8.3   Applying a filter in MATLAB

We can apply a filter to an image in MATLAB with the function `imfilter`, which is part of the image processing toolbox. In the case of the mean filtering operation above, we saw that we had the very simple filter of a 3×3 array with all entries equal to 1/9.

■ Example 11.10

011.A, 011.C   This code shows a filter of uniform values of $\frac{1}{9}$ being created by starting with an array of ones and dividing by 9. This is then applied to a sub-region of the CT image using `imfilter`. The `imfilter` function does the job of "sliding" the filter over each pixel in the image and performing the calculation of the weighted sum. The input and output images are shown below.

```
imData = imread('CT_ScoutView.jpg');
% sub-region
reg = imData(50:120, 400:450);
% Create the mean filter.
myFilt = ones(3) / 9;
% Apply the filter
filtReg = imfilter(reg, myFilt);

figure
imshow(reg);
figure
imshow(filtReg);
```

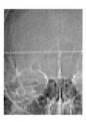

The visualization in Example 11.10 shows the image region before and after applying the filter. Note that the result after the mean filter appears smoother and less sharp than the original. If you look carefully, you should also notice a darker region around the edges of the new image. At these positions, the filter only partly overlapped with the original image.

The step in the commands above, which assigns a filter to the variable `myFilt`, is easily replaced with a different array assignment if we want to use a different filter. The subsequent call to the `imfilter` function would still be the same.

■ Activity 11.11

011.A, 011.C   Make your own 3 × 3 array containing a filter. Apply it to the data array of a gray scale image using the `imfilter` function. Perhaps you can make a filter with a higher weight for the central pixel and lower weights for the remainder. Experiment with filters whose elements sum to one or a value different from one. Visualize or save the results.   ■

■ Activity 11.12

011.C   Write your own implementation of the `imfilter` function, which again takes two arguments: an array of gray scale image data and a square array

containing the filter to apply. You can assume that the filter array is always a
3×3 array to make things simpler. It is possible to carry this out as a series
of nested loops with the following structure:

```
% ...
% pair of loops for whole image array.
for row = 2:nrows - 1
  for col = 2:cols - 1
    % Current pixel has indices 'row' and 'col'

    % Pair of loops for local window
    for rowOffset = -1:1
      for colOffset = -1:1
        % Do stuff here
        % to get a new value
      end
    end % end of window loops

    % May need to assign here.
  end
end % end image loops
```

11.8.4   Filtering and convolution

You may have noticed that the filtering operations described in the previous
sections have a lot in common with the convolution operations described for
1-D signals (Section 11.4). What we have described as a filter can also be called
a *kernel*. The process of "sliding" the filter/kernel over the image to produce
the output is directly analogous to the sliding we looked at in the case of 1-D
signals. The only difference in the case of images is that this sliding operation
needs to take place in two dimensions, along both rows and columns. In the
1-D case, the kernel is a row vector (1-D array), whereas in the 2-D case, the
kernel is a matrix (2-D array).

MATLAB provides a 2-D convolution function called `conv2`, and it can be used
instead of the `imfilter` function. The caveat is that it requires and returns data
of type `double`, so care needs to be taken when passing the data and visualizing
the result.

■ Example 11.11

Recall the `imfilter` function we used to apply a mean filter in Example 11.10. Here, we will use `conv2` instead. Note how the data are converted
to `double` before passing to `conv2`. Note also that the data are scaled to the
0–1 range by dividing by 255 before visualization.

011.A, 011.C

```
imData = imread('CT_ScoutView.jpg');

myFilt = ones(3) / 9;
```

```
reg = imData(50:120, 400:450);

filtReg2 = conv2(double(reg), myFilt, 'same');
imshow(filtReg2 / 255.0)
```

Note also that the call to `conv2` was given the argument `'same'` to ensure that the output array was the same size as the input array. This is again analogous to the similar call in the 1-D case (see Example 11.2). ∎

As was the case with the 1-D convolution, one detail relates to the flipping of the kernel when carrying out the convolution. In the case of `conv2`, the kernel is flipped left-right and up-down before being passed over the image array. In the case of Example 11.11, these flips had no effect as the kernel was symmetric. In general, if the image processing toolbox is available, the use of `imfilter` is a lot easier than the use of `conv2`.

11.9   Summary

In this chapter, we have looked at signals and images and how we can read, process, and visualize them using MATLAB commands. We have examined the convolution operation and how it can be used to process 1-D and 2-D signals. We have also looked at some of the properties of images, such as discrete locations and gray scale versus color images. Focusing on gray scale images, we have introduced some of the basic steps that are used for *image processing*, such as threshold-based operations. These operations rely mainly on the fact that, once an image is loaded into MATLAB, the data it contains can be represented by an array variable that can be manipulated to process the data. Some of the methods have used standard built-in MATLAB functions, and some of the functions we have used are provided by the specialist image processing toolbox. In addition to the fact that these basic steps can be applied to images in general, they form the basis of the important field of *medical image processing*, which has been used in a wide range of clinical settings, since a variety of different types of medical image are now routinely available. The basic operations we have looked at can be "chained together" with the output from one step forming the input for a later step. In this way, image processing "pipelines" can be built up incrementally. Finally, we have discussed image filtering and seen its close relationship to the concept of convolution of 1-D signals.

11.10   Further resources

- The image processing toolbox has a dedicated help section that is distinct from the main MATLAB help. It has sections including
  - *Import, Export, and Conversion*: Reading images into MATLAB and converting/saving them.
  - *Display and Exploration*: Showing an image through the interactive viewer or by a command line call to a function such as `imshow`.

- For readers interested in more details, there is a signal processing toolbox available in MATLAB with its own help section as above. It also provides some interactive tools.
- A number of examples using image processing are also available in the online MATLAB documentation: http://mathworks.com/help/images/examples.html.

11.11   Exercises

These exercises involve use of a number of sample 1-D signals and medical images, which you will need to download from the book's web site before proceeding.

■ Exercise 11.1

In Fig. 11.1 (top), we saw some example ECG signal data. The data for this plot are available on the book's web site, in the file called *example_ecg_data.txt*. Write code to load the data from the file and replicate the plot shown in Fig. 11.1.

011.A, 011.B

The data in the file were sampled at 720 Hz. Write code that down-samples the data to a rate of 60 Hz and plots the down-sampled version of the data overlaid onto the original data. Ensure your plot is annotated and contains a legend.   ■

■ Exercise 11.2

A common kernel for filtering a signal for smoothing is the *Gaussian* kernel. This is based on the Gaussian distribution (also known as the normal distribution). The continuous Gaussian function is shown below (left), and values taken from this function at discrete points may be taken to form discrete Gaussian kernels. The center plot shows three values taken from the function, and the right hand plot shows five values being taken. In each case, the values may be normalized so that they sum to one to generate a kernel.

011.A, 011.B

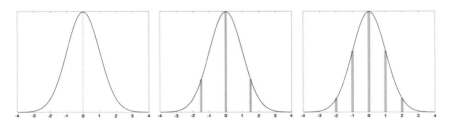

The Gaussian kernels obtained as described above, of lengths three and five are

[0.2, 0.6, 0.2]　　[0.06, 0.24, 0.40, 0.24, 0.06]

Continuing the work of Exercise 11.1, use the `conv` function to smooth the down-sampled (or *sub-sampled*) ECG data using both the Gaussian kernels provided above. Plot the sub-sampled data and both versions of the smoothed data in the same window.　■

■ Exercise 11.3

011.A, 011.B

The data file for one of the sets of heart rate data, shown in Fig. 11.1, is called *heart_rate_series_1.txt* and is available to download.

Write code to read in these measurements, and use the `smooth` function to reduce the level of noise in the data. Write code to find the first time derivative of this heart rate signal (i.e., find the rate of change of the heart rate). See Example 11.4 to get a hint on how to do this.

Make a plot showing the first 15 seconds of the smoothed data in the top half of the window and the time derivative in the bottom half. (*Hint: you can use the* `subplot` *command—see Section 9.2.2.*) Experiment with different values of the span for the `smooth` function to see the effect on the estimated derivative. Remember that the derivative estimates at the end may not contain sensible values, so it is safe to set these to zero.　■

■ Exercise 11.4

011.A, 011.C

Using the *CT_ScoutView.jpg* image, produce, display, and save a binarized image of the head using an appropriate threshold (see the figure below).

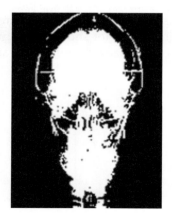

　　　　　　　　　　　　　　　　　　　　　　　　　　　　　■

■ **Exercise 11.5**

Write two functions called `thresholdLow` and `thresholdHigh`, which both take an image and a threshold as their two arguments, and produce a thresholded image as their result (the thresholded images should have zero intensity at any pixel that had a value below/above the threshold in the original image). ■

011.A, 011.C

■ **Exercise 11.6**

Write two functions called `truncateLow` and `truncateHigh`, which both take an image and a threshold as their two arguments, and produce a truncated image as their result (the truncated images should have the intensity of any pixel below/above the threshold set to the threshold). ■

011.C

■ **Exercise 11.7**

Write a pipeline of operations to process *pet-image-liver.png* to highlight the area of high activity in the liver, i.e., the output of the program should be as shown in the figure below.

011.A, 011.C

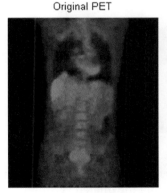

Original PET   Liver activity

From Prof. Dr. Hojjat Ahmadzadehfar.

(Also display a count of the number of pixels that are non-zero in the final image.) ■

■ **Exercise 11.8**

In binary image processing, it is possible to apply a *dilation* operation to an image. You are provided with a simple binary image showing a map of cortical gray matter in the brain. This is saved in the file *cortex-gm.png* and is shown below.

011.A, 011.C

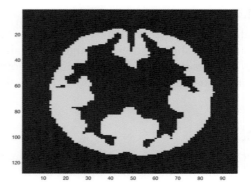

The process of dilating a binary image converts some of the background (blue) pixels to foreground (yellow) pixels. A background pixel is "switched on" if one of its neighbors (up, down, left, or right) is a foreground pixel. Dilating the original image gives the image shown below.

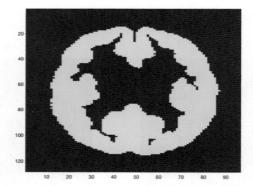

Write a function that accepts an image matrix as an input argument and returns a new matrix as an output containing the dilated version of the input. Write a script *m*-file that loads in the image, calls the function to dilate the image, and then displays both the original and dilated images. ∎

FAMOUS COMPUTER PROGRAMMERS: RICHARD STALLMAN

Richard Stallman is an American software freedom activist and computer programmer, who was born in 1953 in New York. While at high school, he got a summer job at the IBM New York Scientific Centre.

He was tasked with writing a numerical analysis program in FOR-TRAN, an early programming language. He completed this software within 2 weeks and spent the rest of the summer developing a text editor for his own enjoyment. Later, he studied physics at Harvard, and also worked as a programmer at the Massachusetts Institute of Technology (MIT).

While at MIT, he objected to the introduction of passwords to limit access to the laboratory's computers. He wrote a computer program to access and automatically decrypt all users' passwords and sent them messages containing their decrypted password, along with a suggestion to change it to an empty password, i.e., to reinstate free access to computers. Although some users followed his suggestion, eventually password control became the norm in computer access.

Stallman is best known as the originator of the GNU software project (http://www.gnu.org), which aims to promote the concept of free software. Stallman started the GNU project in 1983, and it has gone on to become one of the major influences on software production and distribution in the world today. Stallman has not benefited financially from his important work on the GNU project and professes to care little for material wealth.

Stallman has been a passionate and outspoken advocate for software freedom, and his outspokenness has made him enemies, as well as friends. He has described Steve Jobs, the Apple co-founder, as an "evil genius" for building "computers that were prisons/jails for the user and to make them stylish and chic." Linus Torvalds (see Chapter 13's Famous Computer Programmer), the developer of the LINUX open-source operating system, has criticized Stallman for what he considers "black-and-white thinking" and bringing more harm than good to the free software community. Today, Richard Stallman lives in Cambridge, Massachusetts, and devotes his time to free software advocacy and the GNU project.

"If programmers deserve to be rewarded for creating innovative programs, by the same token they deserve to be punished if they restrict the use of these programs."

Richard Stallman

Graphical user interfaces

12.1   Introduction

In this chapter, we cover how to create graphical user interfaces (GUIs) using MATLAB®. In the preceding chapters, we have been writing MATLAB programs that interact with the user by displaying text in the command window and reading data from the keyboard, but GUIs have an important role to play in making our programs more accessible and usable by a wider range of (non-technical) users. We provide an outline of how to build a simple GUI with basic components, such as text boxes or axes using the built-in `appdesigner` tool. We also look at how to control the behavior of GUI components so that users can interact with the code that is being run.

12.2   Building a graphical user interface in MATLAB

In addition to the fact that a lot of interaction with a software environment can be carried out on a command line or by running a script, it can also often be useful to interact via a *graphical user interface* or GUI.

In MATLAB, we can create our own custom-made GUIs, and we can add objects, including text boxes, for the user to type into, pop-up menus for making choices, images or plots, and buttons, for example, to turn something on or off. We can customize the GUI, laying out the components and sizing them as we want. An example of such a GUI is shown in Fig. 12.1.

295

MATLAB® Programming for Biomedical Engineers and Scientists. https://doi.org/10.1016/B978-0-32-385773-4.00021-6

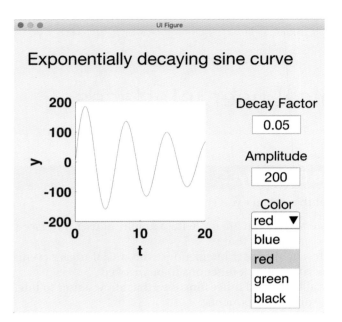

FIGURE 12.1
An example of a GUI. This displays a plot and has some components (text boxes and a drop down menu) to allow the user to interact with the plot.

When a user interacts with a component, such as clicking a button or entering some text, this is called an *event*. When an event happens, MATLAB executes a specific *callback function* that is associated with the event. As developers, we can edit the callback function to control the behavior of the GUI, depending on the user's actions.

12.2.1   Building a GUI with the MATLAB App Designer tool

A custom-made MATLAB GUI is described as an *app*, and it is possible to create, build, and design an app entirely in code using scripts and functions, i.e., programmatically. However, in practice, it is a lot easier to use the *App Designer* tool provided by MATLAB, as it provides an accessible way to select, place, and adjust the layout of components in the app.

We will illustrate the use of the *App Designer* with a series of examples that build up a simple app step-by-step. This app will carry out some basic plotting of data in a way that can be modified by the user.

■ **Activity 12.1 (Creating a blank app and saving it)**

012.A   To start the *App Designer*, go to the *Apps* tab and click the *Design App* button (see Fig. 12.2). Alternatively, simply type `appdesigner` at the command window.

FIGURE 12.2
To start the *App Designer* from the MATLAB environment, click on the *Apps* tab, and then the *Design App* button on the far left.

FIGURE 12.3
The opening page of the *App Designer*.

> This causes a dialog box to be displayed (see Fig. 12.3), in which we can choose the type of app to build. Choose the *Blank App* option. This should open a new *App Designer* window containing a blank canvas for our app, which we can fill in as we like (see Fig. 12.4). ∎

The blank canvas can be overlaid with a grid to help us with the layout of the components (buttons, text boxes, etc.) that are inserted. The controls for toggling the grid and setting its spacing can be found by clicking on the *Canvas* tab at the top (see Fig. 12.4).

There is a set of icons on the left that we can use to construct the app. These can be used to place various components onto the app's canvas. Hovering the mouse over a component's icon gives some text to describe the component. For example, the first icon at the top-left of the set shown in Fig. 12.4 shows a small graph of a function, and this can be used to insert a set of axes into the GUI for plotting graphs or displaying images.

■ **Activity 12.2**

> Our GUI is still untitled, so we will save it first before carrying on. Click on the *Save* icon at the top of the window to open the save dialog box. Because

012.A

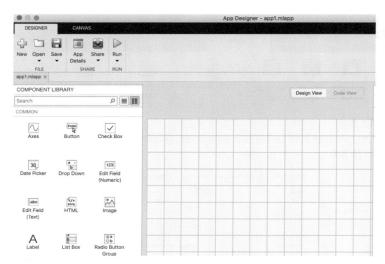

FIGURE 12.4

A blank canvas in the *App Designer* window. In the above, the grid has been turned on to help with the design. Note how the app has been opened in *Design View*. The code underlying the app can be accessed by clicking on the *Code View* button on the right-hand side.

there may eventually be more than one file used by the app, it is a good idea to save it in a folder of its own, where it can be kept together with any other files needed. For example, save it as *myApp.mlapp* in a new folder called *my_apps*. Note how MATLAB uses the suffix *.mlapp* for the file being saved. ∎

After the save step, there will be a single file in the folder, *myApp.mlapp* that contains the information needed to assemble and place the components in our app.

> **Automatic code:** Note that in the background, some code that will be used by the app has been automatically created. This code is not immediately visible, because the *App Designer* will open by default in *Design View*. To see the code that has been generated, click on the *Code View* button at the top right of the canvas (see Fig. 12.4).

It is now possible to run the app that we have just created, even though it does not contain any components yet! This can be done by clicking the *Run* button (▷) at the top of the window (as we would for any MATLAB *m*-file in the *Editor* window). When the GUI runs, as expected, we get a blank window that we cannot interact with yet. We can close the window to continue designing our app.

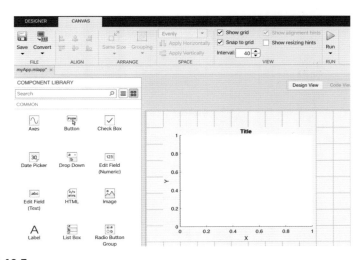

FIGURE 12.5
The canvas of the app after placing a set of axes.

The app can also be run by typing myApp at the command window prompt in the MATLAB environment window (as long as the folder for our app is in the path or we have changed the working directory to that folder).

■ Activity 12.3 (Adding a component to the app)

012.A

Now we can add components to our app. We would like to include a plot, so we start by adding a set of axes. To add the axes, while in *Design View*, you can either click and drag the *Axes* icon onto the blank canvas or click once on the *Axes* icon and drag out a rectangle on the canvas. The GUI canvas should now look something like Fig. 12.5. Move and drag the corners of the axes until they have the size and position required. Leave some space to the right of the axes for some components to be added later.

We can modify the appearance of our axes when working in *Design View*. To do this, click on the axes on the canvas to select the component, then open the *Inspector* tab in the *Component Browser* to the right of the window (see Fig. 12.6). We can modify a number of different fields here to change the appearance of the component. Later on, we will be plotting electrocardiogram (ECG) data, so something suitable might be *ECG* for the title (Title.String), *Time (s)* for the x-axis label (XLabel.String), and *Voltage (mV)* for the y-axis label (YLabel.String). ■

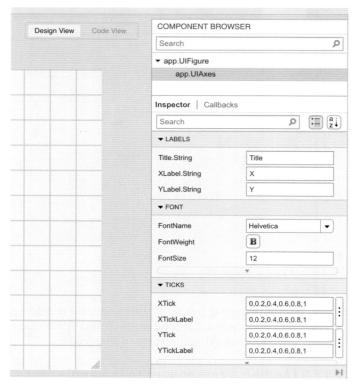

FIGURE 12.6
The inspector for the axes in Fig. 12.5. We can use this to change the appearance of our axes, such as the title, font size, etc.

Re-opening at a later stage: If we close our MATLAB session and later, in a new session, we want to reopen our GUI in *Design View*, we can just type `appdesigner` at the command window or hit the *Design App* button in the *Apps* tab.

We can then open our app from the resulting dialog box by clicking on the *Open ...* button and navigating to our app file (with the name *myApp.mlapp*). Alternatively, our app should appear in the *Recent Apps* list on the left of the *App Designer* window.

12.2.2 Controlling components: events and callback functions

Now that we have a component added to the app (a set of axes), we can start to add code that controls its behavior. In MATLAB, and in some other program-

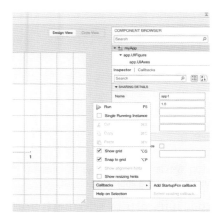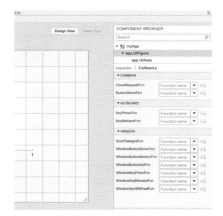

FIGURE 12.7
Accessing callback functions in the *Design View* of the app designer via the context menu (left) or via the *Callbacks* tab in the *Component Browser* (right).

ming languages that allow GUI programming, this can be done using *callback functions*.

A callback is special function that is associated with a GUI or a component within it. The callback is invoked when a particular *event* occurs while the user is interacting with the GUI. For example, an event could be the user clicking on an item in the GUI/app, or resizing the app window. Other less obvious events include the "creation" of components or the "start up" of the app after the components have been created.

When an event occurs, and if code has been written for the associated callback function, then the system will execute that code. We can write and/or modify the callback function, inserting our own code, to control the app's behavior.

■ Activity 12.4 (Accessing a callback function)

Now that we have started to build our app in the *App Designer*, we can have a look at what callback functions it has available. In *Design View*, right click on a blank area outside the app's canvas to bring up a context menu. As shown on the left in Fig. 12.7, we can find a sub-menu called *Callbacks*. Callback functions can also be accessed beneath the *Callbacks* tab in the *Component Browser* on the right of the *App Designer* (this is adjacent to the *Inspector* tab, see the right of Fig. 12.7).

012.B

In this case, we have a few functions available as callbacks for the app. For example, in the left in Fig. 12.7, we can see StartupFcn. This callback is available to the app as a whole, as we clicked outside the canvas area. On the right of Fig. 12.7, we can see further callbacks that are available to the app's figure (see the levels listed above the *Callbacks* pane). The name of each function describes the event that will invoke it. For each function, we

FIGURE 12.8
The argument of the callback function is an instance of our app. We can explore the objects, variables, and functions that it can access by using the dot operator and browsing the auto-complete list.

can either add code for it, or go to that code if it exists by selecting it from the context menu or by clicking the arrow next to its name in the *Component Browser*.

Select the first callback in the context menu (StartupFcn). The *App Designer* will then automatically generate code for this function (if it does not already exist) and will switch focus to this code in the *Code View*. ∎

The automatically generated code will be a function with an empty body and should look like the following:

```
% Code that executes after component creation
function startupFcn(app)

end
```

The code will be run after the components of the app have been created. The name of the function, startupFcn, reflects this, and a descriptive comment is automatically generated for the function. Apart from that, the function is empty, and we can add our own code within it to modify the app's behavior at start up.

In the *Code View* of the *App Designer*, note that some code is grayed out; this code cannot be edited. The sections of the code that can be edited have a white background, for example, inside the body of the callback function we have just created.

Note that the callback function startupFcn has a single argument: app. This argument represents an instance of the class representing our app, the instance that is created when the app is run. The class name will be the one we chose when saving (*myApp*). The running instance can be used to access the functions and components available to the app, and we can explore these by typing its variable name in the code window, within the body of the startupFcn function, and follow it by a single dot. MATLAB will then provide a sub-window to show possible completions, listing the member objects, variables, and functions that the app instance can use. This is illustrated in Fig. 12.8.

Adding code to populate the axes

As described above, the callback function that is invoked when the app starts up is `startupFcn`, and it has an input argument `app` that we can use to access the components of the app. In particular, we can access the plot axes that we added earlier to the app (in *Design View*), as shown in Fig. 12.5. In the next activity, we use the `app` argument to access the axes, as shown in Fig. 12.8.

■ Activity 12.5

We will plot some of the ECG data that we first saw in Fig. 11.1 and later in Exercise 11.1. The data for this plot are available on the book's web site, in the file called *example_ecg_data_b.txt*. Save a copy of this file into the same folder that contains the file saved for the app (i.e., *myApp.mlapp*).

O12.B, O12.C

In *Code View*, add code to the `startupFcn` function so that it looks like the following:

```
% Code that executes after component creation
function startupFcn(app)
    % START ADDED CODE
    % Read data
    ecg = readmatrix("example_ecg_data_b.txt");

    nPts = length(ecg);

    % Sampling frequency
    sampFreq = 720; tStep = 1/sampFreq;

    % Time points at given sampling frequency.
    t = 0:tStep:(nPts-1)*tStep;

    plot(app.UIAxes, t, ecg);
    % END ADDED CODE
end
```

The added code section is marked by comments at the start and the end. Note that the `plot` command takes the app's axes, `app.UIAxes`, as its first argument. ■

Now when we run the app, our axes will contain a plot (see Fig. 12.9). Note that some controls are included just above the plot. If these are not visible, then hovering the mouse over the axes will make them appear. There are controls to pan and zoom the plot with the mouse, and a "home" button to return the view to its original setting. There is also a control to save the plot as a file or to copy it to the clipboard (so that the plot can be pasted into, say, a word processing document).

It is also possible to control the appearance of the plot in code with parameters that the user can vary and control. Note from the plot in Fig. 12.9 how, for

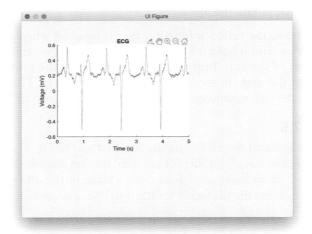

FIGURE 12.9
The app created using the *App Designer*, containing a plot of some ECG data. Note the plot controls just above the plot, to the right of the title.

this ECG data file, we have measurements extending over 5 seconds. As one example of directly controlling the plot, the user may want to set a precise maximum value for the time, and the plot's horizontal axis would be adjusted to reflect this. We will use a text box to get this user input.

Using an edit field to vary the plot

In the *Design View* of the *App Designer*, we can place an *Edit Field* component onto our app's canvas to allow the user to input a time limit. There are two components available for input from the keyboard, *Edit Field (Text)* and *Edit Field (Numeric)*, and we will use the second of these for our user-entered time limit.

■ Activity 12.6

012.A

Drag an *Edit Field (Numeric)* component onto the canvas, and note that it consists of two parts: a white box with a numeric value in it (zero by default) and a descriptive label ("Edit field" by default).

Click on the component that we just added to select it, and use the *Inspector* (see Fig. 12.6) to change the *Label* to "Upper Time Limit" and the (default) *Value* to 5.

It is also possible to move the label relative to the box by clicking it and dragging it. For example, we can position the label above the editable white box. The canvas should now look something like Fig. 12.10. ■

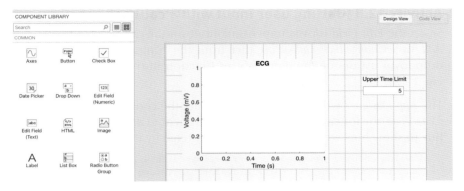

FIGURE 12.10
The app's canvas with a numeric *Edit Field* component for a user-entered upper time limit.

Finding the callbacks for the edit field

Not all components in our app will have the same callback functions, because the behavior of components can vary. As we did earlier for the *Axes* component, we can find out which callbacks are available for our newly added *Edit Field* component, either by using the context menu or by using the *Callbacks* tab adjacent to the *Inspector* (see Fig. 12.7).

■ Activity 12.7

In the *App Designer*, find out which callback functions are available for our numeric *Edit Field* component.   *012.B*

Try temporarily adding an *Edit Field (Text)* component (we can delete it later), and see which callbacks are available for it.   ■

You should find that both types of *Edit Field* (numeric and text) have a callback function named `ValueChangedFcn`. The text *Edit Field* has an additional callback function named `ValueChangingFcn`.

The first callback is generally the most useful. It is called when the user has typed something into the edit box and subsequently hits the *Enter* (or *Return*) key. This signifies that the user has finished typing and now wants to pass information to the app.

The `ValueChangingFcn` of the text *Edit Field* component is invoked *every* time a key is pressed by the user, not just the *Enter* key, i.e., it is called continuously while the user is typing. If you added a text *Edit Field* to the canvas, then delete it now.

Next we will use the `ValueChangedFcn` callback function of the upper time limit *Edit Field* to obtain some user input.

■ Activity 12.8 (Getting user input with a callback function)

012.B, 012.C

Select the *Edit Field* for the upper time limit on the canvas, and then select *Add ValueChangedFcn callback*, either from the context menu or from the dropdown list beside the name of this function in the *Callbacks* pane (see Fig. 12.7).

The *App Designer* will automatically generate the code for the function. In *Code View*, the function will have a brief comment to describe it, and a single line to show how we can access the value entered by the user.

```
% Value changed function: UpperTimeLimitEditField
function UpperTimeLimitEditFieldValueChanged(app, event)
    value = app.UpperTimeLimitEditField.Value;

end
```

Note that the name of the callback function reflects

- the label that we gave to the component: `UpperTimeLimit`
- the type of the component: `EditField`
- the type of event: `ValueChanged`

Now try running the app and entering values into the upper time limit *Edit Field*. Any number can be entered; there will be no effect, as we have not yet written code to respond to this input (the callback function only retrieves the value and assigns it to a variable called `value`).

Try putting non-numeric characters into the *Edit Field*, and note that the app responds with a pop-up message to say that the value must be numeric (because we chose a numeric *Edit Field* component). ■

The automatically generated code for the numeric *Edit Field*'s callback contains a line to show how we can retrieve the value entered by the user. We can edit the line so that the stored variable has a name that is meaningful for our program. It represents the upper time limit, so we can call it `tUpper`, and we can also set a lower limit for the value that the program will accept.

```
%...
tUpper = app.UpperTimeLimitEditField.Value;

if tUpper < 0.01
    return
end
% ...
```

This code retrieves the user input and assigns it to variable `tUpper`. If the number is too small, the function will simply return (recall that we introduced the `return` keyword in Section 4.5). Otherwise, the code goes on to modify the plot as will be shown below.

■ **Activity 12.9**

Place the following code in the *Edit Field*'s callback function so that the user can modify the plotted time range:

O12.B, O12.C

```
% START ADDED CODE
tUpper = app.UpperTimeLimitEditField.Value;

if tUpper < 0.01
    return
end

ecg = readmatrix("example_ecg_data_b.txt");
nPts = length(ecg);

% Sampling frequency
sampFreq = 720; tStep = 1/sampFreq;

% Time points at given sampling frequency.
t = 0:tStep:(nPts-1)*tStep;

plot(app.UIAxes, t, ecg);

% Set time range's upper limit!
xlim(app.UIAxes, [0, tUpper]);
% END ADDED CODE
```

■

This code takes the numeric user input and, if it is below 0.01, the function simply returns. Otherwise, the function re-reads and re-plots the data, and then sets the limits on the horizontal axis to go from zero to the user's given value of `tUpper` (using the `xlim` function, see Section 9.2.5).

Note that the `xlim` function can be given the app's axes (`app.UIAxes`) as its first argument. This is so that the program knows which axes to set the limits for and matches how we specified the axes for the `plot` command that precedes the call to `xlim`.

■ **Activity 12.10**

Run the code above, and experiment by changing the upper time limit in the *Edit Field* box. The field should accept any numeric value that you type. The plot should be updated to reflect the value entered. If an out-of-range number is entered (in this case, below 0.01), the plot should not change. Note what happens if a value greater than 5 is entered.

O12.C

■

12.2.3 Maintaining state: avoiding duplicated code

You may well have noticed some duplicate code in the app that we have built so far. In Activity 12.5, we added some code for reading the data and plotting it on the axes (in the function `startupFcn`). In Activity 12.9, we added code to

FIGURE 12.11
Left: Using the *Property* button in the *Editor* tab to add a property to the app's code. Right: Adding a property using the *Code Browser* pane.

the *Edit Field*'s callback function, also to read and plot the data when the user sets an upper time limit.

This means that we now have similar code repeated in two different places. It would be good to avoid such unnecessary duplication, so we will now take steps towards having a single helper function that plots the data, one that we can call from other parts of the code when it is needed.

To do this, we need to be able to store information that can be accessed in different parts of the code belonging to the app. In particular, we will need to keep track of one or more variables that belong to the app, and we would like the value of any such variable to persist between successive invocations of the callback functions. This can be described as *maintaining state* for our app. We will do this in the next activity by using a *property* for the app and, in the subsequent activity, with a general function that makes use of the property when it carries out the plotting.

■ Activity 12.11 (Add a property to the app)

012.C   We begin by adding a property to the app's code to represent the upper time limit for our plotted data. In the *Editor* tab, click the *Property* button to open a menu. From this menu, we can choose to add a *Private Property* or a *Public Property* (see the left-hand side of Fig. 12.11). We can also add a property from the *Code Browser* pane (see the right-hand side of Fig. 12.11).

Either way, add a private property to the code, and the *App Designer* will automatically generate some code for it that will look something like the following:

```
% ...
properties (Access = private)
    Property % Description
end
% ...
```

The *App Designer* has added a section of properties that, so far, contains a single property called `Property`. Later, we can add to this section if we want further properties.

Edit the code to rename our property to something more meaningful and update the description accordingly:

```
% ...
properties (Access = private)
    % Upper time limit for the plotted data.
    tUpper
end
% ...
```

■

> **The app's components are also properties:** We have actually used properties of the app before. For example, in Activity 12.9, when we called the `plot` function, we passed in `app.UIAxes` as the first argument. This shows that the axes are also a property of the app. The components of the app have their own section of properties in the *Code View*, marked with the comment,
>
> ```
> % Properties that correspond to app components
> ```
>
> and this section of code is in the non-editable (grayed out) area of the code.

We have now made our own property in the app's code to represent the upper time limit. In the next activity, we will add a helper function that will make use of this property when plotting the data.

■ Activity 12.12 (Add a helper plotting function)

We can now add a helper function to the code for the app. Doing this is very similar to adding a property. In the *Editor* tab, click the *Function* button to open a menu with options to add a *Private Function* or a *Public Function* (see the left-hand side of Fig. 12.12). We can also add a function from the *Code Browser* pane (see the right-hand side of Fig. 12.12).

012.C

Add a private function. This will generate a new section in the code that is reserved for private functions. It should look like the following:

```
methods (Access = private)
% ...
% ...
end
```

FIGURE 12.12
Left: Using the *Function* button in the *Editor* tab to add a function to the app's code. Right: Adding a function using the *Code Browser* pane.

Inside this section, the *App Designer* will also place a generic piece of skeleton code for our helper function. It should be similar to the code below:

```
function results = func(app)

end
```

Note how this function expects to receive an `app` instance in the same way that the callback functions did previously (see Activity 12.5 and Activity 12.9). Later, we will use the `app` instance to access our time limit property.

We can edit this skeleton to make it more meaningful for our app. Our function only needs to plot, and does not need to return a value, so we can remove the `results` variable from the above. We can also give it a descriptive name and comment as shown below:

```
function myPlot(app)
    % Helper function to plot the ECG data.

end
```

■

Now we can actually add code inside the helper function's body. Most of this code should be fairly familiar by now!

■ Activity 12.13

O12.C

In the empty helper function we generated in Activity 12.12, add the following code to carry out the plotting task:

```
function myPlot(app)
    % Helper function to plot the ECG data.

    ecg = readmatrix("example_ecg_data_b.txt");
    nPts = length(ecg);

    % Sampling frequency
    sampFreq = 720; tStep = 1/sampFreq;

    % Time points at given sampling frequency.
    t = 0:tStep:(nPts-1)*tStep;
```

```
        plot(app.UIAxes, t, ecg);

        % Set time range's upper limit!
        xlim(app.UIAxes, [0, app.tUpper]);
    end
```

The code that we have added is almost identical to the code that we added previously in the callback functions for starting the app (Activity 12.5) and for handling a value change in the *Edit Field* (Activity 12.9). The key difference is in the last line of the function, which sets the plot limits. In this line, the app's property is accessed using `app.tUpper` in the call to the `xlim` function. ∎

Now that we have made a general plotting function, we need to remove the (now unnecessary) sections that plotted the data in the original code. We will need to replace these with calls to our new function `myPlot`. We need to call our new plotting function from two places in the code in the following two activities:

- Inside the start up function (Activity 12.14);
- Inside the callback function for a changed value in the numeric *Edit Field* (Activity 12.15).

■ Activity 12.14 (Modify the start up callback function)

First, remove the code that performs the plotting inside `startupFcn`, i.e., *delete* the code that was added in Activity 12.5. This should be straightforward, as this code is between the lines `% START ADDED CODE` and `% END ADDED CODE`. Then, replace it so that the body of the function looks like the following:

012.B, 012.C

```
% START ADDED CODE
% Set the initial value for the upper time limit.
app.tUpper = 5;
myPlot(app)
% END ADDED CODE
```
∎

In the next task, we will modify the code for the *Edit Field* callback function.

■ Activity 12.15 (Modify the edit field callback function)

For this callback, we need to delete the code that was added in Activity 12.9. Again this should be easy, as it is between the lines `% START ADDED CODE` and `% END ADDED CODE`. Then, edit it so that it reads as follows:

012.B, 012.C

```
function UpperTimeLimitEditFieldValueChanged(app, event)
    % START ADDED CODE
    val = app.UpperTimeLimitEditField.Value;
```

```
    if val < 0.01
        return
    end
    app.tUpper = val;
    myPlot(app);
    % END ADDED CODE
end
```

As before, if the user enters an invalid time limit in the *Edit Field*, the function returns. But now we assign this user input, if it is valid, to the `app.tUpper` property. We then call our plotting function, which will make use of the updated value of `app.tUpper`.

A further improvement to this code can be made by restoring the last valid time limit to be displayed in the *Edit Field* when an invalid value is entered. This can be done by modifying the `if` clause above to read as follows:

```
if val < 0.01
    app.UpperTimeLimitEditField.Value = app.tUpper;
    return
end
```

Check that the code is running properly by testing it with a range of upper time limits, including non-numeric values. ∎

12.2.4 Keeping track of a property when maintaining state

When we use a property in an app, it is good practice to keep track of the places where it is set and accessed. In particular, a property needs to be set for the first time (i.e., initialized) when the app starts and, while the app is running, any updates to its value need to be checked for validity so that the program continues to work properly. It is worth paying attention to the locations where properties are accessed to help avoid common errors, such as failing to initialize them or setting invalid values.

Note the different places in the code, where our upper time limit property is accessed or set. First, it is declared in the `properties` section:

```
properties (Access = private)
    % Upper time limit for the plotted data.
    tUpper
end
```

There are two places where it is set, the first of these is in the start up callback function.

```
function startupFcn(app)
    % Set the initial value for the upper time limit.
    app.tUpper = 5;
    % ...
```

This is where the value for the upper time limit property is initialized, and this initialization occurs when the app is starting for the first time and its components are being created. Apart from that, the only other place where `app.tUpper` is set is when the *Edit Field* value is changed. Note that the value given in the *Edit Field* is first checked and, if it is not valid, the function returns, and the property is not updated (see the `if` clause in Activity 12.15).

Finally, the property `app.tUpper` is only accessed once, within the plotting function `myPlot`, to set the plot limit.

With regard to property initialization, it is worth noting that this can also be performed in the `properties` section, i.e., we can write the following:

```
properties (Access = private)
    % Upper time limit for the plotted data.
    tUpper = 5
end
```

In this case, we do not need to initialize it in the start up function.

■ Activity 12.16 (Adding feedback text for the user)

In this activity, we will add some text to the app that reports back to the user. Note that the time values loaded from the ECG data file in the previous activities have a maximum value of 5 seconds. Here, we will give a message to the user to indicate whether their chosen upper time limit exceeds the maximum time or not.

O12.B, O12.C

First, in *Design View*, drag a *Label* component onto the canvas. By default, the text it contains will be "Label." Now, we can edit the code to set the initial text contents for the text label. At the end of the start up callback function (see Activity 12.5), we can set this text when the app starts. The function will end up looking like the following:

```
function startupFcn(app)
    % ... previous code here ...
    app.Label.Text = 'Upper time limit within range.';
    % END ADDED CODE
end
```

Now run the app to check it has worked. If the text appears cut off, then the component's size in the canvas is too small. Stop the app, and switch to *Design View*, then resize the label to make it larger.

The next task is to add code that updates the text if the user enters a time limit greater than 5. We can do this in the callback for the *Edit Field* (see Activity 12.8). At the end of the callback we add the following statements:

```
function UpperTimeLimitEditFieldValueChanged(app, event)
    % ... previous code ...
```

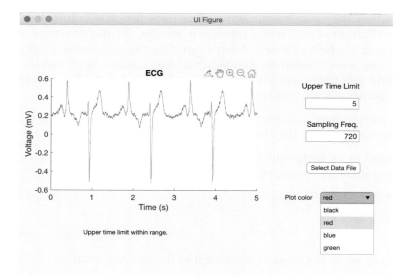

FIGURE 12.13

An illustration of the GUI that has been developed in the activities in this chapter. It also shows further components that will be added in the exercises at the end of the chapter.

```
        if val > 5
            app.Label.Text = 'Time limit out of range.';
        else
            app.Label.Text = 'Upper time limit within range.';
        end
        % END ADDED CODE
    end
```

Now run the app and enter different values for the upper time limit to check that the text gets updated correctly. ∎

12.3   Building an app: recap

So far, we have created a simple app with a graphical user interface. We have added a set of axes to it for plotting ECG data and a text box (or *Edit Field*) that allows the user to set the upper time limit when plotting the data. We have also placed a text label onto the canvas to report back to the user on whether their chosen time limit is within the range of the data. In the exercises, we will add further components to this app so that there are more options for the user to interact with it. An illustration of the final GUI after all components have been included, from the activities above and the later exercises, is given in Fig. 12.13.

12.4   An app for image processing

We now consider a different example of an app that uses some of the techniques described in Section 11.7 on image processing. The components available in the *App Designer* tool allow us to display images easily, and it is possible to use callbacks to control how they are processed or displayed.

■ Activity 12.17 (Display an image in an app)

Use the *App Designer* to create a blank canvas, and add a set of axes to it. Add code in an appropriate place to load an image and, using the `imshow` function, display it on the axes. In this activity, you can load the "Scout" CT image that we first saw in Activity 11.5, and which is available from the book's web site.   *O12.B, O12.C*

It will be a good idea to set a private property for the image array and to initialize it when the app starts. For now, it is fine if the file name for the image is hard-coded in the start up callback function. A separate helper function can access the property and display it on the axes on the canvas.

Hint: The MATLAB function `imshow` *can be given an argument* `'Parent'` *to display an image on a specific set of axes, i.e., a command such as the following can be used:*

```
imshow(image, 'Parent', axes)
```
■

Given that we can load and display an image using an app, it is possible next to carry out operations on the image and display the result. This will be done in the next activity.

■ Activity 12.18 (Process an image in an app)

Following on from Activity 12.17, add a second set of axes to the canvas of the app. We can use the second axes to display the result of image processing operations that are applied to the image loaded and displayed in the first axes.   *O12.B, O12.C*

In this activity, repeat the filtering operation described in Example 11.10. This uses a mean filter, and we can define a private property to represent this (ensure that it is suitably initialized). A second property can be used to store the result of applying the filter to the input image. A good approach would be to define a helper function to apply the filter to the input image and store the result. Another helper function could be used to display both images.

Hints:

- *In the* component browser, *note the names of the different axes on the canvas. They should be something like* `app.UIAxes` *and* `app.UIAxes2`.

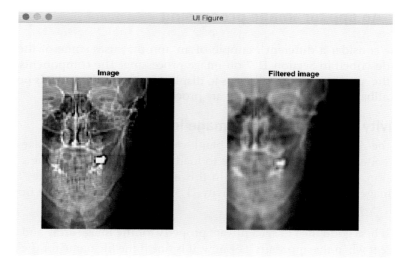

FIGURE 12.14
An app to display an image and the result of applying a filtering operation (see Activity 12.18).

> *Separate calls to the* imshow *function can be used to display the input image and the filtered image in the appropriate axes.*
> ■ *Click on each set of axes, and use the* Inspector *to turn off the labels of the X and Y axes. You can also give a helpful title to each axis, e.g.,* Image *and* Filtered image. ■

After Activity 12.18, the app can be run and the zoom and pan buttons can be used to explore the image data to check the behavior of the code. An illustration of this is given in Fig. 12.14, which shows the input image on the left set of axes and the filtered image on the right.

We will re-visit the image processing app later in the exercises.

12.5 Summary

MATLAB provides a set of tools for creating graphical user interfaces (GUIs). We have seen how we can use the *App Designer* tool to design a GUI by positioning and sizing components such as text boxes and axes. Other components, such as buttons and pop-up menus, can also be added.

In general, when we add a component, we need to control the way it behaves and interacts with the user. This is done using *callback functions*. It is possible to find out which callback functions are available by reading the documentation, but it is generally easy to find out using the context menu when right-clicking on the component that we have added.

The context menu will show which callbacks are available. Selecting one of them from the context menu will cause MATLAB to automatically generate code for the callback along with hints in the usage text. Reading this gives an idea of how to use the component.

The automatically generated skeleton code for the callback functions, plus any code that we write, will control the behavior of the GUI. Each callback is invoked when specific *events* occur, such as the user clicking inside a plot or entering text into an *Edit Field*.

Another way to find out how to add code to a callback function is to place break points in its body. This allows us to inspect the contents of the variables passed in as arguments while the code is running. We can experiment with modifying the code or variables when the debugger is stopped in the callback's function body.

12.6   Further resources

- See the Applications section at the MathWorks File Exchange: http://mathworks.com/matlabcentral/fileexchange.
- The MATLAB documentation has some examples of GUI building under *Documentation → MATLAB → App Building → Develop Apps Using App Designer*.
- The online help lists a set of examples, and these can be found at https://mathworks.com/help/matlab/examples.html (click on *App Building*). It would be a useful exercise to look or work through these, as well as the examples in this chapter.

12.7   Exercises

■ Exercise 12.1

Create a new blank app canvas using the *App Designer* tool, and carry out the following steps:

012.A, 012.B, 012.C

1. Add an *Edit Field (Text)* component. This is for the user to enter text.
2. Add a *Label* component.
3. Save the app, calling it *guiEx1*.
4. Run the app to make sure it works.
5. Close the app, and change the code in the *Edit Field (Text)* callback function so that the characters entered by the user are displayed in the text of the label component.

Hints:

- *Look at how the text for a label was modified in Activity 12.16.*
- *Remember to resize the label component in Design View if the text does not fit.* ■

■ Exercise 12.2

012.A, 012.B, 012.C

Use the *App Designer* to create a blank app, and save it as *guiEx2*.

1. Add two *Edit Field (Numeric)* components. In *Code View*, the property names of these fields should be `EditField` and `EditField1`, which means that we can access them in callbacks through the `app` instance variable using `app.EditField` and `app.EditField1`.
2. Add a *Label* component.
3. Create the start up callback function for the app and add code so that the edit fields both contain a value of 1 when the app starts.
4. Write code so that the *Label* component's text displays the sum of the two numbers shown in the edit fields.

   Hint: Use a separate function called `runOperation` *to actually carry out the adding and the updating of the Label field.*   ■

■ Exercise 12.3

012.A, 012.B, 012.C

Continuing with the app created in Exercise 12.2, use the *App Designer* to drag a *Button Group* onto the app's canvas.

Click on the *Button Group* and use the *Inspector* pane to give it the title *Operation*. Ensure that the *Button Group* has four *Radio Buttons*. If you need to, right click on the group to get the context menu, and choose the option to add another button. We will use each button for one of the four basic arithmetic operations, so click on each one in turn, and use the *Inspector* pane to modify their text properties so that they are *add, multiply, subtract,* and *divide*.

Ensure that the button for the *add* operation is selected by default so that this is the one used on starting the app. To do this, select the *add* button and, in the *Inspector* pane, ensure its *Value* box is checked.

Add code and modify the `runOperation` function so that the correct operation is applied to the current values in the edit fields.

Hints:

- *Create a new (private) property to store the name of the operation that needs to be applied (ensure it is properly initialized).*
- *Add a callback for the Selection Changed event of the Button Group. This will provide a* `selectedButton` *variable, from which we can obtain the name of the operation using the* `selectedButton.Text` *property.*
- *Use a* `switch` *statement in* `runOperation` *to select which operation to apply.*   ■

■ Exercise 12.4

This exercise continues the activities on the ECG data plotting app (Activity 12.1 to Activity 12.16).

012.A, 012.B, 012.C

In this exercise, you will add a push *Button* to the app that, when clicked, will allow the user to select a file of their choice containing the data to plot.

Any file containing a single column of numbers can be loaded and plotted. You can use the file *example_ecg_data_c.txt*, which is available on the web site. Download it, and save it to a folder, perhaps even a different folder from the one containing the code for the app.

Open the *App Designer* tool, and load the last version of the app. You may need to browse to the directory where it was saved. Use the *App Designer* to place a push *Button* component onto the app.

Using the *Inspector* pane, set the text for the push *Button* to *Select Data File*. Even though we have not yet added functionality to the push *Button*, save the app, and test that it runs.

In *Design View*, add a callback for the *Button Pushed* event by clicking on the component to highlight it, and, either use the right click context menu, or the *Callbacks* pane. This will generate an empty function named `SelectDataFileButtonPushed`. Note that the name of the function is derived from the text description that we gave the button.

Add code to the push *Button*'s callback function that allows the user to select a text file. Begin by using the `uigetfile` command as follows:

```
[filename,pathname] = uigetfile({'*.txt'});
```

This opens a file browser for the user, and, if the user chooses a text file, returns the file name and the path to the file in the two output arguments. If the user does not select a file, then the name will equal zero. You can use this information to include a check on the user input. Use the output from `uigetfile` to construct the full file name (see the `fullfile` command).

You will need to maintain state by storing the name of the data file being used. To do this, save the file name in a private property for the app (see Activity 12.11).

The only other part of the code that currently uses the data file is the plotting function `myPlot`. It currently uses a hard-coded string to refer to the original data file (*example_ecg_data_b.txt*). Change its code so that it uses the user set data file via the property that you added.

In the callback function for the push *Button,* after storing the data file in the property, ensure that a call to the plotting function is made so that the display is updated.

Finally, to make the code start up the app correctly, the property is initialized. Do this in the general app start up callback function (near where the upper time limit is initialized and before the first call to plot). ■

■ Exercise 12.5

012.A, 012.B, 012.C

This exercise also continues with the ECG data plotting app. (Activity 12.1 to Activity 12.16 and Exercise 12.4).

Recall that the default sampling frequency of the ECG data was hard-coded in the app. It is set to a value of 720 in the `myPlot` function. In the last exercise, we added a button to allow the user to change the input data file, so it might be useful to allow the user to also change the sampling frequency of the data.

Add a second *Edit Field (Numeric)* component to the canvas using the *App Designer* tool. Give it a text label suitable for the sampling frequency. Use the numeric field to take a number from the user for the sampling frequency of the data. See the earlier activities for adding the upper time limit for guidance. The number will need to be stored in a suitable property, which will need to be initialized to a (default) value of 720. Use this default frequency to set the value of the edit field component in the *Inspector* tab. Ensure the property value of the sampling frequency is used in the plotting function instead of the hard-coded value. ■

■ Exercise 12.6

012.A, 012.B, 012.C

This exercise continues from the previous exercises and also uses the ECG data plotting app.

Insert a pop-up menu onto the app to control the color of the plot. The choice should be between black, red, blue, and green.

Hints:

- *Use a property to store a character array for the plot color. This should contain the single letter that MATLAB uses for plot colors:* `'k'`, `'r'`, `'b'`, `'g'`.
- *Initialize the field appropriately in the start up callback.*
- *Update the* `myPlot` *function so that it uses the plot color property.* ■

■ Exercise 12.7

012.A, 012.B, 012.C

This exercise continues on from the app described in Activity 12.18 for reading and filtering an image.

Add a button on the app to allow the user to browse for and load in their own image. You can use the same approach described in Exercise 12.4. ∎

■ Exercise 12.8

This exercise continues on from the previous exercise on image processing.   *012.A, 012.B, 012.C*

A filtering operation can be repeated any number of times. Add an *Edit Field (Numeric)* component to the canvas of the app so that the user can specify this repetition number. Apply the filter the required number of times before displaying the result. ∎

FAMOUS COMPUTER PROGRAMMERS: BJARNE STROUSTRUP

Bjarne Stroustrup is a Danish computer scientist, born in 1950 in Aarhus, Denmark. He studied at Aarhus University in Denmark and Churchill College, Cambridge, in the UK, where he got his PhD in Computer Science.

Stroustrup is famous for developing the C++ programming language. C++ is responsible for popularizing the concept of object-oriented programming (OOP). Before C++ most programming languages were "procedural," i.e., programs consisted of a sequence of steps to run an algorithm. OOP models problems as a set of objects that contain data along with procedures for operating on that data. In a running OOP program, objects send each other messages and data (see Section 1.2). OOP is now an important paradigm in the world of computer programming. Stroustrup developed C++ in 1978 as an extension to the popular C programming language (see Chapter 9's Famous Computer Programmer). In his own words, he "invented C++, wrote its early definitions, and produced its first implementation ... chose and formulated the design criteria for C++, designed all its major facilities, and was responsible for the processing of extension proposals in the C++ standards committee."

Stroustrup is currently Managing Director in the technology division of Morgan Stanley in New York City and Visiting Professor in Computer Science at Columbia University.

"An organization that treats its programmers as morons will soon have programmers that are willing and able to act like morons only."

Bjarne Stroustrup

Statistics

13.1 Introduction

When writing programs to process biomedical data, it is often desirable to be able to summarize and draw conclusions from the data. This might be, for example, to report numerical summary statistics for a set of measurements, or to identify associations between different variables. Of course, it is always possible to export any data we have produced with MATLAB (see Chapter 7), and then to analyze it with our favorite dedicated statistical package. However, this is not usually necessary, as the MATLAB statistics and machine learning toolbox covers most of the more commonly used statistical techniques. Therefore in this chapter, we give an overview of the statistical capabilities of MATLAB.[1]

Note that, because this book is primarily about computer programming, we will not be discussing statistical theory in detail: for this, we refer the interested reader to one of the excellent textbooks on the theory and application of statistics (see Further Resources at the end of this chapter). We focus instead on the practical side of using statistical techniques in MATLAB and give a number of biomedical examples for illustration.

---

[1]Note that the MATLAB examples, activities, and exercises provided in this chapter require access to the statistics and machine learning toolbox, which may not be installed as part of your standard MATLAB installation.

MATLAB® Programming for Biomedical Engineers and Scientists. https://doi.org/10.1016/B978-0-32-385773-4.00022-8

Table 13.1 MATLAB functions for computing summary statistics of central tendency and dispersion.

Measures of central tendency		Measures of dispersion	
`mean(x)`	Mean value of array	`std(x)`	Standard deviation of array values
`median(x)`	Median value of array	`var(x)`	Variance of array values
`mode(x)`	Mode value of array	`iqr(x)`	Inter-quartile range of array values
		`range(x)`	Range of array values

13.2   Descriptive statistics

We start off with *descriptive statistics*, meaning any technique (numerical or visual) that can be used to describe a set of measurements we have made of one or more variables. We refer to such a set of measurements as a *sample*, and these samples are typically drawn from a larger *population* of such values. Note the distinction here with *inferential statistics* (see Section 13.3), in which we try to *infer*, or draw conclusions about the population based on the sample.

We will separately cover descriptive statistics of *univariate* numerical data (i.e., in which we have one variable per measurement) and *bivariate* data (i.e., two variables per measurement). For each of these, we discuss different numerical tools for summarizing the data and techniques for visualizing the data.

13.2.1   Univariate data

For univariate data, there are two types of numerical statistic that we can use to summarize our measurements: measures of *central tendency* and measures of *dispersion*. Measures of central tendency summarize the average value of the sample, whereas measures of dispersion summarize the spread of the values around this average. Table 13.1 lists the different MATLAB functions for computing these measures.

■ Example 13.1

013.A

The following code illustrates the computation of various measures of central tendency and dispersion from body temperature data (in Celsius) that have previously been saved to a *MAT* file:

```
% load body temperature data
load('temperatures.mat'); % loads in 'temps' variable

% compute measures of central tendency
fprintf('Mean temperature = %.1f\n', mean(temps));
fprintf('Median temperature = %.1f\n', median(temps));
fprintf('Mode temperature = %.1f\n', mode(temps));

% compute measures of dispersion
```

```
fprintf('Std dev of temperature = %.1f\n', std(temps));
fprintf('Variance of temperature = %.1f\n', var(temps));
fprintf('IQR of temperature = %.1f\n', iqr(temps));
fprintf('Range of temperature = %.1f\n', range(temps));
```
■

We have already reviewed MATLAB visualization approaches in Chapters 1 and 9. In this chapter, we introduce some further visualizations that can be useful when analyzing data as part of a statistical analysis. For univariate data, the most common means of visualizing multiple measurements of a variable is the *histogram*. Histograms allow us to visualize the distribution of values in the measurements.

■ **Example 13.2**

Histograms can be displayed for any variable that takes discrete or continuous values. In MATLAB, we use the `histogram` function, as illustrated below.

013.B

```
% load body temperature data
load('temperatures.mat'); % loads in 'temps' variable

% dispay histograms
close all;
figure;
% binning algorithm chooses 7 bins for these data
histogram(temps);
title('Histogram of body temperature values');
xlabel('Body temperature (Celsius)');
figure;
% Explicity set the number of bins to 10
histogram(temps, 10);
title('Histogram of body temperature values');
xlabel('Body temperature (Celsius)');
```

Fig. 13.1 shows the output of this code. As can be seen, histograms can appear quite different, depending on the number of bins used. The `histogram` function uses an algorithm to automatically estimate a good number of bins to use with the data. In the second call, we take control of this and set the number of bins to 10 (using the optional second argument).
■

13.2.2 Bivariate data

For bivariate data that are continuous, discrete, or ordinal, we can measure the *correlation* between the two variables. Commonly applied measures of correlation are the Pearson's correlation coefficient (for continuous or discrete data that are normally distributed) and Spearman's (rank) correlation coefficient (for bivariate data, in which at least one variable is not normally distributed or is ordinal). Both of these measures can be computed in MATLAB using the `corr` function, as illustrated in the following example:

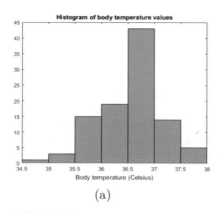
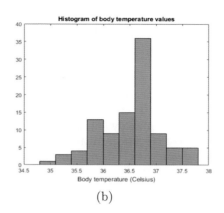

(a)   (b)

FIGURE 13.1

Histograms of body temperature: (a) using 7 bins; (b) using 10 bins.

■ Example 13.3

013.A   Research has suggested that body temperature can reduce with age. Further-more, studies of dementia patients indicate a relationship between body temperature and the decline in cognitive function. The code that follows in-vestigates these possible links using data loaded in from *MAT* files. (Note that these data were not measured from real patients, but were artificially generated based on values reported in the literature.)

Cognitive function is commonly measured using the mini mental state ex-amination (MMSE), a 30-point questionnaire that tests a range of cogni-tive tasks. The data used in this example represent age, body temperature, and MMSE scores for 100 dementia patients. As the histograms shown in Figs. 13.1a and 13.2 illustrate, the temperature and age variables appear to be normally distributed, whereas the MMSE data appear skewed. There-fore to measure the correlation between body temperature and age, we can use Pearson's correlation coefficient, but to measure the correlation between body temperature and MMSE score, we should use Spearman's correlation coefficient. Consider the following code:

```
% load data
load('temperatures.mat');
load('ages.mat');
load('mmse.mat');

% compute Pearson's correlation
[rhoA, pvalA] = corr(ages,temps)

% compute Spearman's correlation
[rhoB, pvalB]=corr(temps,mmse, 'type', 'Spearman')
```

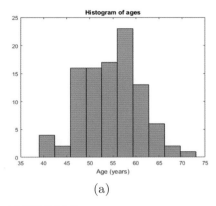
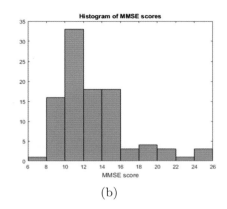

(a)   (b)

FIGURE 13.2
Histograms of (a) age; (b) MMSE score.

Note that, by default, the `corr` function computes Pearson's correlation coefficient. However, if we specify an extra pair of *name-value* arguments (`'type'`, `'Spearman'`), it will compute the Spearman's correlation instead.

The `corr` function returns two values: `rho` represents the correlation coefficient, and `pval` represents the *p-value* of the correlation. This represents the *significance*, i.e., the probability that the correlation observed is due to random variation in the data. A low p-value indicates that the correlation found is likely to represent a real association between the variables in the underlying population.

You can download this code and data from the book's web site. The data have a negative correlation (−0.3409) between body temperature and age, with a low p-value of 5.1832e-04. The correlation between body temperature and MMSE score is 0.1149, with a p-value of 0.2551. Therefore, based upon these results, we can conclude that there is likely to be a real (but weak) negative correlation between body temperature and age. We cannot assume any correlation between body temperature and MMSE score, as there is a 25% chance that we would have seen this pattern of variation in the data even if no correlation between the two variables existed. ■

There are a number of ways to visualize univariate and bivariate data. For bivariate data, perhaps the obvious choice, as the data consist of *corresponding* (or *paired*) observations, is simply to plot one variable against the other for all observations in the sample. This is known as a *scatter plot*; an example is shown in Fig. 13.3a. The scatter plot allows us to visually confirm the negative correlation between body temperature and age that we found in Example 13.3.

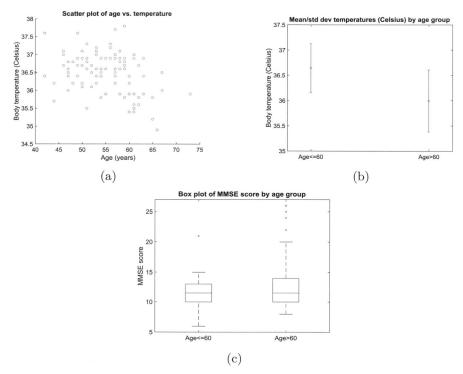

FIGURE 13.3
Bivariate data visualizations: (a) scatter plot of age against body temperature; (b) error bar plot of body temperature for different age groups; and (c) box plot of MMSE score for different age groups.

If we wish to compare the measures of central tendency and dispersion of two symmetrically distributed discrete or continuous variables, we can use an *error bar* plot. An example of this is shown in Fig. 13.3b, which illustrates that the dementia patients who were aged 60 or less had a slightly higher body temperature than those who were over 60. The crosses represent the mean values, and the error bars represent the standard deviations.

Finally, to compare measures of central tendency and dispersion for skewed data, we can use a box plot, which visualizes the medians and inter-quartile ranges of two variables. An example of this is shown in Fig. 13.3c. The red lines represent the medians, and the blue boxes the inter-quartile ranges. The black lines indicate an estimate of the extent of the distribution, whereas the red crosses indicate possible outlier values.

■ Example 13.4

013.B

This example illustrates that MATLAB can be used to generate the visualizations shown in Fig. 13.3. The `scatter` function is used to create a scatter plot and takes two array variables of the same length as arguments (paired data). The `errorbar` function is used to create an error bar plot. Note the use of logical indexing (see Sections 3.4.1 and 10.3.3) when producing this plot. This divides the temperature data into two arrays: one for patients aged 60 or less and one for patients over 60. The `boxplot` function produces a box plot, and takes two arguments: the array of values that are to appear on the y-axis (MMSE scores in our case) and a second array of the same length that will be used to group the values in the first array. We used the same logical array of age classifications for this purpose.

```
% load data
load('temperatures.mat');
load('ages.mat');
load('mmse.mat');

% scatter plot
close all;
figure;
scatter(ages, temps);
xlabel('Age (years)');
ylabel('Body temperature (Celsius)');
title('Scatter plot of age vs. temperature');

% error bar plot
age_class = ages<=60; % logical array
means = [mean(temps(age_class)) mean(temps(~age_class))];
stdevs = [std(temps(age_class)) std(temps(~age_class))];
figure;
errorbar(means, stdevs, 'x');
ylabel('Body temperature (Celsius)');
set(gca, 'XTick', 1:2);
set(gca, 'XTickLabel', {'Age<=60','Age>60'})
title('Mean/std dev temperatures (Celsius) by age group');

% box plot
figure;
boxplot(mmse, age_class);
ylabel('MMSE score');
set(gca, 'XTick', 1:2);
set(gca, 'XTickLabel', {'Age<=60','Age>60'})
title('Box plot of MMSE score by age group');
```
■

Table 13.2 provides a summary of the different MATLAB functions for working with bivariate data.

■ Activity 13.1

The enlargement of brain ventricles has been proposed as an indicator of Alzheimer's disease progression. The change in volume of the brain ventricles can be measured from three-dimensional (3-D) medical images (e.g.,

013.A, 013.B

Table 13.2 MATLAB functions for working with bivariate data: computing numerical measures and producing visualizations.

Numerical measures	
`[rho, pval] = corr(x,y)`	Correlation (`rho`) between variables `x` and `y` and the associated p-value (`pval`)
Visualization functions	
`scatter(x,y)`	Scatter plot between variables `x` and `y`
`errorbar(means, stdevs)`	Error bar plot showing means and standard deviations of multiple variables
`boxplot(x,y)`	Box plot of variable `x` broken down by variable `y`

MR and/or CT scans). The file *Alzheimers_data.mat* contains the result of a study in which 17 Alzheimer's patients had their disease *progression* measured using two methods: the change in MMSE score and the change in ventricle size. Each patient was measured at two time points, and the file contains the changes in MMSE score (first column, positive values show symptoms becoming worse) and the changes in ventricle volume (second column, in cm^3). Analyze these data as follows:

1. Produce histograms of the two variables, i.e., change in MMSE score and change in volume. What do the histograms suggest?
2. Visualize the data using a scatter plot. What does the scatter plot show?
3. Calculate an appropriate measure of correlation between the two variables. Also determine the p-value of the correlation. ∎

13.3 Inferential statistics

When working with biomedical data, as well as visualizing and summarizing, we often want to try to answer a specific question using the data we have as evidence. For example, based on data we might collect on the response to a new treatment, can we say that the treatment works better than an existing treatment or a placebo? This type of question brings us into the realm of *inferential statistics*, in which we try to *infer* something about the population of interest based on a sample we have drawn. In this section, we review some of the functions that MATLAB provides to help us answer such questions, which are also known as *hypothesis tests*. Hypothesis tests help us to decide whether a given hypothesis can be rejected (or not), given a set measurements we have in our data. Typically, the hypothesis that we actually test is known as the *null hypothesis*, and the alternative to it is known as the *alternative hypothesis*.

13.3.1 Testing the distributions of data samples

Before we look at specific questions, such as whether a treatment works or not, it is important to know something about the *distribution* of our data. We have already seen how histograms can be used to visually inspect data distributions (see Section 13.2.1), but in inferential statistics, we want to be more precise than this. To be able to choose an appropriate hypothesis test for answering questions using the data, it is first important to be able to test the hypothesis that the data sample was drawn from a population with a particular distribution. The MATLAB function `chi2gof` is one way to do this. It performs a *chi square (χ^2) goodness-of-fit test* on the data. This tests the null hypothesis that the data sample comes from a particular distribution, for a given level of confidence (e.g., 95%). The alternative hypothesis is that the data do not come from such a distribution. A common distribution to test for in inferential statistics is the *normal*, or *Gaussian* distribution, as the following example illustrates:

■ Example 13.5

An experiment is performed to test the efficacy of a new dietary supplement intended to lower cholesterol levels in humans. Total cholesterol level data in mmol/L were gathered from 125 subjects who suffered from high cholesterol. The readings were taken before starting to take the new supplement and 3 months afterward. The data are available in the files *cholesterol_pre.txt* and *cholesterol_post.txt*. We want to test if the distributions of the two variables (i.e., the total cholesterol levels before and after) are normally distributed. Examine the following code:

013.C

```
% Null Hypothesis = data drawn from a normal distribution
%   with same mean and variance as sample
% Alternative Hypothesis = data does not come from such a
%   distribution

% load data
chol_pre = load('cholesterol_pre.txt');
chol_post = load('cholesterol_post.txt');

% always a good idea to visualise data first ...
subplot(2,1,1);
histogram(chol_pre);
xlim([2 9]);
title('Histogram of cholesterol levels before treatment');
subplot(2,1,2);
histogram(chol_post);
xlim([2 9]);
title('Histogram of cholesterol levels after treatment');

% Apply chi square goodness-of-fit tests
% with default 95% confidence
%   h==1 Reject null hypothesis
%   h==0 Do not reject
[h_pre, p_pre] = chi2gof(chol_pre)
[h_post, p_post] = chi2gof(chol_post)
```

Note that the `chi2gof` function returns two arguments: `h` refers to the result of the hypothesis test (1 means reject the null hypothesis, 0 means do not reject), whereas p is the p-value of the test, i.e., the probability of seeing this level of variation in the data if the null hypothesis were true. The default confidence level to work to is 95%, but we can change this by using optional *name-value* pairs of arguments, as shown below (an alpha value of 0.01 corresponds to 99% confidence).

```
[h_pre,p_pre] = chi2gof(chol_pre, 'alpha', 0.01)
```

In this example, for both variables, the return value h was 0, meaning that the null hypothesis could not be rejected. Therefore we assume, for the purposes of our statistical analysis, that both variables are normally distributed ∎

See Section 13.5 at the end of this chapter for details of alternative ways of using MATLAB to test if our data are normally distributed.

13.3.2   Comparing data samples

The main reason for testing whether a data sample was drawn from a population with a normal distribution is to help us to decide which type of test to apply for further questions based on the data. Two categories for such tests are *parametric* and *nonparametric* tests. Broadly speaking, we can apply parametric tests if all of our data are normally distributed, and we must apply nonparametric tests otherwise.

There are three common types of test for comparing data samples, depending on the number of samples we have and their relationship to each other:

- A one sample test against an expected value.
- A two sample *paired data* test, in which there exist one-to-one correspondences between values in the first sample and values in the second sample.
- A two sample *unpaired data* test in which no such correspondences exist.

Each of these tests can be either *one-tailed* or *two-tailed*. With a one-tailed test, we are trying to decide whether the values in the population from which one sample was drawn are *larger* or *smaller* than either a specific value or the set of values in another sample. With a two-tailed test, we are looking for *any* difference, regardless of whether it is larger or smaller.

The most commonly applied parametric hypothesis test is the Student's t-test. MATLAB functions for applying a Student's t-test for the three different situations are summarized below.

- `[h,p] = ttest(x,m,'alpha',`α`,'tail',t)`: Performs a one sample t-test between the dataset x and the expected value m. The null hypothesis is that there is no difference between the mean of x, and m. The alternative hypothesis is that there is such a difference. An h value of 1 is returned if the null hypothesis can be rejected, and an h of 0 means it cannot be rejected. The variable p contains the p-value, which represents the probability that the null hypothesis is true, and that the sample data occurred by chance. The extra optional pairs of arguments, `'alpha'`, α, and `'tail'`, t, specify the significance level and type of the test. The default value for the significance level α is 0.05, representing 95% confidence. The default value for the type t is `'both'`, meaning a two-tailed test. If t is `'right'`, it will perform a one-tailed test for the mean of x being greater than m. If t is `'left'`, it will perform a one-tailed test for the mean of x being less than m.

- `[h,p] = ttest(x,y,'alpha',`α`,'tail',t)`: Performs a two sample paired data t-test between the datasets x and y. The null hypothesis is that the mean difference between pairs of x and y values is zero. The alternative hypothesis is that this mean is not zero. An h value of 1 is returned if the null hypothesis can be rejected, and an h of 0 means it cannot be rejected. The extra optional argument pairs specifying α and t have similar meanings to those described above. A `'tail'` of `'left'` tests if the mean of x is less than the mean of y, whereas `'right'` tests if the mean of x is greater than the mean of y.

- `[h,p] = ttest2(x,y'alpha',`α`,'tail',t)`: Performs a two sample unpaired data t-test between the datasets x and y. The null hypothesis is that x and y come from normal distributions with the same mean. The alternative hypothesis is that their means are different. An h value of 1 is returned if the null hypothesis can be rejected. The extra optional argument pairs specifying α and t have the same meanings, as described above.

Note that we use the same function (`ttest`) to perform both a one sample t-test against an expected mean, and a two sample paired data t-test. The difference lies in the nature of the second argument supplied: if it is a scalar value, a one-sample t-test will be performed, whereas if it is an array of the same length as the first argument, a two-sample paired data t-test will be performed.

■ Example 13.6

In this example, we extend the cholesterol case study we introduced in Example 13.5. Now that we know the data are normally distributed, we want to use parametric tests to answer the following questions:

013.C

- Is the total cholesterol level *before* taking the supplement significantly different to the "healthy" level of 5 mmol/L?
- Is the total cholesterol level *after* taking the supplement significantly different to the "healthy" level of 5 mmol/L?
- Is the total cholesterol level after taking the supplement significantly lower than the total cholesterol level before taking the supplement?

We will work to a 95% confidence level. The following code illustrates how to answer these questions using MATLAB:

```
% load data
chol_pre = load('cholesterol_pre.txt');
chol_post = load('cholesterol_post.txt');

% Test both datasets against expected value of 5mmol/L
% 2-tailed, 95% confidence
% Null Hypothesis = there is no difference between the
%  sample mean and 5
% Alternative Hypothesis = there is such a difference
% (h=1 means reject null hypothesis, h=0 means do not reject)
[h_pre5, p_pre5] = ttest(chol_pre, 5)
[h_post5, p_post5] = ttest(chol_post, 5)

% Test for decrease in cholesterol after taking supplement
% 1-tailed, 95% confidence
% Null Hypothesis = mean difference between pre and post
%  data not < 0
% Alternative Hypothesis = mean difference between pre
%  and post data < 0
% (h=1 means reject null hypothesis, h=0 means do not reject)
[h_cmp, p_cmp] = ttest(chol_pre, chol_post, 'tail', 'right')
```

Note that for the first two questions we perform a two-tailed test, as we are looking for *any* difference, but for the third question, we are looking for one variable to be less than the other, so we perform a one-tailed test. We use a paired data t-test as the pre-supplement and post-supplement cholesterol levels were measured from the same subjects. In this example, we are able to reject the null hypothesis for all three questions, meaning that both variables are different to 5 mmol/L, and that the post-supplement cholesterol level is lower than the pre-supplement level. ∎

If at least one of our data samples was drawn from a population which is not normally distributed, we cannot use a Student's t-test. In this case, a nonparametric test should be used. MATLAB also provides functions for common nonparametric tests that broadly correspond to the three types of parametric test we introduced above. These are summarized below:

- `[p,h] = signrank(x,m,'alpha',`α`,'tail',t)`: Performs a one sample Wilcoxon signed rank test for a single sample x against a specified population median m. The null hypothesis is that there is no difference between the median of x, and m. The alternative hypothesis is that there is such a difference. An h value of 1 is returned if the null hypothesis can be rejected. The extra optional argument pairs specifying α and t have similar meanings to those described earlier for the `ttest` and `ttest2` functions. If the value of t is `'both'` (the default value), the test will be two-tailed. If t is `'right'`, it will perform a one-tailed test for the sample median being greater than m. If t is 'left', it will perform a one-tailed test for the sample median being less than m.[2]

- [p,h] = signrank(x,y,'alpha',α,'tail',t): Performs a two sample paired data Wilcoxon signed rank test between the datasets x and y. The null hypothesis is that the median of the differences between x and y is zero. The alternative hypothesis is that there is a difference. An h value of 1 is returned if the null hypothesis can be rejected. The extra optional argument pairs specifying α and t have similar meanings to those described earlier: a 'tail' of 'left' tests if the median of x is less than the median of y, whereas 'right' tests if the median of x is greater than the median of y.

- [p,h] = ranksum(x,y,'alpha',α,'tail',t): Performs a two-tailed unpaired Mann–Whitney U test between the datasets x and y. The null hypothesis is that x and y come from identical distributions. The alternative hypothesis is that their medians are different, but the distributions are otherwise identical. An h value of 1 is returned if the null hypothesis can be rejected. The extra optional argument pairs specifying α and t have the same meanings as they do for the signrank function.

In the above, note the reversed ordering of the output arguments p and h when compared to the ttest and ttest2 functions.

■ Example 13.7

To illustrate the use of non-parametric hypothesis tests, we introduce a new case study. A company develops a new surgical technique for inserting hip implants. The angular accuracy of the placement can be assessed post-operatively using X-ray imaging. Accuracy data (in degrees) have been gathered from a cohort of 100 patients. Half of the cohort (randomly selected) underwent the new surgery, and the other half underwent traditional surgery. The data are available in the file *implant_data.mat*. This contains two MATLAB variables: trad_errors and new_errors, which represent the angular errors for the traditional surgery group and the new technique's group, respectively.

013.C

Assuming that at least one of the two samples was drawn from a population that is not normally distributed, we want to perform an appropriate hypothesis test to determine, with 95% confidence, if the new surgical technique results in lower errors than the traditional technique. Examine the following code:

```
% load data
load('implant_data.mat');
```

---

[2]N.B. The one sample Wilcoxon signed rank test assumes symmetrically distributed differences. If this is not the case, a sign test can be performed with the function signtest.

```
% Test for decrease in error with new technique
% Unpaired data, so Mann Whitney U test
% 1-tailed, 95% confidence
% Null Hypothesis = median of new errors not less than
%  median of trad errors
% Alternative Hypothesis = median of new errors less than
%  median of trad errors
% (h=1 means reject null hypothesis, h=0 means do not reject)
[p,h] = ranksum(trad_errors, new_errors, 'tail', 'right')
```

We use a Mann–Whitney U test. Even though the two samples we are comparing have the same length (50), there is no pairing between the patients who underwent traditional surgery and those who underwent the new surgical technique. Therefore we have *unpaired* data. We are looking for a *decrease* in errors, so we apply a one-tailed test. For these data, ranksum returns an h value of 1, meaning that we reject the null hypothesis and conclude that the new technique has lower errors. ∎

For all of the aforementioned MATLAB functions for hypothesis testing, note that we can always find the p-value of the test as one of the return arguments. The p-value is effectively the probability that we would have seen the level of difference that we have (or a stronger difference) *if the null hypothesis were true*. If this value is sufficiently low (less than the significance level of the test, e.g., 0.05 for 95% confidence), then we reject the null hypothesis. Therefore you should notice that these MATLAB functions will return h=1 (i.e., reject the null hypothesis) whenever the p-value is less than the significance level. Table 13.3 provides a summary of the different MATLAB functions we can use when performing inferential statistics.

■ Activity 13.2

O13.B, O13.C

Bilirubin is a substance found in the blood, which is the breakdown product of the clearance of aged red blood cells. A study is investigating a possible link between high bilirubin levels and the risk of developing cardiovascular disease.

A group of subjects were randomly selected 10 years ago, and data have been gathered from them over the past 10 years. For each subject, the data consist of the average serum bilirubin level over the 10 year period (in mg/dL) and whether or not the subject developed a cardiovascular disease during the 10 year period (1=yes, 0=no). The data are available in the file *bilirubin_data.mat*, and consists of two variables: bilirubin and cardiovascular_disease.

Write code to split the bilirubin level data into two groups: one for those subjects who had cardiovascular disease, and one for those who didn't.

Table 13.3 MATLAB functions for performing inferential statistics: testing sample distributions and comparing samples. A return value of h=1 means reject the null hypothesis, and p is the associated p-value.

Testing sample distributions	
`histogram(x)`	Display a histogram of variable x
`[h,p] = chi2gof(x)`	Perform a chi square goodness-of-fit test on variable x, with the null hypothesis that x comes from a normal distribution
Comparing samples	
`[h,p] = ttest(x,m)`	Perform a one sample Student's t-test between variable x and expected mean value m, with the null hypothesis that the mean of x is not significantly different to m
`[h,p] = ttest(x,y)`	Perform a paired Student's t-test between variable x and variable y, with the null hypothesis that the mean difference between x and y is zero
`[h,p] = ttest2(x,y)`	Perform an unpaired Student's t-test between variable x and variable y, with the null hypothesis that x and y have the same means
`[h,p] = signrank(x,m)`	Perform a Wilcoxon signed rank test between variable x and expected median value m, with the null hypothesis that the median of x is not significantly different to m
`[h,p] = signrank(x,y)`	Perform a Wilcoxon signed rank test between variable x and variable y, with the null hypothesis that the median difference between x and y is zero
`[h,p] = ranksum(x,y)`	Perform a Mann–Whitney U test between variable x and variable y, with the null hypothesis that x and y come from identical distributions

Visualize the distributions of the data values in these two groups. What are your comments?

Perform an appropriate hypothesis test to determine, with 95% confidence, if the bilirubin levels of those who had cardiovascular disease are from a different distribution to the bilirubin levels of those who didn't. Explain why you chose the test you did, and comment on the result. ∎

13.4 Summary

MATLAB provides functions for most common statistical tasks, either through its core functionality or through the statistics and machine learning toolbox.

Table 13.1 summarizes the functions available to numerically summarize either the central tendency or the dispersion of a univariate data sample. The correlation between two variables can be measured using the corr function.

The distribution of a univariate sample can be visualized using a histogram of its values, which, in MATLAB, can be produced using the histogram function. Multivariate data visualizations can be produced using the `scatter`, `errorbar`, and `boxplot` functions.

As well as numerically summarizing and visualizing data (descriptive statistics), MATLAB also provides functions for answering questions about the population from which the data were drawn (inferential statistics). A chi square goodness-of-fit test can be used to test if a univariate sample fits a specific distribution, such as the normal distribution. The MATLAB function `chi2gof` can be used for this purpose. The results of such a test help us to decide whether subsequent tests should be parametric or nonparametric.

It is possible to carry out parametric hypothesis tests using MATLAB. The Student's t-test can be performed using the `ttest` and `ttest2` functions. The `ttest` function is used for one sample comparisons against an expected mean value, and also for two sample comparisons using paired data. The `ttest2` function is used for two sample comparisons using unpaired data.

For nonparametric tests, the MATLAB functions `signrank` and `ranksum` perform, respectively, a Wilcoxon signed rank test (either one sample or two sample paired data) and a Mann–Whitney U test (two sample unpaired data).

All hypothesis tests can be applied using different confidence levels and as either one-tailed or two-tailed tests.

13.5   Further resources

- "Statistics for Biomedical Engineers and Scientists: How to Visualize and Analyze Data," by Andrew King and Robert Eckersley [7] provides a much more in-depth and complete coverage of statistical theory and its application to biomedical problems using MATLAB.
- "Vital Statistics: An Introduction to Health Science Statistics," by Stephen McKenzie [8] is an accessible introduction to the theory and application of statistics with a focus on the health sciences.
- We introduced `chi2gof` as one way of checking to see if our data were drawn from a normally distributed population. The following are alternatives:
  - Kolmogorov–Smirnov test: see the MATLAB documentation for `kstest`.
  - Lilliefors test: see the MATLAB documentation for `lillietest`.
  - Z-test: see the MATLAB documentation for `ztest`.
  - Shapiro–Wilk test: there is no built-in MATLAB function, but implementations are available from the Mathworks File Exchange (http://mathworks.com/matlabcentral/fileexchange).
- See the MATLAB documentation for more details on the functions we have covered, and for other useful functions:
  - Descriptive statistics: http://mathworks.com/help/stats/descriptive-statistics.html.

- Statistical visualization: http://mathworks.com/help/stats/statistical-visualization.html.
- Hypothesis testing: http://mathworks.com/help/stats/hypothesis-tests-1.html.

13.6 Exercises

■ Exercise 13.1

A new drug has been developed to suppress the desire for drinking alcohol among alcoholic patients. Data have been gathered of the number of units of alcohol consumed weekly by a group of 73 alcoholics before and after treatment with the drug. The data are in the file *alcohol.mat*, which contains two variables: `units_before` and `units_after`.

013.A, 013.B, 013.C

In this exercise, you will use MATLAB to determine if the alcohol intake figures are significantly lower after treatment compared to before treatment. This will involve analyzing the data to determine if they are normally distributed, numerically summarizing the data, and then choosing and applying an appropriate hypothesis test.

1. Load in the data, and visualize the distributions of the alcohol consumption data, both before and after taking the drug.
2. Perform appropriate hypothesis tests to determine if the before treatment and after treatment data are normally distributed. Work to a 95% confidence level.
3. Choose and compute appropriate numerical summary statistics of the before and after treatment data. Justify your choices.
4. Choose an appropriate hypothesis test and apply it to determine if the after treatment data are significantly lower than the before treatment data. Work to a 95% confidence level. Explain your choice of hypothesis test, and comment on the result. ■

■ Exercise 13.2

A large scale study of the UK population has gathered data from volunteers through questionnaires, a variety of tests, and imaging studies. A research team would like to use these data to investigate a potential link between diet and cardiac health.

013.A, 013.C

Data gathered from 5000 subjects is available to you in the file *heart.mat*. The file contains two variables. The `diet` variable is a classification of each subject's diet into one of three categories: good ($=2$), average ($=1$), or poor ($=0$). The `ef` variable is a measurement of each subject's *ejection fraction*, as measured using dynamic magnetic resonance imaging. Ejection fraction represents a measure of how much blood is pumped out of the heart in each beat, and is considered to be a good indicator of cardiac health. Typical (healthy) values are between 50% and 65%, but people with poor cardiac health are likely to have lower values.

Use MATLAB to perform the following tasks:

1. Load the data, and visualize them using four histograms included in the same figure. These should show respectively:
   - All ejection fraction values.
   - Ejection fraction values for subjects with a poor diet.
   - Ejection fraction values for subjects with an average diet.
   - Ejection fraction values for subjects with a good diet.

   Comment on the distributions that the histograms show.
2. Compute an appropriate measure of correlation between diet and ejection fraction. Also, compute the probability that this correlation is due to random variation in the data. Comment on the results.
3. Perform appropriate hypothesis tests to determine, with 95% confidence, if the ejection fraction data for subjects with poor, average, and good diets were drawn from normal distributions.
4. Based on the results of these three tests, choose and apply appropriate hypothesis tests to determine if:
   - The ejection fractions of subjects with poor diets are significantly different to the ejection fractions of subjects with good diets.
   - The ejection fractions of subjects with poor diets are significantly less than 50%.

   In the above, as well as calling the appropriate MATLAB functions, interpret the outputs to give your conclusions in plain English. ■

■ Exercise 13.3

013.A, 013.B, 013.C

Positron emission tomography (PET) imaging enables in-vivo quantitative measurements of radiotracer concentration and is commonly applied to monitor the progress of cancer. A common measure is the standardized uptake value (SUV). If the SUV of a tumor increases over time, this means the cancer is becoming more active.

1. SUV values can be affected by the presence of motion artifacts in the PET images, so correcting for motion, such as that caused by breathing, is of interest. A research group has developed a new technique for respiratory motion correction of PET images. PET images have been acquired from 100 patients and SUV values calculated for their tumors. These data are in the file *suv_pre_correction.txt*. The images were then motion corrected using the new algorithm, and the SUV

values recalculated. The recalculated values are in the file *suv_post_correction.txt*. Using MATLAB, read in the data, visualize them, and choose and apply an appropriate hypothesis test to determine with 95% confidence if the new technique changes the SUV values.

2. SUV values for clinical use are sometimes computed by either taking the mean or maximum SUV value within a region of interest. These approaches may be subject to noise in the images. A better way may be to use a statistical test based on all values within the region. A patient undergoing chemotherapy has had PET images acquired before and after treatment. SUV values for all voxels within a tumor region for the two images can be found in the files *suv_pre_chemo.txt* and *suv_post_chemo.txt*. Because the tumor may have changed size, the two files may contain different numbers of SUV values, and the values do not represent corresponding voxels. Doctors wish to know if the activity of the tumor has changed as a result of the treatment. First, use MATLAB to compute mean and maximum SUV values for before and after chemotherapy to answer this question. Next, perform an appropriate statistical hypothesis test to answer the same question with 95% confidence. ■

FAMOUS COMPUTER PROGRAMMERS: LINUS TORVALDS

Linus Torvalds is a Finnish computer scientist, born in 1969 in Helsinki, Finland. He is famous for being the developer of the LINUX open-source operating system. As a child, Torvalds was always interested in home computers. He owned several and enjoyed writing software for them, including games and a text editor. He went on to study computer science at the University of Helsinki.

In 1991, while still a student at Helsinki, Torvalds purchased a personal computer (PC), which ran the Microsoft DOS operating system. But Linus preferred the UNIX operating system (see Chapter 9's Famous Computer Programmer) that he had used on the university's computers. Therefore he set himself the task of writing a version of UNIX for PCs. After he had produced the core of the new operating system, he made the source code freely available on the internet and asked other programmers around the world to contribute to the project. Many took up the offer, and the LINUX project was born. Torvalds initially wanted to call his open-source operating system FREAX (free UNIX), but his friend Ari Lemmke, who administered the FTP server where the software was hosted for download, didn't like this name and so called it LINUX (Linus' UNIX). These days, LINUX is one of the most widely used operating systems in the world. Versions of LINUX are run by many PCs, servers, and smart phones.

The fact that the source code of the LINUX kernel was released freely under the GNU software license (see Chapter 11's Famous Computer Programmer) meant that Torvalds never benefited financially from the success of LINUX. However, in 1999, two leading commercial developers of LINUX-based software, Red Hat and VA Linux, presented Torvalds with stock options in gratitude for his creation, which made him a wealthy man.

Torvalds currently lives in Portland, Oregon, USA, and works full-time for the LINUX Foundation on improving LINUX.

"Most good programmers do programming not because they expect to get paid or get adulation by the public, but because it is fun to program."

Linus Torvalds

CHAPTER 14

Machine learning

LEARNING OBJECTIVES

At the end of this chapter you should be able to:

O14.A Explain the meaning of the term machine learning, outline the difference between unsupervised and supervised machine learning and give examples of each type

O14.B Use MATLAB® to perform cluster analysis on data

O14.C Use MATLAB to reduce the dimensionality of high dimensional data and visualize the transformed representation

O14.D Use the MATLAB classification learner app to train, evaluate, and deploy a supervised classification model

O14.E Use the MATLAB regression learner app to train, evaluate, and deploy a supervised regression model

O14.F Use the MATLAB deep network designer app to train and evaluate a pre-defined deep learning model using transfer learning

14.1 Introduction

In this chapter, we introduce the fundamental concepts of machine learning and give an overview of the functionality provided by MATLAB to develop and apply machine learning models.[1] Machine learning has a strong relationship with the field of statistics. Therefore many of the ideas discussed in this chapter build upon content introduced in Chapter 13. You may wish to refer back to parts of Chapter 13 as you work through this chapter. Machine learning is a big topic, and many books have been written about it. In this chapter, we attempt to provide an overview of the key concepts and practical skills needed to get started with analyzing biomedical data using machine learning. We refer the reader to one of the more specialist text books on machine learning for a more detailed coverage of the field.

---

[1] Note that the MATLAB examples, activities, and exercises provided in this chapter require access to the statistics and machine learning toolbox, which may not be installed as part of your standard MATLAB installation. The later example on deep learning also requires the deep learning toolbox.

MATLAB® Programming for Biomedical Engineers and Scientists. https://doi.org/10.1016/B978-0-32-385773-4.00023-X

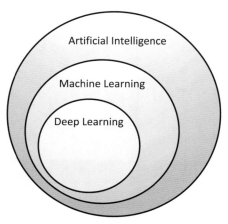

FIGURE 14.1

The relationship between artificial intelligence, machine learning, and deep learning.

14.2 Artificial intelligence, machine learning, and deep learning

The terms *artificial intelligence* (AI), *machine learning*, and *deep learning* are often heard these days, both in the academic literature and the popular media. The terms are often used interchangeably, but in reality they have specific meanings, and so we commence by introducing these meanings.

As illustrated in Fig. 14.1, AI, machine learning, and deep learning are progressively more precise terms. AI refers to any technique that aims to enable computers to exhibit "intelligent" behavior.[2] Machine learning refers to AI techniques that work by learning new behaviors from data, rather than having these behaviors explicitly programmed. Deep learning methods are one way of implementing machine learning, and work by defining a deep network of artificial "neurons," whose operation in some way mimics that of real neurons in biological systems, such as the human brain. In recent years, machine learning and deep learning techniques have achieved significant success, and they are often referred to as AI.[3] In this chapter, we focus on machine learning, and we use this term, rather than the more general term "AI." We also include a brief taster of deep learning methods toward the end of the chapter.

One of the main differences between traditional machine learning and deep learning techniques lies in their use of *features* that describe and/or are extracted from the data. The original data to be analyzed will typically consist of a number of variables, for example, patient age, height, weight, blood pressure, or intensities of images acquired from the patient. We will normally measure

---

[2]Clearly defining the term "intelligence" is actually quite hard, but we gloss over this difficulty here.
[3]This is not incorrect, but it is imprecise.

these variables from multiple individuals (e.g., patients). We refer to such data as *multivariate*, as they contain multiple variables for each individual (cf. univariate and bivariate data, see Chapter 13). In general terms, a feature refers to a characteristic of the individuals we are studying and gathering data from. A feature can be as simple as one of the variables we have measured (e.g., the blood pressure of a patient), but features can also be more complex, and represent transformations of combinations of several variables. It is the features of the data that are used by machine learning models to perform the tasks we give them. A common distinction to make here is between *hand-crafted features* and *learned features*. Hand-crafted features are those identified as being important by the algorithm developer, typically using some domain knowledge of the application being studied. Learned features are identified by the machine learning model itself, often as being useful for some specific task. As a rough distinction, traditional machine learning models use hand-crafted features, whereas deep learning models use learned features.

14.3 Types of machine learning

Machine learning models can be used for different purposes. Sometimes we have a specific task in mind, for example, we may wish to train a model to diagnose a disease from imaging and/or blood test data. In this case, we have what we call *labels* or *annotations* for each individual. For example, for disease diagnosis, we might have a binary label (disease/no disease) for each individual. In such cases, we can use a *supervised* machine learning model, and the supervision comes in the form of the provided labels. The task of the machine learning model is then to predict the label of new individuals based only on the input data features. Alternatively, we may not have any labels, or we may not even know what task we want to perform with the data. We just want to examine the data and learn something about their structure and the relationships between individuals. In such cases, we can use an *unsupervised* machine learning model. We discuss each of these two types of machine learning in separate sections below.

14.4 Evaluating machine learning models

Supervised machine learning models need to be trained, and they need data to drive this training. However, they also need to be evaluated before they can be deployed on new data, and this evaluation typically also requires data. Therefore it is common to divide the total amount of data we have into *training* and *test* data. The training data and their labels will be used to learn a model that can predict the label given the data. The test data and their labels are not available to the model during training, and will only be used to evaluate the model's performance after training has finished. This approach is known as *holdout validation*, since we "hold out" a portion of the total data for testing purposes. In

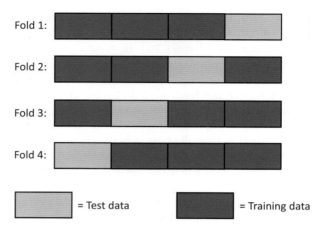

Fold 1:

Fold 2:

Fold 3:

Fold 4:

☐ = Test data   ☐ = Training data

FIGURE 14.2

Illustration of a 4-fold cross validation scheme. Data are divided into training and test data four times, with each individual appearing in the test set once. The overall evaluation is based on a combination of the test set performances over all four folds.

holdout validation, in addition to the training and test data, a separate set of *validation* data is sometimes used. The validation data are not used for optimizing the model, but they can be used for selecting the final model used for deployment. For example, validation data can be used for determining when to terminate an iterative optimization process or for selecting optimal model *hyperparameters*.[4]

Holdout validation is a common strategy in machine learning when we have a large amount of data. However, sometimes data are more limited, due to expense, ethical considerations, or difficulties with annotation, amongst other reasons. When data are limited, a common approach to training and evaluation is *cross validation*. With cross validation, the training and evaluation processes are performed several times with different splits of the data into training and test sets. The number of different splits used is known as the number of *folds* of the cross validation. For example, a simple 4-fold cross validation is illustrated in Fig. 14.2. We can see that four different splits into training and test are made, and we ensure that each individual appears in the test set once. Four different models are trained using different subsets of the data, and the resulting models evaluated on the "left out" test data. The overall evaluation is based on a combination of the test results of the four folds. A specific type of cross validation is known as *leave-one-out* validation. In this case, the left-out test sets consist of a single individual, and we have as many cross validation folds as there are individuals.

---

[4]In machine learning, a hyperparameter is a parameter whose value is used to control the learning process, but they do not form a part of the final trained model. For example, the learning rate used to update model parameters is a hyperparameter.

Finally, in addition to the test set that was set aside from our original dataset, it is important to also evaluate our trained model on *external* test data. These are data that might come from a different *distribution* to our original data. For example, a model may be tested on medical images acquired on a different make of scanner from the images with which the model was trained. Testing models on such external test sets is important to ensure that the model can generalize well to the potentially diverse types of data that will be encountered after deployment. A good review of the evaluation of machine learning models in medicine can be found in [9].

14.5   Overfitting and underfitting

One design choice that we must make when choosing a machine learning model is the complexity of the model. By complexity, we refer to the number of parameters that must be optimized to train the model. The more parameters there are, the more (potentially) powerful the model is. However, we need to be aware of potential dangers that can come with models with many parameters. Two relevant terms that we should introduce and explain here are *underfitting* and *overfitting*.

We illustrate these concepts using a simple regression problem. Consider Fig. 14.3. These three plots show the same data points with three different polynomial curves fitted to them (see Section 1.11). Fig. 14.3a shows a first-order polynomial (straight line) fitted to the data. It is clear that this model is not powerful enough to capture the variation visible in the data, and we describe this as the model underfitting the data. Fig. 14.3b shows a fifth-order polynomial fitted to the same data. The model is clearly powerful enough to fit to the data because of its extra parameters. But it is unlikely that this model would *generalize* well. For example, if we wanted to test how good its fit was at other x locations, there would probably be significant errors. We refer to this as overfitting the data. Fig. 14.3c shows a third-order polynomial, which, in this case, probably represents a good fit. The curve fits well to the known data, but does not overfit to the noise in the data, and is therefore more likely to generalize well.

Although this is a simple example, the same principle applies when selecting more complex machine learning models. There are many models to choose from, and they vary in their complexity. It is always a good idea to start simple, and only use more complex models if we think it is required for the problem being studied. To help us to make these decisions, it is essential to validate the models we choose on data not used for model optimization (i.e., the validation and test sets referred to in Section 14.4 above).

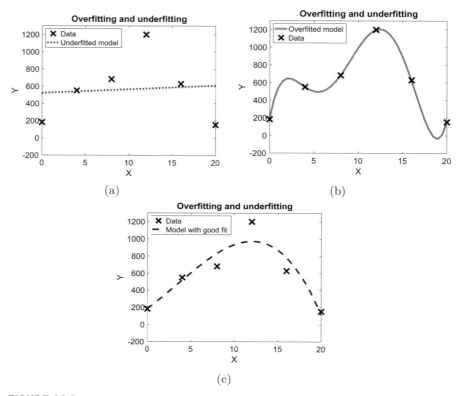

FIGURE 14.3

Illustration of overfitting and underfitting on a simple regression problem. The same data points are shown in all three plots. (a) Shows a first-order polynomial (straight line) fit to the data (underfitting), (b) shows a fifth-order polynomial (overfitting), (c) shows a third-order polynomial (good fit).

14.6 Unsupervised learning

As noted above, machine learning models can be either unsupervised or supervised. Recall that the aim of an unsupervised model is to learn about the underlying structure of the data and the relationships between individual data points.

There are many different unsupervised learning techniques, but a useful way to classify some of the more commonly used ones is into *clustering* techniques and *dimensionality reduction* techniques. We review these two types of approach below.

14.6.1 Clustering

Clustering techniques aim to group individual data points together, according to their similarity. The groups of similar data points are referred to as clusters.

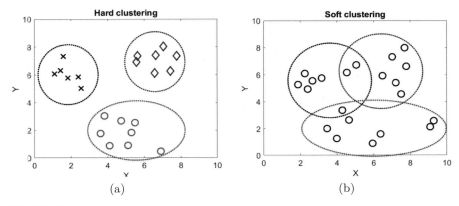

FIGURE 14.4

Hard and soft clustering techniques. (a) Hard clustering, in which each data point is in only one cluster; (b) soft clustering, in which data points can be in multiple clusters.

For example, in Fig. 14.4a, we can see that a cluster analysis of the 2-D data points has grouped them into three distinct clusters, based on the similarity of their coordinates. We can also see that each point is only in one of the clusters. This is known as *hard clustering*. An alternative approach to hard clustering is shown in Fig. 14.4b, in which some points are in more than one cluster. This type of approach is known as *soft clustering*.

There are many different algorithms for both hard and soft clustering, and we do not attempt to review all of them here. We illustrate the use of clustering in MATLAB using one of the most widely applied hard clustering techniques: *k-means clustering*. We refer the reader to Section 14.10 at the end of this chapter for information about alternative algorithms.

■ Example 14.1

We illustrate the application of k-means clustering using a realistic biomedical example from cardiology. The thickness of the left ventricular myocardium (the muscle that provides the pumping action of the heart) is known to provide useful information about the heart's function. Abnormal thickening of the myocardium, either globally throughout the myocardium, or locally in a small region, is often a sign of pathology. Myocardium thickness can be measured from imaging data, such as magnetic resonance (MR) or ultrasound.

014.B

A group of 163 patients have had their myocardium thicknesses measured from imaging data. The measurements were made radially from the center of the left ventricular blood pool in 10 degree steps, resulting in 36 measurements for each subject, each in millimeters. The data are in the file *myothicknesses.mat*, in a 163×36 array of numerical values. We wish to investigate the relationships between the thickness measurements of different subjects, and try to identify clusters of patients with similar measurements.

To begin with, we do not know how many clusters to look for. One way of estimating this is by trial-and-error, but to assess how good a particular number of clusters is, we need ways of quantifying and visualizing clusterings. (In the examples in Fig. 14.4, it was easy to visualize the clusters, because the data were 2-D. But for high dimensional data it is more challenging.) The code shown below illustrates how to apply k-means clustering on the myocardium thickness data (using 3 clusters as an illustration) and visualize the results using a *silhouette plot*.

```
%% Load data
load('myothicknesses.mat');

%% Apply and visualise clustering
% k—means clustering
[clusters centroids sumD] = kmeans(myocardium,3);

% produce silhouette plot
silhouette(myocardium,clusters);
xlabel('Distance to other clusters')
ylabel('Cluster')
title('Silhouette Plot of Myocardium Thickness Clusters');
```

The code uses the MATLAB `kmeans` function, and its two arguments represent the matrix of data points and the number of clusters. In this case, the return argument `clusters` is a 163×1 array containing cluster numbers for each patient. The other two return arguments contain:

- `centroids`: A 3×36 array, with each row representing a cluster centroid;
- `sumD`: A 3×1 array containing the sums of within-cluster distances.

Silhouette plots are shown in Fig. 14.5 for 2, 3, and 4 clusters. A silhouette plot displays a normalized measure of the distances between each data point in a cluster and its neighboring clusters. The distances range between 1 (points that are very far from neighboring clusters) and -1 (points that are probably in the wrong cluster). A distance of 0 indicates that a point is not clearly in any cluster. Therefore a good clustering should have mostly high numbers for each cluster in its silhouette plot. We can see from these plots that for both 2 and 3 clusters, there is good separation between the clusters (i.e., large distances to neighboring clusters), but for 4 clusters, we have some negative values in the fourth cluster. Therefore we conclude that either 2 or 3 clusters is probably a good number.

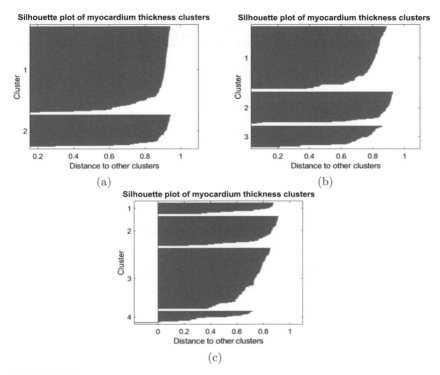

FIGURE 14.5
Silhouette plots of k-means clustering results: (a) 2 clusters; (b) 3 clusters; (c) 4 clusters. The plot for 4 clusters shows some negative distances for cluster 4, indicating poor separation.

We can perform some further investigation by visualizing the average thickness measurements within each cluster. The code below illustrates how to achieve this, and the resulting plots (for 2 and 3 clusters only) are shown in Fig. 14.6.

```
%% Compute and plot average thickness curves for each cluster
curve1 = mean(myocardium(clusters==1,:), 1);
curve2 = mean(myocardium(clusters==2,:), 1);
curve3 = mean(myocardium(clusters==3,:), 1);
figure;
plot(1:36, curve1, '-k');
hold on;
plot(1:36, curve2, ':r');
plot(1:36, curve3, '.-b');
title('Average thickness curves for 3 clusters');
xlabel('Measurement number');
ylabel('Thickness (mm)');
legend('Cluster 1', 'Cluster 2', 'Cluster 3');
```

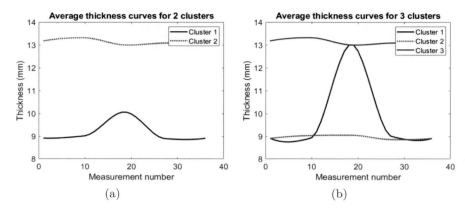

FIGURE 14.6
Average myocardium thicknesses after clustering: (a) 2 clusters; (b) 3 clusters.

> We can see that, using 2 clusters, one cluster has uniformly thicker myocardium, and the other is thinner, but with a local thickening. However, when we extend the analysis to 3 clusters, a third category emerges with uniformly thinner myocardium. This suggests that 3 clusters is an appropriate number for this application. ∎

14.6.2 Dimensionality reduction

The data we used in the previous example were high dimensional (there were 36 measurements for each patient, so the data were 36-dimensional). Visualizing high dimensional data is difficult, so it is desirable to be able to reduce the dimensionality of such data. A family of approaches collectively known as *dimensionality reduction* techniques can be used for this purpose.

The simplest way to reduce the dimensionality of data is to exclude some of the dimensions. For example, in the myocardium thickness example, we could select a small number of thickness measurements, and work only with those. This approach is known as *feature selection*. However, care must be taken when selecting features to ensure that important information is not lost. For example, if we had excluded the measurements at the local thickening of the myocardium shown in Fig. 14.6b, then we would not have been able to distinguish between clusters 1 and 2. Feature selection is most commonly performed manually, using some domain knowledge and after some initial inspection of the data, but automatic methods do exist (see Section 14.10).

An alternative approach to dimensionality reduction is *feature transformation*. Feature transformation techniques apply a transformation to all features of the input data, resulting in a new, lower-dimensional representation of the data. The idea is that the important aspects of the structure of the data are preserved in this new representation, but that it can be visualized or further processed more easily, due to its reduced dimensionality.

A wide range of techniques have been proposed for feature transformation, involving both linear and nonlinear transformations. We will illustrate the concept using one of the most commonly used techniques, *principal components analysis* (PCA). PCA works by determining the linear transformation that results in a new representation, in which the first dimension "explains" or contains as much of the data's variance as possible, then the second dimension explains as much as possible of the remaining variance, i.e., variance that has not been accounted for by the first dimension, and so on.

■ Example 14.2

To demonstrate the use of dimensionality reduction techniques, we return to the myocardium thickness example introduced in Example 14.1. We will apply PCA to these data to reduce their dimensionality using a linear transformation. Consider the code below.

014.C

```
%% Load data
load('myothicknesses.mat');

%% PCA
% input matrix: rows = observations, cols=variables
[coeff,score,latent,tsquared,explained,mu] = pca(myocardium);
% coeff: principal component coefficients
% score: representations of data in principal component space
% latent: variances within dimensions
% tsquared: Hotelling's t squared statistics
% explained: percentage of variation explained by components
% mu: mean of observations

fprintf('PCA variance explained:\n');
fprintf('Mode 1 explains %.1f%%\n', explained(1));
fprintf('Mode 2 explains %.1f%%\n', explained(2));
fprintf('Mode 3 explains %.1f%%\n', explained(3));

% We find that the first two components explain ~98%
% of the variation, so let's plot the data using these
% two dimensions only
plot(score(:,1), score(:,2), 'xk');
xlabel('Component 1');
ylabel('Component 2');
title('First two PCA dimensions of myocardium data');
```

We can see that the MATLAB function `pca` can be used to perform PCA, and that it takes a matrix as input. In this matrix, the rows represent observations (i.e., patients), and the columns represent variables (myocardium thickness measurements). The function returns a number of arguments, but the most

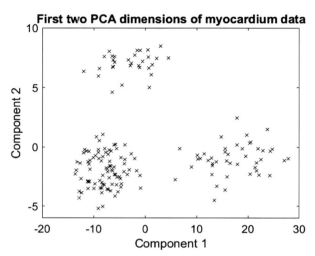

FIGURE 14.7

Plot of the first two principal components of a principal components analysis (PCA) of the myocardium data. Three distinct clusters are clearly visible.

important ones are:

- coeff: The *principal components*, which for our example is a 36 × 36 matrix, in which the columns represent the principal component vectors. We can think of these as vectors in the original representation, which indicate the directions of dimensions in the transformed representation. The first principal component represents the direction in the 36-dimensional space containing the largest variance. Although there are still 36 dimensions in the new representation, we can choose to ignore dimensions that contain a low amount of variance.
- explained: The percentage of total variance explained by each principal component (in our example this is a 36 × 1 array). This can be used to help decide which dimensions to ignore.
- mu: The mean of the observations, i.e., the average myocardium thickness vector over all patients.
- score: The principal component *scores*. These are the original data transformed to the new representation, i.e.,
  ```
  score = (myocardium - mu) * coeff.
  ```

As the code shows, we can investigate how much of the variance is explained by the different components. For our myocardium data, we find that approximately 98% of the variance is explained by the first two components, so we produce a plot of the transformed myocardium data using only these two components. The result is shown in Fig. 14.7, and we can clearly see three distinct clusters in the data, confirming our findings from the cluster analysis. ∎

PCA is probably the most commonly applied feature transformation technique. But the linear nature of its transformation can sometimes be too limiting, and alternatives, many of which employ nonlinear transformations, do exist. Please see Section 14.10 for more information on alternative feature transformation techniques.

From our examples, we can see that dimensionality reduction and clustering are sometimes complementary techniques. Dimensionality reduction and subsequent visualization is often a good first step for investigating high dimensional data, even if the final aim is to perform clustering. It also offers a flexible way to transform the representation of the data for subsequent tasks. For example, dimensionality reduction techniques have been applied in biomedical engineering applications, such as image registration, image fusion, and image denoising.

■ **Activity 14.1**

Radiomics refers to a set of methods for extracting a large number of features (e.g., textural, appearance) from medical images using automated algorithms. The radiomics features can be used for further study of tissue characteristics and for discriminating between patients. *014.B, 014.C*

Radiomics features have been derived from tumors segmented from MR scans acquired from 300 cancer patients. Researchers wish to investigate if the radiomics features can be used to characterize the tumors, with a view to optimizing treatment. The radiomics data are in the file *radiomics.mat*, in which the rows represent patients, and the columns represent features.

Use unsupervised machine learning techniques to analyze the data, and decide if the patients could potentially be divided into groups based on the radiomics features. What future investigations would you recommend? ■

14.7 Supervised learning

Recall that with supervised learning our aim is to predict a label for some new data, based on some features of the input data. We can identify two distinct cases of supervised learning, depending on the nature of the label. If the label is categorical,[5] we train a *classification* model; and if it is numerical, we train a *regression* model. We discuss each of these separately below. In both cases, our task is to predict an output variable from some input variables. When discussing supervised learning, we commonly refer to the input variables as *predictors* and the output variable as a *response*.

---

[5]Categorical data can take one of a limited number of possible values. For example, a gender variable could only take the values "male" or "female." (See Section 6.4.)

14.7.1 Classification

MATLAB features a dedicated app for training and evaluating supervised classification models: the classification learner app. This can be started by clicking on *Classification Learner* in the *Apps* tab. This app can be used to explore data, select or transform features, define validation schemes, train different models, and evaluate their results. The app allows many different types of classification model to be trained and their performances visualized and compared. Once the preferred model has been chosen, it can be exported to the MATLAB workspace for subsequent use.

■ Example 14.3

O14.D

To illustrate the use of the classification learner app, we introduce a new example. A team of researchers is interested in using machine learning to predict the 1-year survival of patients admitted to a hospital accident and emergency department. Data have been gathered from a cohort of 150 patients. From each patient, four variables were recorded at admission time:

- `iss`: injury severity score (ISS): A measure of injury severity ranging between 0 and 75, with high numbers indicating severe injuries (see Activity 1.11).
- `gcs`: Glasgow coma scale (GCS): A measure of the patient's level of consciousness between 3 (unconscious) and 15 (fully conscious).
- `age`: Age in years.
- `sbp`: Systolic blood pressure in mmHg.

The survival or otherwise of each patient one year after admission was also recorded in the variable `survive1` (1 = yes, 0 = no). All data are available in the file *survival.mat*.

The classification learner app requires the predictors to be in a single variable in the workspace. Therefore we first prepare our predictor variable by combining the four variables we have been provided with.

```
% load data from workspace
load('survival.mat');

% combine predictors into a single matrix
predictors = [iss gcs sbp age];
```

Next, in the classification learner app window, we click on *New Session* to begin our analysis. The dialog box shown in Fig. 14.8 will appear. In this dialog, we should select our predictor (the *Data Set Variable*) and response variables. We can also specify a validation scheme (holdout validation, cross validation or none—see Section 14.4). In this example, we choose 5-fold cross validation. Next, we click on *Start Session*.

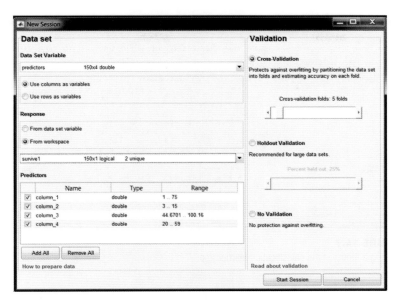

FIGURE 14.8
Defining the predictors, response, and validation scheme in the classification learner app.

In the next dialog box, we are able to explore the data and train different models. In Fig. 14.9 we have used the *Scatter Plot* button in the *Plots* tab to visualize the first two predictor variables (ISS and GCS), with the data points color coded by their response. (Recall that we introduced scatter plots in Section 13.2.2.) We can see that there is a reasonably good separation between patients who survived at least one year and those who did not. We have also trained a *Quadratic SVM* model by selecting this button in the *Model Type* tab and clicking on *Train*. The *History* panel shows that this trained model has an accuracy of 87.3%. In the scatter plot, the points are plotted as circles if they were correctly classified by the model and as crosses if not.

Among the other visualizations available, the *Confusion Matrix* can be particularly informative when assessing model performance. This summarizes the numbers of correctly and incorrectly predicted classifications of all data points. In binary classification cases, such as our example, we refer to the correct predictions as *true positives* and *true negatives* (depending on their label), and the incorrect predictions as false positive and false negatives. A perfect classifier would have non-zero values only on the leading diagonal of the confusion matrix, i.e., there would be only true positives and true negatives.

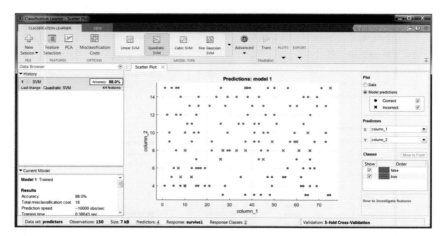

FIGURE 14.9
Training a model in the classification learner app.

Before we export our trained model, we can perform some further experimentation with the predictor variables. Under the *Features* tab we have the option to deselect certain predictor variables and then retrain the model. For example, if we believe that ISS and GCS are most predictive of survival we could try training a model that uses only these variables. We can also apply PCA to reduce the dimensionality of the predictors before using them to train the model. This can be sometimes be useful for very high dimensional data.

Once we are happy with our model, we can export it to the MATLAB workspace for subsequent use. Here, it is useful to distinguish between two types of trained model: the *validated model* and the *full model*. The validated model is the one that has been trained and evaluated using our specified validation scheme. So, if we specified holdout validation with 25% of the data held out, this model will only be trained on 75% of the data. However, at the same time as training the validated model, MATLAB also trains a full model, which is identical to the validated model, except that it is trained using all data. It is important to remember that when we investigate different models in the classification learner app, it is the validated models that we investigate. But when we export a model from the app, it is always the full model that gets exported.

To export a model from the app, we click on *Export Model* under the *Export* tab. When we do this, we must specify a variable name for the exported model, which will be a `struct` type (see Section 6.3).

To use an exported model to make predictions on new data, in the command window we need to access the element of the `struct` that contains the prediction model. For example, the following code makes a 1-year survival prediction for a new patient with the following measurements: ISS = 54, GCS = 10, systolic blood pressure = 90mmHg, age = 40.

```
p = [54 10 90 40];
surv = trainedModel.predictFcn(p)
```

■

■ Activity 14.2

Cardiac resynchronization therapy (CRT) is a common treatment for heart failure, which involves the implantation of a pacemaker. Patients are routinely selected for CRT using a range of metrics of cardiac health. For example, one set of selection criteria is the following:

014.D

- A NYHA class of between 2 and 4. This is an integer categorization of symptom severity defined by the New York Heart Association, ranging between 1 (mild) and 4 (severe).
- A QRS duration of more than 120 ms. The QRS duration is the length of the QRS complex in an ECG signal.
- An ejection fraction of less than 35%. The ejection fraction is the proportion of blood pumped out of the left ventricle in one heart beat.

However, approximately one-third of patients selected in this way do not respond positively to the treatment.

Researchers are interested in using machine learning and a wider range of metrics to improve the selection criteria for CRT. A cohort of 100 patients who underwent CRT have been studied over time, and classified as either responders or non-responders to the treatment. The following data were recorded for each patient at the time of selection for treatment:

- NYHA class
- QRS duration
- Ejection fraction
- Age
- Quality of life questionnaire score. This is a questionnaire that measures the quality of an individual's life across a broad range of areas, including health.
- Six-minute walk distance: This is the distance in meters that the subject can walk in six minutes.

These data are available in the file *CRT.mat* as six separate variables: `age`, `nyha`, `qrs`, `ef`, `qol`, and `sixmwd`. An additional variable `response` indicates whether or not the patients responded to treatment (1 = responder, 0 = non-responder).

Use the classification learner app to train and validate a range of models to predict CRT response from the provided data. Use a 4-fold cross validation to compare their performances. Export the best model, and use it to classify an additional test set of 25 patients available in the file *CRT_test.mat*. What classification accuracy do you achieve on the test set?   ■

14.7.2   Regression

Regression is a type of supervised learning in which the response is a numerical value. For example, predicting a patient's age from a brain scan would require a regression model. We have already seen some simple examples of regression in Section 1.11, in which we fitted polynomial models to observed data. More powerful and flexible machine learning regression models are available through the MATLAB regression learner app.

To start the regression learner app, click on *Regression Learner* in the *Apps* tab. In many ways, this app is similar to the classification learner app. The exceptions all stem from the fact that we must specify a numerical response variable, rather than a categorical one. This entails different types of data visualization, different trainable models, and different ways of measuring performance.

■ Example 14.4

014.E

To illustrate the use of the regression learner app, we return to the Accident and Emergency patient project from Example 14.3. The researchers are now interested in predicting the length of hospital stay (in days) for each patient. The same ISS, GCS, age, and systolic blood pressure data are available in the file *stay.mat*, along with an extra variable, called `stay`, which indicates the hospital stay in days for each of the 150 patients.

In a similar way to the previous classification examples, our first step is to prepare the data by combining all predictors into a single variable in the MATLAB workspace.

```
% load data from workspace
load('stay.mat');

% combine predictors into a single matrix
predictors = [iss gcs sbp age];
```

Next, in the regression learner app dialog box, we click on *New Session*, and select our predictor and response variables. As shown in Fig. 14.10, we do this in the same way as we did in the classification learner app, but we choose the numerical variable `stay` as the response. A validation scheme is chosen in the same way as before; we choose a 5-fold cross validation. We can then click on *Start Session*.

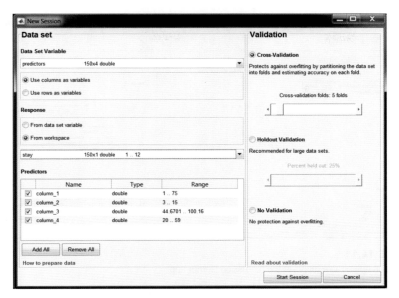

FIGURE 14.10
Defining the predictors, response, and validation scheme in the regression learner app.

Both the classification learner and regression learner apps allow models to be chosen manually, or for a range of models to be trained and compared. In addition, they allow some more complex *optimizable* models to be trained. Optimizable models are those whose operation depends upon certain hyperparameters (see Section 14.4). Both apps allow such models to be automatically trained several times and the values of these hyperparameters optimized. For example, in Fig. 14.11, an optimizable *Gaussian process regression* model has been trained and the *minimum MSE plot* shows the errors of the model for different hyperparameter settings. The best set of hyperparameters is used for training the final model, and its performance is quantified by the *RMSE*, which stands for the root mean square error of the predicted responses compared to the true responses. In our case, we have a RMSE of 1.1056 days.

The *minimum MSE plot* is only available for optimizable models. For all models, the following visualizations are also available:

- *Response Plot*, in which the *x*-axis is used for different data points, and both predicted and actual responses are plotted on the *y*-axis.
- *Predicted versus Actual Plot*, which shows the true response on the *x*-axis and the predicted response on the *y*-axis.
- *Residuals Plot*, which plots the true response on the *x*-axis against the prediction error (or *residual*) on the *y*-axis. This can be used to assess if

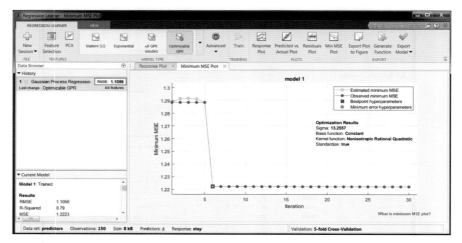

FIGURE 14.11
Training an optimizable model in the regression learner app.

there is any bias in the model's performance, i.e., if the size of the error depends on the size of the value being predicted.

Finally, we can export our chosen model in the same way as we did in the classification learner app by choosing *Export Model* from the *Export* tab.

Now suppose that we wish to predict the hospital stay in days of the new patient from Example 14.3; recall the measurements: ISS = 54, GCS = 10, systolic blood pressure = 90mmHg, age = 40. We can use the following code:

```
new_data = [54 10 90 40];
new_pred = trainedModel.predictFcn(new_data);
fprintf('Expected stay for patient = %f days\n', new_pred);
```

■

■ Activity 14.3

014.E   A doctor's surgery wants to select a list of its patients, who may be at higher risk of health problems, for further tests and screening. To this end, they wish to use machine learning to estimate each patient's "heart age," based upon self-reported data from a questionnaire. To train the model, they have gathered data from 500 patients. The following data are available in the file *heartage.mat*:

- `age`: The subject's age in years.
- `children`: Their number of children.
- `married`: Married (=1) or single (=0).

- `gender`: Female (=1) or male (=0).
- `height`: The subject's height in meters.
- `weight`: The subject's weight in kilograms.
- `lifestyle`: A measure of how active their lifestyle is (1 = sedentary, 2 = medium, 3 = active).
- `income`: Household income.
- `heartage`: The actual heart age of the subject, based upon a physical examination and ECG recording.

Use the regression learner app to train and validate a model to predict heart age from the other variables. Use holdout validation with 25% of the patients set aside for validation. First, use all predictor variables, and experiment with different models to produce the model with the lowest RMSE. Is there any bias in the model? Next, use feature selection to train a model using only age as a predictor; does it have a significantly higher RMSE than the model which uses all predictors? Which other variable(s) can you add as predictors (i.e., in addition to age) to improve the model's performance? ∎

14.8 Deep learning

As described in Section 14.2, deep learning is a type of machine learning, in which the model consists of interconnected layers of artificial "neurons." Deep learning models are potentially very powerful and have achieved impressive results in a range of applications in recent years. In this section, we provide an introduction to the training and evaluation of a deep learning model in MATLAB, using a realistic biomedical example.

∎ Example 14.5

In cardiac MR imaging, it is common to acquire a *short axis* stack of dynamic images showing the movement of the heart as it beats. Short axis images are acquired at a series of planes perpendicular to the long axis of the left ventricle. Planes acquired at the top of the left ventricle are known as *basal* slices, those at the bottom are *apical* slices, and those in the middle are *mid* slices. Since the precise coverage of clinical short axis acquisitions can vary, it is useful to be able to classify such images as basal, mid, or apical.

014.F

In this example, we will train a deep learning classifier to identify short axis images as either apical or mid slices. We will use an existing classification architecture, the VGG-16 model [10], which has been pre-trained on a computer vision classification task. We will then retrain the model for our task. In machine learning, this process of updating a model trained for a different (but related) task is known as *transfer learning*.

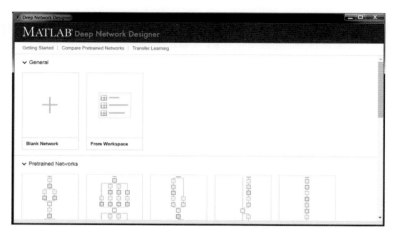

FIGURE 14.12
Choosing a model in the deep network designer app.

Although deep learning functionality can be used by calling functions in a script or function, the easiest way to get started with deep learning in MAT-LAB is to use the deep network designer app. When we start the app, we are presented with the dialog box shown in Fig. 14.12. We can either choose to design our own model or we can use one of the pre-defined models. We will scroll down and select the VGG-16 model. (The first time you do this, you will be asked to install the add-on for the model. Once you have done this once, you do not need to do it again.)

The resulting dialog box (Fig. 14.13) allows us to alter the architecture of the pre-defined model. In this dialog, the center pane shows the model, and we can zoom in to view different layers of the network by holding *Ctrl* and moving the mouse wheel. The original VGG-16 model was trained on an image classification task with 1000 classes. Therefore we need replace the third-last and last layers of the model to reflect the fact that we only have 2 classes in our problem (i.e., apical and mid slices). The third-last block is a fully connected layer: we can drag a new *fullyConnectedLayer* into the center pane of the dialog box from the left pane, under the subheading *Convolution and Fully Connected*. We click on the new layer and set the *OutputSize* to 2 to match our number of classes. The last layer is a classification layer, so in the same way we can drag a *classificationLayer* (under the *Output* subheading) into the center pane. Next, we delete the original third-last and last layers by clicking on them and pressing the *Delete* key, then drag connections to join our new layers into the model. Fig. 14.13 shows what the model should look like now.

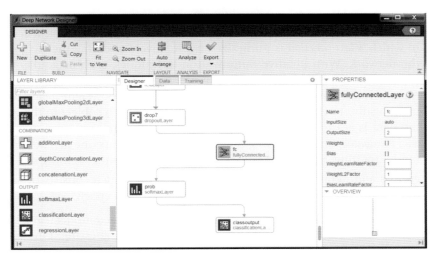

FIGURE 14.13
Altering a model in the deep network designer app.

The next stage is to import our data. The data we use come from the publicly available "ACDC" dataset [1]. Apical and mid short axis slices from 100 subjects have been manually cropped and organized into two folders, one for each class. The names of these folders will be used for the class names. These data can be downloaded from the book's web site. To import the data, we click on the *Data* tab in the center pane and choose *Import Data*. In the next dialog box, we select the folder *above* the two folders containing the image data. We can also select the proportion of the data that will be held out as test data; we keep the default value of 30%. Since our training data consists of 100 subjects, we have 70 subjects (two images each, one apical, and one basal) for training and 30 subjects for testing. To import all data, we click on *Import*. (Note that the VGG-16 model was designed to work with color images, and so the inputs to the model are expected to have 3 color channels. Cardiac MR data are gray scale images, and so could not normally be used as input to the VGG-16 model. We have avoided this problem by duplicating the gray scale data in the three channels.)

We can now check to see if the app identifies any problems with our model and/or data. Click on the *Designer* tab, and then click on *Analyze*. If we get no errors it is OK to proceed, so we can close the Analyzer window and return to the main dialog.

Now we click on the *Training* tab. We can either train with default training parameters or modify these parameters by clicking on *Training Options*. Here it will be useful to give a brief description of how the training of deep learning models works and what these parameters mean. The most common algorithm for training deep learning models is some form of *gradient descent*, in which a *loss function* is minimized by iteratively updating the weights of the network. With iterative optimization techniques, such as gradient descent, training a machine learning model consists of a number of *epochs*. An epoch refers to a number of updates of the network weights, each based upon a subset of the total training data, such that each subject in the training data is seen once. Each of these subsets of the training data is known as a *minibatch*, and processing one minibatch represents one *iteration* of the model training. So, if the minibatch consists of the entire training set, then epochs and iterations are the same thing. If a minibatch consists of $\frac{1}{n}^{\text{th}}$ of the training set, then there will be n iterations in each epoch.

As training proceeds, it is common to validate the learned model using some test data, so that we can visualize how training is progressing. One of the parameters that we can alter (*ValidationFrequency*) refers to how often this validation occurs, i.e., how many iterations between each validation? Finally, the *InitialLearningRate* is a parameter that specifies how much the network weights will be changed in each learning update. If this is too high, learning may not be effective at all. If it is too low, learning may be too slow.

For our example, we alter the parameters slightly as shown in Fig. 14.14 and summarized below:

- *Optimizer* = SGDM
- *InitialLearnRate* = 0.0001
- *ValidationFrequency* = 1
- *MaxEpochs* = 10
- *MiniBatchSize* = 140

All other parameters have default values. Note that "SGDM" stands for stochastic gradient descent with momentum, which is a common gradient descent algorithm. Because we have a training set of 140 images (70 of each class) and our minibatch size is 140, we process all training data in each iteration, so the number of epochs we specify will be the number of iterations.

To start the training of our model, we can click on *Train*. Depending on the speed of the computer being used, this may take some time, as training deep learning models can be computationally expensive. But the progress of the training will be displayed as it takes place, by plotting the training loss and the classification accuracy on the test set. Fig. 14.15 shows the completed training for our model, for which the final classification accuracy on the held out test set was 91.67%.

The deep network designer app also allows us to export our trained model by clicking on *Export*, then *Export Trained Network and Results*. The following

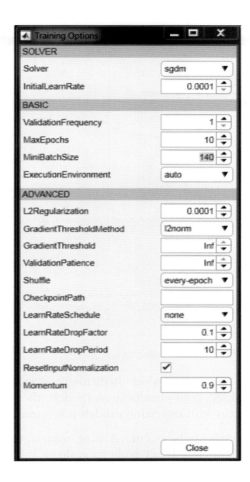

FIGURE 14.14
Altering training parameters in the deep network designer app.

code illustrates how to use the exported model to classify an image:

```
im = imread('apical/patient001.jpg');
im=imresize(im,[224 224]);
[pred,p] = classify(trainedNetwork_1,im);
fprintf('Image is %s with probability %.2f%%\n', ...
        string(pred), max(p)*100);
```

14.9 Summary

Machine learning techniques make use of features of data which are often high dimensional. Traditional machine learning techniques use hand-crafted features, which are defined by the algorithm developer, whereas deep learning techniques learn features for a specific task.

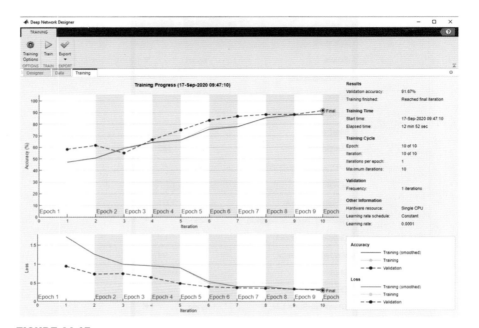

FIGURE 14.15
Visualizing model training in the deep network designer app.

Supervised machine learning involves predicting a response variable, given some predictor variables. With classification models, the response is a categorical variable, whereas with regression models it is numerical.

Machine learning models can be evaluated using holdout validation, in which a subset of the data available is set aside to evaluate the model trained on the rest of the data. An alternative to holdout validation is cross validation, in which multiple splits into training and test data are made, such that each individual is part of the test set once.

Overfitting of a machine learning model refers to using a model that is too complex for the provided training data, so that it ends up fitting to noise in the data and generalizing poorly to new data. Underfitting occurs when the model is not complex enough to capture the underlying structure of the data.

Unsupervised learning includes both clustering and dimensionality reduction techniques. k-means clustering is a common clustering algorithm and can be run in MATLAB using the `kmeans` function. PCA is a common dimensionality reduction technique, which is implemented by the `pca` function in MATLAB.

The classification learner and regression learner apps can be used to train and evaluate supervised classification and regression models, respectively. Each allows multiple models to be trained and their performances visualized and quantitatively compared. Trained models can be exported to the MATLAB workspace for deployment.

A quick way to get started with deep learning in MATLAB is by using the deep network designer app. This allows pre-defined models to be retrained and evaluated using new data.

14.10 Further resources

- MATLAB documentation:
  - Dimensionality Reduction and Feature Extraction/Selection: https://uk.mathworks.com/help/stats/dimensionality-reduction.html.
  - Cluster Analysis: https://uk.mathworks.com/help/stats/choose-cluster-analysis-method.html.
  - Classification Learner App: https://uk.mathworks.com/help/stats/classificationlearner-app.html.
  - Regression Learner App: https://uk.mathworks.com/help/stats/train-regression-models-in-regression-learner-app.html.
  - Deep Network Designer App: https://uk.mathworks.com/help/deep-learning/ref/deepnetworkdesigner-app.html.
- The dimensionality reduction toolbox, developed by Laurens van der Maaten, contains implementations of a wide range of techniques for dimensionality estimation and reduction: https://lvdmaaten.github.io/drtoolbox/.

14.11 Exercises

■ Exercise 14.1

Blood tests can be analyzed to produce a wide range of measurements of blood chemistry. Researchers investigating a newly discovered virus would like to investigate if carriers of the virus have significantly different levels of blood chemistry measurements to non-carriers. They have analyzed blood tests from 500 volunteers. It is not known how many of the volunteers were carriers of the virus, but it is believed that a number of them were. The blood test data are available in the file *blood.mat*, in which rows represent volunteers and columns represent blood chemistry readings.

O14.B, O14.C

Use unsupervised machine learning techniques in MATLAB to analyze these data, and decide if you think the volunteers can be divided into two groups (carriers and non-carriers), based upon their blood test results.   ■

■ Exercise 14.2

Continuing with the blood chemistry case study from Exercise 14.1, the researchers now have potential diagnoses of the virus, based upon reported

O14.D

symptoms. The original blood chemistry data and the diagnoses are contained in the file *blood_diagnoses.mat*. The `diagnosis` variable has a value of 1 for positive diagnoses and 0 otherwise.

Use a supervised machine learning technique in MATLAB to train and validate a model to diagnose the virus from blood test results. ■

■ Exercise 14.3

014.D

For this and the next exercise, we return to the radiomics case study that we introduced in Activity 14.1. The researchers now have classifications of the tumors as benign or malignant, based on biopsies. The file *radiomics_class.mat* contains the original radiomics data and a new variable called `tumour_type`, which is equal to 0 for benign tumors and 1 for malignant tumors.

Use the MATLAB classification learner app to train and validate a classifier to predict whether a tumor is benign or malignant from the radiomics data. ■

■ Exercise 14.4

014.E

The researchers now wish to determine the *grade* of the tumor from the radiomics data. The file *radiomics_grade.mat* contains the original radiomics data and a new variable called `tumour_grade`. The tumor grades are integers between 1 and 4 indicating tumor development, with 4 being the most advanced stage.

Use the MATLAB regression learner app to train and validate a regression model to predict tumor grade from the radiomics data. ■

■ Exercise 14.5

014.F

Download the file *ACDC_cropped.zip*, which contains the image data used in Example 14.5. Work through the example, and try to reproduce similar training results to those shown in Fig. 14.15. Next, try the following different settings:

1. Try adding data augmentation to the training scheme. Data augmentation involves adding new training data, based on applying transformations to the original data, and in some cases can improve model generalizability. In the deep network designer app you can specify data augmentation settings when you import the data. Experiment with different transformations, and see if they improve the final test accuracy.

2. Try reducing the minibatch size to 14 instead of 140. This will use $\frac{1}{10}$th of the training data in each iteration, so there will be 10 iterations in an epoch.

3. Try using a different optimizer such as the *Adam* optimizer.
   (Hint: You will probably need to alter the InitialLearningRate parameter to get the Adam optimizer to work.)

Note that by using the 30% of held out data to tune parameters in this way, these data can no longer be considered independent test data, since the chosen model will be biased toward these data. They are now considered *validation* data in the terminology introduced in Section 14.4.   ∎

FAMOUS COMPUTER PROGRAMMERS: LARRY PAGE AND SERGEY BRIN

For this chapter, we have two famous computer programmers: Larry Page and Sergey Brin. Page and Brin are the co-founders of Google. Through founding Google and developing much of the core software upon which it is based, these two American computer scientists have revolutionized the way in which we use the internet.

Initially Google was solely an internet search engine, although it has now come to include a wide range of other functionality, such as email and social networking. The Google search engine started off as a research project in 1996 whilst Page and Brin were both PhD students at Stanford University. Its main innovation was a technology known as PageRank, which used much more sophisticated ways of determining a web page's relevance to a search term. Most search engines at that time used relatively simplistic algorithms that just counted how many times a search term appeared in the page. PageRank also incorporated information about the network of hyperlinks between relevant pages.

Page and Brin did most of the programming for PageRank in their dormitory room, which was crammed with computers. The project soon began to place a big load on Stanford's computing infrastructure, but by this time Page and Brin had decided to end their studies and focus solely on the search engine. They scraped together funds from friends and family, bought some servers and rented a garage to host them. Shortly after this they founded the Google company and gained further investment. Today Google has an estimated value of US$280 billion.

"Solving big problems is easier than solving little problems."

Sergey Brin

Engineering mathematics

15.1 Introduction

This chapter covers some of the basic mathematics that is needed for engineering applications. The focus will be on a small handful of topics: scalars and vectors, complex numbers, matrices, and calculus. Simple examples are given to illustrate the MATLAB functions that are relevant for each topic. Additional areas for further reading and investigation are given at the end of the chapter.

15.2 Scalars and vectors

Mathematically, we make a distinction between scalars and vectors. Informally, scalars can be specified with a single number, for example, a temperature t, whereas vectors have a magnitude and an associated direction, for example, a measurement of velocity $\mathbf{v}$. In general, a vector needs an ordered list of numbers to describe it, and these are the vector's *components*.

■ Example 15.1

The basic numeric types in MATLAB are stored as arrays. A scalar can simply be represented as a 1×1 array.

O15.A

```
>> t = 32;
>> whos
  Name      Size        Bytes  Class     Attributes

  t         1x1             8  double
```

■

MATLAB® Programming for Biomedical Engineers and Scientists. https://doi.org/10.1016/B978-0-32-385773-4.00024-1

Dimension

A vector **v** has d dimensions, with real-valued components, and this can be written $\mathbf{v} \in \mathbb{R}^d$. In MATLAB, the vector can be represented using an array that is either $1 \times d$ or $d \times 1$ to store its components.

■ Example 15.2

015.A   A two-dimensional (2-D) vector expressed as a 1×2 array:

```
>> v = [-3, 4];
>> whos
  Name        Size              Bytes  Class     Attributes

  v           1x2                  16  double
```

■

Magnitude

As described earlier, a vector has a magnitude, and a common way to define the magnitude is to use the Euclidean norm:

$$|\mathbf{v}| = \sqrt{\sum_{k=1}^{d} v_k^2} \text{ where } \mathbf{v} = (v_1, v_2, \ldots, v_d).$$

The MATLAB `norm` function can be used to compute vector magnitude as shown below.

■ Example 15.3

015.A   Calculate the magnitude of a 2-D vector:

```
>> v = [-3, 4];
>> norm(v)
ans =
     5
```

■

If we consider **v** to be a $1 \times d$ array, then $\sum v_k^2$ can be obtained by carrying out the *matrix* multiplication $\mathbf{v}\mathbf{v}^T$, where $\mathbf{v}^T$ indicates the transpose of **v**. This means that the magnitude can be obtained using the MATLAB $*$ operator as follows:

```
>> sqrt(v * v')
ans =
     5
```

See Section 15.4.2 for more details on matrix multiplication.

Direction in 2-D

In two dimensions, the direction of a vector can be given by the angle between the vector and the x-axis (measured anti-clockwise). This can be obtained using the `atan2` function, which takes two arguments representing the y and x coordinates and returns the angle measured in radians (between $-\pi$ and π).

■ Example 15.4

Obtain the direction of a 2-D vector in radians, then convert it to degrees: *015.A*

```
>> v = [-3, 4];
>> dir_v_rad = atan2(v(2), v(1)); % dir_v_rad = 2.2143
>> dir_v_deg = dir_v_rad * 180 / pi; % dir_v_deg = 126.8699
```
■

■ Activity 15.1

Use MATLAB to find the magnitude of the vector $\mathbf{w} = (5, -1.5)^T$ and the   *015.A*
angle that it makes with the *positive y* axis.   ■

Adding and subtracting vectors

For a pair of vectors $\mathbf{u}, \mathbf{v} \in \mathbb{R}^d$, addition and subtraction are carried out component-wise, and the standard MATLAB addition and subtraction operators work as we would expect.

■ Example 15.5

Adding and subtracting vectors: *015.A*

```
>> u = [1, 2, 3]; v = [-1, 1, 4];
>> u + v; % ans =    0    3    7
>> u - v; % ans =    2    1    -1
```
■

Unit vectors

Unit vectors are vectors with magnitude equal to one. For a vector $\mathbf{v}$, we can define a unit vector $\hat{\mathbf{v}}$ by $\hat{\mathbf{v}} = \mathbf{v}/|\mathbf{v}|$, as long as the vector has non-zero magnitude, and, in this case, $\hat{\mathbf{v}}$ and $\mathbf{v}$ will both have the same direction.

■ Example 15.6

Normalize a vector to obtain a unit vector in the same direction: *015.A*

```
>> v = [-3, 4];
>> unit_v = v / norm(v)
unit_v =
    -0.6000    0.8000
```
■

The unit vectors along each of the main x, y, and z axes in three dimensions are often written as $\hat{\mathbf{i}}$, $\hat{\mathbf{j}}$, and $\hat{\mathbf{k}}$, i.e., $\hat{\mathbf{i}} = (1, 0, 0)$, $\hat{\mathbf{j}} = (0, 1, 0)$, and $\hat{\mathbf{k}} = (0, 0, 1)$.

Scalar (dot) product

For a pair of vectors, $\mathbf{u}, \mathbf{v} \in \mathbb{R}^d$, the dot (or scalar) product is the sum of the products of their components:

$$\mathbf{u} \cdot \mathbf{v} = \sum_{k=1}^{d} u_k \, v_k,$$

which can be carried out in MATLAB using the `dot` function.

■ Example 15.7

015.A

Use the `dot` function to evaluate the scalar product of two three-dimensional (3-D) vectors:

```
>> u = [1, 2, 3]; v = [-1, 1, 4];
>> dot(u, v)
ans =
     13
```

■

Viewing $\mathbf{u}$ and $\mathbf{v}$ as matrices, the dot product can be written as a matrix multiplication $\mathbf{u} \cdot \mathbf{v} = \mathbf{u}\mathbf{v}^T$, so, in MATLAB, we can use the matrix multiplication operator $*$ to achieve the same result:

```
>> u * v'
ans =     13
```

See Section 15.4.2 for more details on matrix multiplication.

Angle between vectors

For a pair of vectors, $\mathbf{u}$ and $\mathbf{v}$, we can find the cosine of the angle between them by dividing the dot product by the norms of each vector. Written formally, this is

$$\cos\theta = \frac{\mathbf{u} \cdot \mathbf{v}}{|\mathbf{u}|\,|\mathbf{v}|}.$$

■ Example 15.8

015.A

Find the angle between the vector $\mathbf{u} = (1, 2, 3)^T$ and the vector $\mathbf{v} = (-1, 1, 4)^T$.

We can calculate this as follows:

```
>> u = [1, 2, 3]; v = [-1, 1, 4];
>> c = dot(u, v) / ( norm(u) * norm(v) );
>> theta = acos(c) * 180 / pi; % theta =    35.0229
```

■

Direction in 3-D

For a 3-D vector, it is possible to define its direction based on the angles it makes with each of the axis vectors $\hat{\mathbf{i}}$, $\hat{\mathbf{j}}$, and $\hat{\mathbf{k}}$. Again, we can use the dot product to find these angles.

■ **Example 15.9**

Find the angle between the vector $\mathbf{u} = (1, 2, 3)^T$ and each of the x, y, and z coordinate axes.

015.A

First, we define $\mathbf{u}$ and the axes:

```
>> u = [1, 2, 3];
>> norm_u = norm(u);
>> ax_1 = [1,0,0]; ax_2 = [0,1,0]; ax_3 = [0,0,1];
```

Next, we calculate the angles in radians:

```
>> c1 = dot(u, ax_1) / norm_u;
>> c2 = dot(u, ax_2) / norm_u;
>> c3 = dot(u, ax_3) / norm_u;
```

Finally, we convert to degrees and display:

```
>> ang_1 = acos(c1)*180/pi;
>> ang_2 = acos(c2)*180/pi;
>> ang_3 = acos(c3)*180/pi;
>> disp([ang_1, ang_2, ang_3])
   74.4986   57.6885   36.6992
```

■

■ **Activity 15.2**

Find the angles between the vector $w = (-1, 0, 7)^T$ and each of the coordinate axes. Note that the vector is contained in the $x - z$ plane, so you should know what value to expect for one of these angles. ■

015.A

Cross product

A pair of 3-D vectors, $\mathbf{u}, \mathbf{v} \in \mathbb{R}^3$, can define a third vector that is perpendicular to each vector in the pair, i.e., a vector that is at right angles to the plane that contains them. This vector can be found by evaluating the cross product of $\mathbf{u}$ and $\mathbf{v}$:

$$\mathbf{u} \times \mathbf{v} = (u_2 v_3 - u_3 v_2)\hat{\mathbf{i}} + (u_3 v_1 - u_1 v_3)\hat{\mathbf{j}} + (u_1 v_2 - u_2 v_1)\hat{\mathbf{k}},$$

and the magnitude of the cross product gives the area of the parallelogram defined by the pair of vectors.

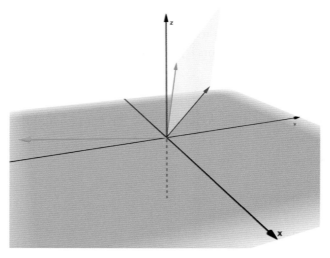

FIGURE 15.1
A visualization of the cross product of vectors **u** (red) and **v** (green) to give **w** (cyan). The area of the parallelogram is equal to the length of **w**.

■ Example 15.10

015.A

Find the area of the parallelogram defined by the vectors $\mathbf{u} = (1, 2, 3)^T$ and $\mathbf{v} = (-1, 1, 4)^T$:

```
>> u = [1, 2, 3]; v = [−1, 1, 4];
>> w = cross(u, v) % perp. to u and v
w =
     5     −7     3

>> norm(w) % area of parallelogram
ans =
     9.1104
```

An illustration of the vectors above is given in Fig. 15.1.   ■

15.3   Complex numbers

In some settings, it is necessary to use complex numbers to represent values, for example, when analyzing radio-frequency data. In this case, MATLAB is equipped to carry out operations on such data.

A complex number consists of the sum of a *real* and an *imaginary* part, where the imaginary part is a multiple of the square root of -1, which, in mathematics, we normally denote as $i = \sqrt{-1}$. The following can all be considered as complex numbers:

$2 + 3i$ 3.14 $\sqrt{-16}$

The first number above is clearly a complex number; it is the sum of a *real* part, 2, and an *imaginary* part, $4i$. The second and third numbers in the list can be made to look more like a complex number by making their real and imaginary parts explicit:

$$3.14 = 3.14 + 0i \qquad \sqrt{-16} = 0 + 4i$$

The number 3.14 can be considered a *pure real* number, but it should be noted that the real numbers are a subset of the complex numbers.

> As described above, the mathematical convention is to use the letter i to denote $\sqrt{-1}$. However, the *engineering* convention is to write $\sqrt{-1} = j$. MATLAB allows for both conventions.

> **Tip:** It is common for `i` and `j` to be used as loop variables, but, because they can be used to represent the square root of -1 in complex and imaginary numbers, it is good advice to *avoid* using these letters as loop index variables when complex numbers are being used.

Such examples can be entered directly at the command window to explicitly show them as complex numbers in MATLAB.

■ Example 15.11

Define some complex numbers: *015.B*

```
>> 2 + 3i
ans =
   2.0000 + 3.0000i

>> 3.14 + 0i
ans =
    3.1400

>> sqrt(-16)
ans =
   0.0000 + 4.0000i
```

When displaying the answer for the second number above, MATLAB does not show the imaginary part (which is zero). ■

Basic arithmetic operations are available for complex numbers. For example, addition and subtraction are straightforward.

■ Example 15.12

015.B

Calculate $z_1 + z_2$ and $z_1 - z_2$ for $z_1 = 2 + 3i$ and $z_2 = 5 - 2i$:

```
>> z1 = 2 + 3i;
>> z2 = 5 - 2i;
>> z1 + z2
ans =
   7.0000 + 1.0000i

>> z1 - z2
ans =
  -3.0000 + 5.0000i
```

■

We can also multiply two complex numbers as follows:

$$
(a + bi)(c + di) = a \cdot c + bi \cdot c + a \cdot di + bi \cdot di
$$
$$
= ac + bci + adi + bdi^2
$$
$$
= (ac - bd) + (bc + ad)i
$$

■ Example 15.13

015.B

Find the product of $2 + 3i$ and $-1 + 2i$.

First, we work out the answer mathematically:

$$
(2 + 3i)(-1 + 2i) = 2 \cdot (-1) + 3i \cdot (-1) + 2 \cdot 2i + 3i \cdot 2i
$$
$$
= -2 - 3i + 4i + 6i^2
$$
$$
= -8 + i
$$

Compare this answer against the value obtained at the MATLAB command window:

```
>> (2+3i) * (-1+2i)
ans =
  -8.0000 + 1.0000i
```

■

It is important to note the use of brackets in the above example. If they are not used, then this can lead to the incorrect result as shown below:

```
>> 2+3i * -1+2i % Not using brackets ...
ans =
   2.0000 - 1.0000i % Not what we expect!
```

The reason for the difference in the results is due to the precedence of operators in MATLAB (see Table 2.2). Because multiplication (∗) takes precedence over addition (+), the above expression is evaluated mathematically as $2 + (3i \cdot (-1)) + 2i = 2 - 3i + 2i = 2 - i$. This shows that some care should be taken when carrying out arithmetic on complex numbers.

More advanced functions, such as trigonometric and exponential functions, also allow complex numbers as arguments. We can, for example, demonstrate Euler's formula for exponentials of pure imaginary values ix (where x is real-valued):

$$e^{ix} = \cos x + i \sin x.$$

■ Example 15.14

Choosing $x = \pi/3$, for example, we can carry out the following to demonstrate Euler's formula:

015.B

```
>> x = pi / 3;
>> exp(1i * x)
ans =
    0.5000 + 0.8660i

>> [cos(x), sin(x)]
ans =
    0.5000    0.8660
```

■

Further illustrations of the use of complex values in polynomials, trigonometric functions, and exponentials are shown below.

■ Example 15.15

Polynomials, trigonometric functions, and exponentials with complex numbers:

015.B

```
>> z = 1i
z =
    0.0000 + 1.0000i

>> 1 + z + z^2 + z^3
ans =
    0

>> sin(i * pi)
ans =
    0.0000 +11.5487i

>> exp(1 + 2i)
ans =
    -1.1312 + 2.4717i
```

■

We can confirm the last example mathematically using the basic rules of complex exponentials:

$$e^{a+bi} = e^a \, e^{bi} = e^a \, (\cos b + i \sin b).$$

Substituting $1 + 2i$ for $a + bi$,

$$e^{1+2i} = e^1 (\cos 2 + i \sin 2) = e \cos 2 + i \, e \sin 2.$$

■ Example 15.16

015.B

We can confirm the calculated result above by evaluating the real and imaginary parts separately:

```
>> exp(1 + 2i)
ans =
  -1.1312 + 2.4717i

>> e = exp(1); % e =     2.7183
>> [e * cos(2), e * sin(2)]
ans =
  -1.1312    2.4717
```

■

The built-in functions `real` and `imag` can be used to return the real and imaginary parts of a complex number.

■ Example 15.17

015.B

Find the real and imaginary parts of $z = e^{1+2i}$.

```
>> z = exp(1 + 2i);
>> real(z)
ans =
  -1.1312

>> imag(z)
ans =
   2.4717
```

■

■ Activity 15.3

015.B

Define $z_1 = e^{\pi/6}$ and $z_2 = 3 + 4i$. Find the real and imaginary parts of the sum and difference of z_1 and z_2.

Change z_2 so that $z_2 = e^{-\pi/6}$, and repeat the above. What do you notice? ■

Complex numbers written as $a + bi$ are said to be written in *Cartesian form* and can be represented in the complex plane by a point at (a, b), where the horizontal axis represents the real part, and the vertical axis represents the imaginary part (see Fig. 15.2). The distance of a point $z = a + ib$ in the complex plane from the origin is known as the *modulus* of z. The angle between the positive real axis and the line from the origin to z is known as the *argument*. A complex number described by its modulus and argument is said to be in *polar* form.

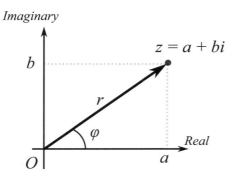

FIGURE 15.2
The complex plane showing the number z in Cartesian form $a + bi$ and in polar form $re^{i\phi}$.

The modulus r of $z = a + bi$ is given by Pythagoras's theorem: $r = \sqrt{a^2 + b^2}$, and the argument ϕ can be obtained using trigonometry rules: $\phi = \arctan(b/a)$. Conversely, the real and imaginary parts of z can be obtained from r and ϕ using the trigonometric rules: $a = r \cos \phi$ and $b = r \sin \phi$.

As stated above, the number z can be written in Cartesian form as $a + bi$ and in polar form $z = re^{i\phi}$. These can be shown to be equivalent, again using Euler's formula:

$$re^{i\phi} = r\,(\cos \phi + i \sin \phi) = r \cos \phi + ir \sin \phi = a + ib.$$

In MATLAB, the functions `abs` and `angle` can be used to find the modulus and argument of a complex number.

■ Example 15.18

Find the modulus and argument of $z = e^{1+2i}$.   *O15.B*

```
>> z = exp(1 + 2j);
>> abs(z)
ans =
    2.7183

>> angle(z)
ans =
    2
```

We can confirm the relationship between the Cartesian and polar forms in MATLAB. In the following, we derive the latter from the former:

■ Example 15.19

015.B

Show how the real and imaginary parts of $z = e^{1+2i}$ can be used to calculate its modulus and argument:

```
>> a = real(z)
a =
   -1.1312

>> b = imag(z)
b =
    2.4717

>> sqrt(a*a + b*b)
ans =
    2.7183

>> atan2(b, a)
ans =
    2
```

■

In the following, we go the other way and derive the Cartesian form from the polar form:

■ Example 15.20

015.B

Show how the modulus and argument of $z = e^{1+2i}$ can be used to calculate its real and imaginary parts:

```
>> r = exp(1); % r =    2.7183
>> phi = 2;
```

Define $z = re^{i\phi}$:

```
>> z = r * exp( i * phi ); % z =   -1.1312 + 2.4717i
>> a = real(z); % a =   -1.1312
>> b = imag(z); % b =    2.4717
```

■

Writing complex numbers in polar form can help with multiplication and division. For example, consider $z_1 = re^{i\phi}$ and $z_2 = se^{i\theta}$. Their product and quotient can be expressed as follows:

$$z_1 z_2 = re^{i\phi} se^{i\theta} = rs\, e^{i(\phi+\theta)}$$

$$\frac{z_1}{z_2} = \frac{re^{i\phi}}{se^{i\theta}} = \frac{r}{s}e^{i(\phi-\theta)}$$

In other words, the modulus of the product is the product of the moduli, with an equivalent statement holding for the quotients. Also, the argument of the product is the sum of the individual arguments ($\phi + \theta$), whereas the argument of the quotient is their difference.

Polar form is also useful for representing powers of complex numbers. Consider $z = re^{i\phi}$, then we have

$$z^n = \left(re^{i\phi}\right)^n = r^n\, e^{in\phi}, \tag{15.1}$$

which shows that the modulus of z^n is the original modulus raised to the same power, whereas its argument is n times the original argument.

■ Activity 15.4

For $z_1 = 3 + 2i$ and $z_2 = 1 - 3i$, confirm that the product of the moduli is the same as the modulus of the product. Then confirm that the argument of the sum equals the sum of the arguments.   ■

015.B

15.3.1   Time-varying signals

Time-varying signals are often encountered in a range of biomedical engineering applications, and complex arithmetic can be helpful in this context. One example is given by alternating current (AC) electrical signals, where, say, the voltage V in a conductor varies according to a function of time:

$$V(t) = 10\cos\pi t.$$

This AC voltage has an amplitude of $10V$, i.e., it varies between 10 and -10 sinusoidally. It has a wavelength of 2 seconds (because πt varies between 0 and 2π, as t varies between 0 and 2). The value that is observed, in this case, the voltage, can be viewed as the real part of a complex number:

$$10\cos\pi t + 10i\sin\pi t.$$

As we have seen above, this can be represented by a complex exponential:

$$10\cos\pi t + 10i\sin\pi t = 10\left(\cos\pi t + \sin\pi t\right) = 10e^{\pi it}.$$

The signal $V(t)$ starts at its maximum value at $t = 0$, and we can define its *phase* to be zero. An AC signal might be shifted along the time axis, say, $V_2(t) = 10\cos\pi(t - 0.5)$, i.e., with a phase of -0.5 s. Again, we can view this signal as the real part of a complex number that can be written as $10e^{\pi i(t-0.5)}$.

When processing such signals, it is possible to work with these complex representations and extract the real part at the end to derive our final answer. For instance, if we want to find the sum of the two AC signals, V and V_2, we can add the complex forms:

$$10e^{\pi it} + 10e^{\pi i(t-0.5)}.$$

Using the various rules of complex arithmetic relating to arguments, modulus, etc., we can simplify the above expression.

$$
\begin{aligned}
10e^{\pi it} + 10e^{\pi i(t-0.5)} &= 10\left(e^{\pi it} + e^{\pi it}e^{-0.5\pi}\right)\\
&= 10e^{\pi it}\left(1 + e^{-0.5\pi}\right)\\
&= 10e^{\pi it}\left(1 - i\right)\\
&= 10e^{\pi it}\sqrt{2}\left(\frac{1}{\sqrt{2}} - i\frac{1}{\sqrt{2}}\right)\\
&= 10e^{\pi it}\sqrt{2}\left(\cos\left(-\frac{\pi}{4}\right) + i\sin\left(-\frac{\pi}{4}\right)\right)\\
&= 10\sqrt{2}\,e^{\pi it}e^{-\frac{\pi}{4}i}\\
&= 10\sqrt{2}\,e^{\pi i(t-0.25)}
\end{aligned}
$$

Using the rules of complex arithmetic, we have shown that the sum of the two signals results in a signal with an amplitude of $10\sqrt{2} \approx 14V$ and a phase of -0.25 s.

15.4   Matrices

We already introduced the fundamentals of matrices in MATLAB in Section 1.7. Matrices are very commonly used in a wide range of applications, and calculations involving matrices of numbers, i.e., rectangular arrays, are one of the main tasks that MATLAB was designed to perform. A matrix A with m rows and n columns is termed an $m \times n$ matrix, and, if its entries are real-valued, we can write $A \in \mathbb{R}^{m \times n}$. We can use lower case letters to denote the element in row i and column j of the matrix A as a_{ij}. If $m = n$, we say that A is a *square* matrix.

15.4.1   Matrix structure

The diagonal

The diagonal of a matrix $A \in \mathbb{R}^{m \times n}$ consists of the elements a_{jj} for $1 \leq j \leq \min(m, n)$. In MATLAB, this can be obtained using the diag function. The following shows how we can retrieve the diagonal of a matrix.

■ **Example 15.21**

015.C

Extract the elements on the main diagonal from a 2×3 matrix, in this example containing the digits 1–6:

```
>> B = reshape(1:6, 2, 3)
B =

     1     3     5
     2     4     6

>> diag(B)
```

```
ans =
    1
    4
```

■

The `diag` function can also be used to *construct* a square matrix if it is given a list of elements. The given elements will then appear on the diagonal of the new matrix, and the remaining elements will be zero.

■ Example 15.22

Create a matrix with the digits 1–4 on the diagonal: *O15.C*

```
>> diag(1:4)
ans =
    1    0    0    0
    0    2    0    0
    0    0    3    0
    0    0    0    4
```

In the above, we placed the values $\{1, 2, 3, 4\}$ on the main diagonal of the resulting 4×4 matrix. We might want to place values on a diagonal above or below the main diagonal; this is possible by passing a second non-zero integer argument that is positive for above and negative for below the main diagonal. For example, to place values in the first diagonal below the main one, we can make the call

```
>> diag([0.1, 0.5, 0.2], −1)
ans =
         0         0         0         0
    0.1000         0         0         0
         0    0.5000         0         0
         0         0    0.2000         0
```

■

Identity matrix

A square matrix that has ones on the diagonal and zeros elsewhere is known as the *identity* matrix, and the specific MATLAB function `eye` can be used to generate an identity matrix.

```
>> eye(3)
ans =
    1    0    0
    0    1    0
    0    0    1
```

If a non-square matrix is required with ones on the main diagonal, it is possible to give two arguments to specify the rows and columns:

```
>> eye(2,3)
ans =
    1    0    0
    0    1    0
```

Triangular matrices

An upper triangular matrix is one for which the elements *below* the diagonal are zero, for example,

$$\begin{pmatrix} 1 & 2 & 3 \\ 0 & 4 & 5 \\ 0 & 0 & 6 \end{pmatrix},$$

and vice-versa for the definition of a lower triangular matrix. Formally, we say that matrix A is upper triangular if we have $a_{ij} = 0$ for $i > j$, and that it is lower triangular if $a_{ij} = 0$ for $i < j$.

■ Example 15.23

015.C

For a given matrix, we can extract a lower (upper) triangular matrix using the functions `tril` (`triu`). In this example, we take the lower triangular part from a 4×4 magic square.

```
>> A = magic(4)
A =
    16     2     3    13
     5    11    10     8
     9     7     6    12
     4    14    15     1

>> tril(A)
ans =
    16     0     0     0
     5    11     0     0
     9     7     6     0
     4    14    15     1
```

■

■ Activity 15.5

015.C

Construct a 5×7 matrix A containing the numbers from 1 to 35 inclusive.

Find the diagonal of the matrix A, and create a matrix equal to A, but with zero values below the diagonal.   ■

15.4.2   Matrix operations

Addition and subtraction

Two matrices of the same size can be added or subtracted, and this operation is carried out element-wise. For example, if $A, B \in \mathbb{R}^{m \times n}$, then $C = A \pm B$ has elements c_{ij}, where $c_{ij} = a_{ij} \pm b_{ij}$ for $1 \leq i \leq m$ and $1 \leq j \leq n$. In MATLAB, this works as expected using the standard addition and subtraction operators.

■ **Example 15.24**

Simple matrix addition: *015.C*

```
>> A = ones(2,3)
A =
      1      1      1
      1      1      1

>> B = reshape(1:6, 2, 3)
B =
      1      3      5
      2      4      6

>> A + B
ans =
      2      4      6
      3      5      7
```

Take care to ensure that matrices added/subtracted have the same shape:

```
>> B + eye(2)
Matrix dimensions must agree. % Error!
```
■

Matrix multiplication

For a given pair of matrices, A and B, we can carry out the *matrix multiplication* $C = A B$. This can only be done if the number of columns in A equals the number of rows in B, because matrix multiplication is carried out by evaluating element-wise products of pairs taken from the rows of A and the columns of B, the sums of which then form the elements of C. For example,

$$A = \begin{pmatrix} 1 & 2 \\ 1 & 0 \\ 2 & -1 \end{pmatrix} \qquad B = \begin{pmatrix} 1 & 2 & 0 \\ 3 & 0 & 1 \end{pmatrix}.$$

We have

$$AB = \begin{pmatrix} 1\cdot1+2\cdot3 & 1\cdot2+2\cdot0 & 1\cdot0+2\cdot1 \\ 1\cdot1+0\cdot3 & 1\cdot2+0\cdot0 & 1\cdot0+0\cdot1 \\ 2\cdot1+-1\cdot3 & 2\cdot2+-1\cdot0 & 2\cdot0+-1\cdot1 \end{pmatrix} = \begin{pmatrix} 7 & 2 & 2 \\ 1 & 2 & 0 \\ -1 & 4 & -1 \end{pmatrix}.$$

■ **Example 15.25**

For a pair of matrices in MATLAB, the standard multiplication operator $*$ *015.C*
will carry out matrix multiplication:

```
>> A = [1 2; 1 0 ;2 -1];
>> B = [1, 2 , 0; 3, 0, 1];
>> C = A * B
C =
```

$$
\begin{array}{rrr}
7 & 2 & 2 \\
1 & 2 & 0 \\
-1 & 4 & -1
\end{array}
$$

We can check the sizes of the above matrices:

```
>> size(A)
ans =
        3    2

>> size(B)
ans =
        2    3

>> size(C)
ans =
        3    3
```

◼

Multiplying square matrices

For a pair of square matrices, matrix multiplication can be carried out if they are the same size.

◼ Example 15.26

O15.C   The product of two 2×2 matrices to give another:

```
>> M = [2, 1; 1 3]
M =
        2    1
        1    3

>> N = [0, -1; 2, 0]
N =
        0   -1
        2    0

>> M * N
ans =
        2   -2
        6   -1
```

◼

Multiplying by the identity matrix

When carrying out matrix multiplication, if one of the matrices is the identity matrix I, then the other matrix is unchanged. In other words, for $A \in \mathbb{R}^{m \times n}$, we have

$$A I_n = A \qquad \text{and} \qquad I_m A = A,$$

where I_n is an $n \times n$ identity matrix, and I_m is an $m \times m$ identity matrix.

■ Example 15.27

Show that a 2×2 matrix is unaffected by multiplication by the identity matrix.   *015.C*

```
>> A = [2, 3; 3, 5];
>> A * eye(2)
ans =
     2     3
     3     5

>> eye(2) * A
ans =
     2     3
     3     5
```
■

Element-wise or Hadamard multiplication

It is worth noting that, for two matrices M and N of the same size, it is also possible to carry out *element-wise multiplication*. Mathematically, this is sometimes called Hadamard (or Schur) multiplication and denoted $H = M \odot N$, where $h_{ij} = m_{ij}\, n_{ij}$.

In MATLAB, element-wise multiplication is an example of an *array* operation (as opposed to a matrix operation). To do this, we precede the multiplication operator with a dot.

■ Example 15.28

Show that matrix multiplication differs from element-wise multiplication   *015.C*
for a pair of 2×2 matrices.

```
>> M = [2, 1; 1 3];
>> P = [1, 2; 3, 4];
>> M .* P % element—wise multiplication
ans =
     2     2
     3    12

>> M * P % matrix multiplication
ans =
     5     8
    10    14
```
■

Matrices acting on vectors

We can treat column vectors as a special case of matrices, where the number of columns is one, e.g.,

$$\mathbf{v} = \begin{pmatrix} 1 \\ 3 \\ 5 \end{pmatrix}.$$

Matrices can be used to act on column vectors by pre-multiplication:

$$M = \begin{pmatrix} 1 & 0 & 1 \\ 0 & -2 & 1 \end{pmatrix}, \quad \mathbf{v} = \begin{pmatrix} 1 \\ 3 \\ 5 \end{pmatrix} \quad \Rightarrow \quad M\mathbf{v} = \begin{pmatrix} 6 \\ -1 \end{pmatrix}.$$

When viewing a matrix as acting on a vector by pre-multiplication, we mean that the matrix is on the left of the vector when the operation is applied. In the above case, the 3-D vector was pre-multiplied by M to give a 2-D output.

Matrices represent linear transformations

In general, a $m \times n$ matrix A can therefore be seen to represent a linear transformation from the space of n-dimensional vectors to the space of m-dimensional vectors. If x is a n-dimensional vector, i.e., $\mathbf{x} \in \mathbb{R}^n$, and $\mathbf{y} = A\mathbf{x}$, then $\mathbf{y} \in \mathbb{R}^m$, and we can view pre-multiplication by matrix A as an operation that transforms the n-vector $\mathbf{x}$ to the m-vector $\mathbf{y}$. We can denote the operation as f, and then f is a function from $\mathbb{R}^m$ to $\mathbb{R}^n$, and we can write

$$f(\mathbf{x}) := A\mathbf{x} = \mathbf{y} \qquad f : \mathbf{x} \to \mathbf{y} \qquad A \in \mathbb{R}^{m \times n}, \ \mathbf{x} \in \mathbb{R}^n, \ \mathbf{y} \in \mathbb{R}^m.$$

Transformations for square matrices

If a matrix represents a transformation between vector spaces with the same dimension, n say, then A is a square matrix, and, in this case, the matrix represents a transformation f from $\mathbb{R}^n$ to $\mathbb{R}^n$:

$$f(\mathbf{x}) := A\mathbf{x} = \mathbf{y} \qquad f : \mathbf{x} \to \mathbf{y} \qquad A \in \mathbb{R}^{n \times n}, \ \mathbf{x}, \mathbf{y} \in \mathbb{R}^n.$$

The matrix determinant

For a square matrix, it is possible to capture some information about the corresponding transformation in a single numeric value, known as the *determinant*. One way to think about the determinant is that it represents the factor by which the size of a region changes under the transformation. For example, in the case of a 2×2 matrix A, we can ask what effect applying it would have on the area of a unit square. To illustrate, we take the unit square with a corner on the origin, i.e., the points $(0, 0)$, $(1, 0)$, $(1, 1)$, and $(0, 1)$. These are shown by a black line in Fig. 15.3.

■ Example 15.29

015.A, 015.C

We define the points and a matrix to represent the transformation:

```
>> points = [0 1 1 0; 0 0 1 1]
points =
     0     1     1     0
     0     0     1     1
```

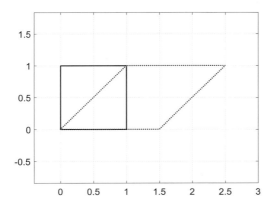

FIGURE 15.3
A set of points defining a square (black line) and the points after being transformed by matrix multiplication (dotted blue line). The change in area is equal to the determinant of the matrix. See Example 15.29.

```
>> A = [1.5, 1 ; 0 1];
```

Now, we apply the transformation to all of the points:

```
>> A * points
ans =
         0    1.5000    2.5000    1.0000
         0         0    1.0000    1.0000
```

After transformation, the points represent a parallelogram with a base length of 1.5 and a height of 1. This is shown by a dotted blue line in Fig. 15.3. The area of the parallelogram is 1.5, which means that the area of the unit square has been multiplied by a factor of 1.5. We can confirm this by finding the determinant of the original matrix:

```
>> det(A)
ans =
    1.5000
```

This property of the determinant extends to higher dimensions. For example, the determinant of a 3×3 matrix tells us the scale factor that would be multiplied by the volume of a unit cube in 3-D when it is transformed by the matrix.

■ **Activity 15.6**

Construct the matrix below in MATLAB.   *015.A, 015.C*

$$M = \begin{pmatrix} 0 & 0 & 2 \\ \frac{1}{6} & 0 & 0 \\ 0 & 3 & 0 \end{pmatrix}$$

Show that, when applying M to the points defining a volume, the overall volume will not change after transformation. ∎

Invertibility

When a transformation is represented by a square matrix A, we can ask whether the transformation is invertible, i.e., if we are given any $\mathbf{y} \in \mathbb{R}^n$, is it possible to find a vector $\mathbf{x}$ that solves $A\mathbf{x} = \mathbf{y}$?

Matrix inverse

If the transformation for a square matrix A is invertible, then it is possible to obtain the *inverse matrix* of A. This is denoted by A^{-1} and is a matrix of the same size such that $A A^{-1} = A^{-1} A = I$, where I is the $n \times n$ identity matrix. For such an invertible transformation, it is possible to use the inverse to solve for $\mathbf{x}$ in the equation $A\mathbf{x} = \mathbf{y}$ as follows:

$$A\mathbf{x} = \mathbf{y} \quad \Rightarrow A^{-1} A\mathbf{x} = A^{-1}\mathbf{y} \Rightarrow I\mathbf{x} = A^{-1}\mathbf{y} \Rightarrow \mathbf{x} = A^{-1}\mathbf{y}.$$

∎ Example 15.30

015.C   This is a simple example of obtaining the inverse of a 2×2 matrix:

```
>> A = [6, 5; 7, 6];
>> inv_A = inv(A)
inv_A =
      6.0000    -5.0000
     -7.0000     6.0000
```
∎

We can demonstrate the property of the inverse described above.

∎ Example 15.31

015.C   Show that $A A^{-1} = A^{-1} A = I$ for a 2×2 matrix.

```
>> A * inv_A
ans =
      1.0000          0
     -0.0000     1.0000

>> inv_A * A
ans =
      1.0000          0
     -0.0000     1.0000
```
∎

When a linear transformation is *not* invertible, the corresponding matrix is called *singular*, and MATLAB raises a warning when we try to calculate its inverse.

■ Example 15.32

Any matrix where one row (column) is a multiple of another row (column) *015.C*
is singular. We can use this property to easily create a non-invertible matrix.
In this case, the last row is double the first:

```
>> B = [1, 2, -1; 2, 0, 3; 2, 4, -2]
B =
     1     2    -1
     2     0     3
     2     4    -2
```

Now, when we try to find the inverse; the result contains infinity values.

```
>> inv(B)
Warning: Matrix is singular to working precision.
ans =
    Inf   Inf   Inf
    Inf   Inf   Inf
    Inf   Inf   Inf
```

■

Matrices and systems of linear equations

We have shown before (see Section 1.7) how matrix inverses can be used to
solve a system of linear equations. For example, if we have

$$A = \begin{pmatrix} 5 & 1 \\ 6 & 3 \end{pmatrix} \qquad \mathbf{x} = \begin{pmatrix} x_1 \\ x_2 \end{pmatrix} \qquad \mathbf{y} = \begin{pmatrix} 5 \\ 9 \end{pmatrix},$$

the matrix equation $A\mathbf{x} = \mathbf{y}$ can be seen as equivalent to a linear system of
equations

$$\begin{pmatrix} 5 & 1 \\ 6 & 3 \end{pmatrix}\begin{pmatrix} x_1 \\ x_2 \end{pmatrix} = \begin{pmatrix} 5 \\ 9 \end{pmatrix} \qquad \text{or} \qquad \begin{matrix} 5x_1 + x_2 = 5 \\ 6x_1 + 3x_2 = 9 \end{matrix} \tag{15.2}$$

Solving linear systems of equations using the inverse

The above means that we can use the inverse in MATLAB to solve such equations (which we did before in Example 1.1).

■ Example 15.33

Solve the linear system in Eq. (15.2) using the inverse matrix: *015.C*

```
>> A = [5, 1; 6, 3];
>> inv(A)
ans =
    0.3333   -0.1111
   -0.6667    0.5556
```

We can use the inverse to obtain the solution for **x**:

```
>> y = [5; 9];
>> x = inv(A) * y
x =
    0.6667
    1.6667
```

And we can confirm that the vector **x** satisfies the equation:

```
>> A * x
ans =
    5.0000
    9.0000
```

■

Solving linear systems of equations using `mldivide`

In general, the use of the inverse to solve systems of equations is not advisable, especially for a large number of variables (which means the corresponding matrix is large). There are better ways of solving equations that are more numerically stable and less computationally demanding, and these are available in MATLAB. One useful function is `mldivide`, which can be used to solve systems of equations of the form $A\mathbf{x} = \mathbf{y}$.

■ Example 15.34

015.C   Solve the linear system in Eq. (15.2) using the `mldivide` function:

```
>> A = [5, 1; 6, 3];
>> y = [5; 9];
>> x = mldivide(A, y)
x = 0.6667
    1.6667
```

This provides the same solution previously obtained using the matrix inverse in Example 15.33.

■

There is also a shorthand way of using the `mldivide` function, using the backslash operator as follows:

```
>> A \ y    % Same as mldivide(A, y)
ans =
    0.6667
    1.6667
```

■ Activity 15.7

015.C   Use `mldivide` to solve the linear equations:

$$-x_1 + 3x_2 = -6.5$$
$$2x_1 + x_2 = -1$$

■

Under-determined systems of equations

In the previous simple examples of systems of equations, we have only looked at the case where the number of equations equals the number of unknowns. In some cases, it is possible to have a system where there are *fewer* equations than unknowns, for example,

$$x_1 + 3x_2 + x_3 = 5$$
$$2x_1 \qquad -x_3 = 4$$

(15.3)

and this represents an *under-determined* system, which can be written in the form $A\mathbf{x} = \mathbf{y}$ if we set

$$A = \begin{pmatrix} 1 & 3 & 1 \\ 2 & 0 & -1 \end{pmatrix} \qquad \text{and} \qquad \mathbf{y} = \begin{pmatrix} 5 \\ 4 \end{pmatrix}.$$

Note that, because there are fewer equations than unknowns, there are multiple solutions that can satisfy the system. Here are two examples:

$$\begin{pmatrix} 1 & 3 & 1 \\ 2 & 0 & -1 \end{pmatrix} \begin{pmatrix} -3 \\ 6 \\ -10 \end{pmatrix} = \begin{pmatrix} 5 \\ 4 \end{pmatrix} \quad , \quad \begin{pmatrix} 1 & 3 & 1 \\ 2 & 0 & -1 \end{pmatrix} \begin{pmatrix} 4 \\ -1 \\ 4 \end{pmatrix} = \begin{pmatrix} 5 \\ 4 \end{pmatrix}.$$

(15.4)

We can set up such an under-determined system in MATLAB by assigning to the matrix A and the vector $\mathbf{y}$:

```
>> A = [1, 3, 1; 2, 0, -1]
A =
     1     3     1
     2     0    -1

>> >> y = [5; 4]
y =
     5
     4
```

■ Example 15.35

Confirm that there are multiple solutions to an under-determined system of equations by performing the calculations for the examples given in Eq. (15.4).

015.C

```
>> A * [-3; 6; -10]
ans =
     5
     4

>> A * [4; -1; 4]
ans =
```

```
        5
        4
```

In this example, A is a 2×3 matrix. Therefore as it is a non-square matrix, we cannot use the `inv` function to find a solution. However, the `mldivide` function can still be used to provide a solution when the matrix of the system is non-square.

■ Example 15.36

015.C   For the system in Eq. (15.3), use `mldivide` to find a solution, and ensure that it is indeed a solution.

Here, we use the shorthand \ notation to obtain a solution, then check it.

```
>> x = A \ y
x =
    2
    1
    0

>> A * x
ans =
    5
    4
```

Over-determined systems of equations

In the case when there are *more* equations than unknowns, we say the system is *over-determined*. In the following example, we have three equations and two unknowns:

$$
\begin{aligned}
x_1 + 2x_2 &= 5 \\
-3x_1 + x_2 &= -2 \\
2x_1 - x_2 &= 1
\end{aligned}
\tag{15.5}
$$

Because there are more constraints than variables, it is possible, indeed likely, that there is *no* solution that can satisfy all three equations simultaneously. In other words, we may not be able to find choices for x_1 and x_2 in Eq. (15.5) that lead to the exact values on the right hand side of each of the three equations, which represent the vector **y** when the system is written in the form $A\mathbf{x} = \mathbf{y}$. We can, however, look for solutions that are approximately correct. For instance, we can seek a solution that gives a vector that is as close as possible to the vector **y**.

In this case, the `mldivide` function can still be used to find such an approximate solution. In the next example, we use `mldivide` to calculate an approximate solution vector **x** for the over-determined system in Eq. (15.5).

■ **Example 15.37**

Find an approximate solution to an over-determined system of equations.   *O15.C*

First, we set up the matrices:

```
>> A = [1, 2; -3, 1; 2, -1];
>> y = [5; -2; 1];
```

Next, use `mldivide`:

```
>> x = A \ y
x =
    1.3200
    1.8267
```

■

Note that the above obtains an *approximate* solution, and we can inspect what result is obtained by substituting **x** into the original equation.

```
>> A * x
ans =
    4.9733
   -2.1333
    0.8133
```

Recall that the target output was the vector $\mathbf{y} = (5, -2, 1)^T$. We can look at the error, or difference, between our result and the target vector, as well as its magnitude:

```
>> e = A * x - y
e =
   -0.0267
   -0.1333
   -0.1867

>> norm(e)
ans =
    0.2309
```

For this over-determined system of equations, the `mldivide` function has given us a solution that, while not exact, is as close as possible to the required output.

Equation solving: summary

To summarize, MATLAB can be used to solve systems of linear equations if we can represent them in the form of a matrix multiplication acting on vectors, i.e., in the form $A\mathbf{x} = \mathbf{y}$. The number of equations in a system may equal the number of unknowns, it may be under-determined (fewer equations than unknowns), or it may be over-determined (more equations than unknowns). In all cases, the `mldivide` function can be used to provide a solution (or at least an approximate one). The `mldivide` function is actually a wrapper function for various methods, and the method used will depend on the properties of the system of equations, and is selected at run-time. For more information on how the function works and how it chooses which solutions to provide for the user, further details can be found in the MATLAB help, under *Mathematics* → *Linear Algebra* → *Systems of linear equations*.

■ Activity 15.8

015.C   Decide whether the following system of equations is over- or under-determined:

$$3x_1 - x_2 + 2x_3 = 11$$
$$x_1 + 2x_2 \quad\quad = -4$$

Confirm that the system can be solved by $x = (2, -3, 1)^T$, and find a second solution using `mldivide`. How does the second solution differ from the one given?   ■

15.5 Calculus

The basic operations of calculus, namely, differentiation and integration, are a routine part of many problems in engineering, and MATLAB offers a variety of tools for estimating derivatives and integrals from data.

15.5.1 Differentiation

The `diff` function: estimating derivatives from differences

We begin with estimating derivatives numerically for 1-D data. One of the simplest functions to use in this case is the `diff` function.

Assume that measurements of a function $f(x)$ are obtained at a *regular* set of points $\{x_1, \ldots, x_N\}$, i.e., where the N points are equally spaced. We can denote the distance between successive x values by h, i.e., $x_{k+1} - x_k = h$. The measurements can be written $\{f(x_1), \ldots, f(x_N)\}$ or, if we want to save space, $\{f_1, \ldots, f_N\}$.

We can make an estimate of the derivative of f at location x_k by using the *forward difference* method:

$$\left. \frac{df}{dx} \right|_{x=x_k} = f'(x_k) \approx \frac{f_{k+1} - f_k}{x_{k+1} - x_k} = \frac{f_{k+1} - f_k}{h}.$$

For a concrete example, consider a set of x values with a spacing of 0.2 between zero and one. Including the endpoints, there will be six sample points altogether, as shown below:

```
>> x = linspace(0, 1, 6)
x =
        0    0.2000    0.4000    0.6000    0.8000    1.0000
```

We can obtain values for the function $f(x) = \sin 2x$ at the chosen sample locations. In the following, we assign these values to a variable, called f:

```
>> f = sin(2 * x)
f =
        0    0.3894    0.7174    0.9320    0.9996    0.9093
```

■ Example 15.38

Use the diff function to estimate the derivatives for the data above.

015.D

First, we obtain the raw values of the forward differences, then an estimate for the derivative can be found by dividing these values by the spacing:

```
>> diff_f = diff(f); % 0.389 0.328 0.214 0.0675 −0.0903
>> h = x(2) − x(1) % 0.2000
>> dfdx_est = diff(f) / h
dfdx_est =
        1.9471    1.6397    1.0734    0.3377    −0.4514
```

■

For this simple function, we can also calculate the exact derivatives and compare these with the estimates found above. The function $f(x) = \sin 2x$ has a derivative $f'(x) = 2\cos 2x$:

```
>> dfdx = 2 * cos(2 * x)
dfdx =
        2.0000    1.8421    1.3934    0.7247    −0.0584    −0.8323
```

There are a couple of things to note here. First, the estimated values of the derivative have errors compared to the exact derivative. This is expected when we estimate gradients numerically. Second, the derivative estimates only cover the sample points up to but not including the last one. This will be true in general. For N points, we can only calculate $N-1$ differences: $x_2 - x_1, \ldots, x_N - x_{N-1}$.

Derivatives using the gradient *function*

An alternative to estimating derivatives with the diff function is to use the gradient function. Continuing with the same x sample locations and f values defined above for $f(x) = \sin 2x$, we can obtain a new set of estimates for the derivative as follows:

```
>> dfdx_est2 = gradient(f, h)
dfdx_est2 =
        1.9471    1.7934    1.3566    0.7055    −0.0569    −0.4514
```

Note that the gradient function takes the f values for the function as the first argument and the spacing between x sample locations (h) as the second argument. Note also that this function provides estimates for all sample values.

We can compare the exact values we expect (`dfdx`) with the values obtained from the gradient function (`dfdx_est2`) and the values estimated using the `diff` function earlier (`dfdx_est`):

```
>> dfdx
dfdx =
    2.0000    1.8421    1.3934    0.7247   -0.0584   -0.8323

>> dfdx_est2
dfdx_est2 =
    1.9471    1.7934    1.3566    0.7055   -0.0569   -0.4514

>> dfdx_est
dfdx_est =
    1.9471    1.6397    1.0734    0.3377   -0.4514
```

Note that the estimates provided by the `gradient` function are generally more accurate. This, combined with the fact that it provides an estimate for all samples, gives it an advantage over the `diff` function.

How the `gradient` function estimates derivatives

The `gradient` function uses a forward difference estimate for the first sample location, and this is why its estimate agrees with the `diff`-based estimate there. For the last point, `gradient` uses a *backward* difference estimate, i.e., if there are N points, then the gradient at the last (N^{th}) position uses the function values at positions $N - 1$ and N, i.e.,

$$f'(x_N) \approx \frac{f_N - f_{N-1}}{h}$$

For intermediate points x_k with $1 < k < N$, there are sample points to the left and to the right, which means that a *central* difference estimate of the gradient can be made:

$$f'(x_k) \approx \frac{f_{k+1} - f_{k-1}}{2h}.$$

■ Example 15.39

015.D

Confirm that the different methods (forward, central, backward differences) are provided by `gradient` for the data above.

```
>> dfdx_est2 % Values given by gradient function.
dfdx_est2 =
    1.9471    1.7934    1.3566    0.7055   -0.0569   -0.4514
```

```
>> (f(2) - f(1)) / h % Manually get first value
ans =
    1.9471

>> (f(3:end) - f(1:end-2)) / 2 / h % Intermediate values
ans =
    1.7934    1.3566    0.7055    -0.0569

>> (f(end) - f(end-1)) / h % Last value
ans =
    -0.4514
```

∎

∎ Activity 15.9

Consider the function $y = -\log\cos^2 x$. Define a suitable interval around $x = 2$, and use both `diff` and `gradient` to estimate $\frac{dy}{dx}$ for $x = 2$. Calculate the value we expect analytically, and compare this with the numerical results.

015.D

∎

Comparing `gradient` with convolution

Recall Example 11.4, which used convolution to estimate the derivative of time-series data representing the position of a sensor. The approach used was to read in the data, smooth them, and convolve the smoothed data with the kernel $(-1/2\Delta_t, 0, 1/2\Delta_t)$, where Δ_t is the time step in the data. Note that for a series of measurements $f_1, \ldots, f_k, \ldots, f_N$, the convolution by the above kernel at the kth sample will be

$$\frac{-1}{2\Delta_t} f_{k-1} + 0 f_k + \frac{1}{2\Delta_t} f_{k+1} = \frac{f_{k+1} - f_k}{2\Delta_t},$$

which represents a central difference estimate for the derivative at x_k. We recall the code used for the convolution method below (it is condensed and slightly modified):

```
delta_t = 0.4;
z_data = readmatrix('pos-z.csv');
n_samp = numel(z_data);

t_vals = 0:delta_t:(n_samp-1)*delta_t;

kernel = [-1, 0, 1] / (2 * delta_t);
kernel = fliplr(kernel);

deriv_est1 = conv(z_data, kernel, 'same');
```

The variable `deriv_est1` contains the convolution estimate of the derivatives. Recall also that we arbitrarily set the estimate values at the end points to zero,

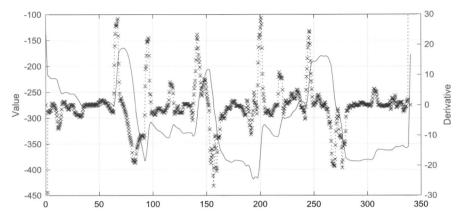

FIGURE 15.4
A set of sensor data and two estimates for the derivative. The original data (solid blue line) correspond to the left-hand y-axis. The derivative estimates correspond to the right-hand y-axis. The estimates were obtained using the convolution method (red crosses) and the `gradient` method (black dotted line).

as there was insufficient data for a good estimate there, and the `'same'` option was passed to the `conv` function:

```
deriv_est1(1) = 0;
deriv_est1(end) = 0;
```

■ Example 15.40

015.D Obtaining a derivative estimate using the `gradient` function is a lot easier:

```
deriv_est2 = gradient(z_data, delta_t);
```
■

We can compare the two estimates for the derivative using a plot. The code below plots the original data and both estimates in a figure with two y axes:

```
yyaxis('left')
plot(t_vals, z_data)
yyaxis('right')
plot(t_vals, deriv_est1, 'xr') % red crosses
ylabel('Derivative')
hold on
plot(t_vals, deriv_est2, ':k') % black dotted
grid on
```

The resulting plot is shown in Fig. 15.4, which shows that both methods of estimating the gradient agree quite closely with each other. Note that the command `grid on` was used above to show an overlaid grid of lines on the plot that is aligned with the tick marks on the axes. This can be used to aid reading a plot.

The location of derivative estimates

When using the `gradient` function, we provide a set of measured function values $f_1, f_2, \ldots, f_N$ at a set of N sample locations, $x_1, x_2, \ldots, x_N$. In this case, the function returns N estimates of the gradient $f_1', \ldots, f_N'$, and these correspond unambiguously with the sample locations.

The `diff` function, however, returns a number of estimates, one fewer than the number of samples given, and we have some choice over how to make the estimates match up to the sample locations. For example, consider a sequence of equally spaced sample locations, $x_1, x_2, \ldots, x_N$, with corresponding function values of $f_1, f_2, \ldots, f_N$. The individual estimate

$$\frac{f_{k+1} - f_k}{h}$$

can be viewed as a *forward* estimate for the gradient at location x_k, as discussed earlier. Alternatively, it can be viewed as a *backward* estimate for the gradient at the next sample point, x_{k+1}.

When the function `diff` is given an argument f containing all the function values, i.e., $(f_1, f_2, \ldots, f_N)$, then the sequence returned will contain $N - 1$ values,

$$f_2 - f_1, f_3 - f_2, \quad \cdots \quad , \quad f_N - f_{N-1}.$$

Dividing these values by the spacing h, which, in MATLAB, means calculating `diff(f)/h`, will give a set of gradient estimates that can be viewed as forward estimates for $x_1, x_2, \ldots, x_{N-1}$ or backward estimates for $x_2, x_2, \ldots, x_N$.

With a slight abuse of notation, the same set of derivatives can be viewed as *central* estimates for a set of sample locations that can be written $x_{1.5}, x_{2.5}, \ldots, x_{N-0.5}$, which are the points that are half-way between the original sample locations.

Returning to the earlier example,

```
>> x = linspace(0, 1, 6)
x =
        0    0.2000    0.4000    0.6000    0.8000    1.0000

>> f = sin(2 * x);
>> diff_f = diff(f)
diff_f =
    0.3894    0.3279    0.2147    0.0675   -0.0903
```

The estimates contained in `diff_f` can be viewed as central estimates for the set of x locations $\{0.1, 0.3, 0.5, 0.7, 0.9\}$ that are mid-way between the locations originally obtained.

Higher derivatives

Numerical estimates of higher derivatives can be obtained by repeatedly applying the `diff` function. To obtain a higher-order difference, we can apply the `diff` function to get a first-order difference, then apply `diff` to the result to get a second-order difference, and so on. Dividing by the correct spacing will then provide estimates of first, second, and higher-order derivatives.

As an example, given a sequence of measurements of a function, illustrated by the top row of numbers below, the first application of `diff` provides the numbers in red on the second row. Applying `diff` to these values, gives the second-order differences, which are in blue on the third row.

$$\begin{array}{ccccccccc} \ldots & 12 & & 9 & & 11 & & 15 & \ldots \\ & \ldots & -3 & & 2 & & 4 & \ldots & \\ & & \ldots & 5 & & 2 & \ldots & & \end{array}$$

The numbers on the last row can be obtained with a MATLAB call to `diff(diff(f))`, where f represents the list of numbers in the top row. Alternatively, a second argument can be passed to the function to indicate the order required. This can be done by calling `diff(f, 2)`. We can confirm that the call `diff(f, 2)` gives the same result as would be obtained by calling `diff` twice:

```
>> a = diff(diff(f));
>> b = diff(f, 2);
>> all( a == b )
ans =
     1
```

Here, the built-in MATLAB `all` function returns a logical value indicating if all elements of its array argument are nonzero or true.

We can represent the sequence of function values, and the first- and second-order differences symbolically, as shown below:

$$\begin{array}{ccccccc} \ldots & f_{k-1} & & f_k & & f_{k+1} & \ldots \\ \ldots & & f_k - f_{k-1} & & f_{k+1} - f_k & & \ldots \\ \ldots & & & f_{k+1} - 2f_k + f_{k-1} & & \ldots & \end{array}$$

Dividing by the spacing between samples h, the first-order differences can provide estimates for the derivative of the function to the left and the right of x_k:

$$\frac{f_k - f_{k-1}}{h} \quad \text{and} \quad \frac{f_{k+1} - f_k}{h}.$$

Using these two estimates of the first derivative, we can obtain a central estimate for the second derivative at x_k, in which, once again, h is the spacing between the samples:

$$\frac{1}{h}\left[\frac{f_k - f_{k-1}}{h} - \frac{f_{k+1} - f_k}{h}\right] = \frac{f_{k+1} - 2f_k + f_{k-1}}{h^2}. \tag{15.6}$$

The above shows that we can obtain an estimate for the second derivative from diff by dividing the second difference by h^2. It also shows how the estimate at x_k is obtained using the values of the function at x_{k-1}, x_k, and x_{k+1}.

■ Example 15.41

Find the second derivative estimates for $f(x) = \sin 2x$ at x values between zero and one with an equal spacing of 0.2.   *015.D*

```
>> x = linspace(0, 1, 6); % samples
>> h = x(2) - x(1); % spacing
>> f = sin(2 * x); % function values
```

Note that there are six sample locations between zero and one inclusive. Now we can use diff to estimate the second derivative:

```
>> est_d2fdx2 = diff(f, 2) / (h * h)
est_d2fdx2 =
   -1.5370   -2.8314   -3.6787   -3.9453
```
■

Note that in Example 15.41, we know the analytic expression of the second derivative:

$$\frac{d^2 f}{dx^2} = -4\sin 2x,$$

so we can obtain these values and compare against the estimate:

```
>> d2fdx2 = -4 * sin(2 * x)
d2fdx2 =
        0   -1.5577   -2.8694   -3.7282   -3.9983   -3.6372
```

There are four derivative estimates obtained via the diff call, which is two fewer than the number of sample locations. This is because the second-order differences can correspond to the sample locations $x_2, \ldots, x_5$ (see Eq. (15.6)).

■ Example 15.42

Use the analytically obtained values for the derivatives to assess the error and relative error of the estimates obtained using diff.   *015.D*

```
>> est_d2fdx2
est_d2fdx2 =
   -1.5370   -2.8314   -3.6787   -3.9453
```

```
>> d2fdx2(2:5)
ans =
    -1.5577    -2.8694    -3.7282    -3.9983

>> err = abs(est_d2fdx2 - d2fdx2(2:5))
err =
     0.0207     0.0381     0.0494     0.0530

>> rel_err = err ./ abs(d2fdx2(2:5))
rel_err =
     0.0133     0.0133     0.0133     0.0133
```

■

In general, we can obtain an estimate of the Mth derivative of a set of measurements f, with a sample spacing of h, via the `diff` function using the call

```
>> diff(f, M) / (h ^ M);
```

and, as shown above, some care must be taken to identify a good correspondence between the estimates provided and the sample locations. In the case of the second derivative estimates shown above, the locations for which estimates are provided excludes the first and the last sample locations.

Gradients for functions with more than one variable

When we consider functions with more than one variable, we refer to derivatives with respect to each of their variables as *partial derivatives*. For example, a function can represent temperature t, and this can depend on two factors: latitude l and the month of the year m. We can write this as $t = T(l, m)$, and we can find how the function T varies with respect to each of its input variables separately. That is, we can measure how much T varies for a change in l, and we can separately measure how much T varies with respect to m. These two rates of change can be written as

$$\frac{\partial T}{\partial l} \quad \text{and} \quad \frac{\partial T}{\partial m}.$$

We do not use a straight d, as would be the case for a function of a single variable, such as df/dx. The curly ∂ indicates that we had a choice of which variable we could have differentiated with respect to; the variable chosen is written at the bottom part of each expression.

In another example, electrical charge q can depend on location, i.e., it can depend on x, y, and z coordinates. This leads to three possible partial derivatives:

$$q = Q(x, y, z) \quad \rightarrow \quad \frac{\partial Q}{\partial x} \quad \frac{\partial Q}{\partial y} \quad \frac{\partial Q}{\partial z}.$$

Partial derivatives using the `gradient` *function*

In MATLAB, we can use the `gradient` function to obtain partial derivatives. Previously, we applied the `gradient` function to 1-D data and estimated the derivative by making a call such as `gradient(f, h)`. The second argument, in this case, specifies the spacing between samples along the single dimension.

Consider an example where the value of a function z depends on two variables, x and y, say. In this case, we can still use the `gradient` function to estimate the partial derivatives $\frac{\partial z}{\partial x}$ and $\frac{\partial z}{\partial y}$. The measurements for z must be contained in an array, with each measurement corresponding to a sample location (x_j, y_k), where j and k index the locations of a regular grid of sample points. We can also pass the spacings of the sample points to the `gradient` function, as second and third arguments, to obtain the estimate. Below we run through an example of this.

First, we need to define the grid of sample points in the xy plane. 1-D linear grids are first defined in the x and y directions, and these are then passed to the `meshgrid` function (see Section 9.2.3) to obtain a 2-D grid represented by a pair of arrays, one containing the x coordinates and the other containing the y coordinates.

■ Example 15.43

Make a function $z(x, y)$ representing two Gaussian blobs, one centered at $(-0.5, 0.5)$ and another centered at $(2, -1)$. Define them on a mesh grid between -4 and 4 in each dimension.

O15.C

```
>> x = linspace(-4, 4, 51);
>> y = linspace(-4, 4, 51);
>> [x,y] = meshgrid(x, y);
>> z = exp( - (x+0.5).^2 - (y-0.5).^2 ) + ...
    2 * exp( - (x-2).^2 - (y+1).^2 );
```
■

We can visualize the function as an image using the `imagesc` function. The result of this is shown in Fig. 15.5:

```
>> imagesc(z), axis xy, colormap gray
```

In the above, the function `imagesc` visualizes an array as an image, with a suitable scaling so that it is properly displayed. The `axis xy` command ensures that the row dimension of the array is on the horizontal axis, and the column dimension is vertical. The `colormap gray` command ensures the image is displayed with gray scale coloring.

■ Example 15.44

Use the `gradient` function to estimate the partial derivatives of the function z in Example 15.43.

O15.D

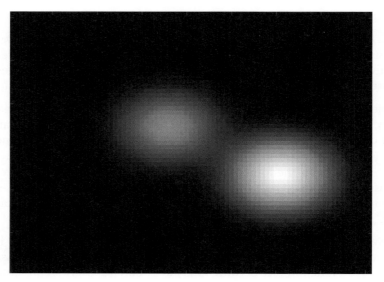

FIGURE 15.5
A function z of two variables (x and y). This function represents two Gaussian blobs.

First, we calculate the spacing of the sample grid in each of the x and y directions:

```
>> dx = x(1,2) - x(1,1);
>> dy = y(2,1) - y(1,1);
```

Now, we call the `gradient` function:

```
>> % Estimate partial derivatives in x and y directions.
>> [dzdx, dzdy] = gradient(z, dx, dy);
```

In the above, as the data are 2-D, we collect two return values. The first represents $\frac{\partial z}{\partial x}$, and the second represents $\frac{\partial z}{\partial y}$.

Finally, we can visualize the two partial derivatives. Each one can be passed to the `imagesc` function separately, and the resulting visualizations for each partial derivative are shown in Fig. 15.6.

```
>> % Plot partial derivative w.r.t. x and y
>> figure, imagesc(dzdx), axis xy
>> figure, imagesc(dzdy), axis xy
```

■ Activity 15.10

015.D The equation of a hyperbolic paraboloid is given by $p = x^2 - y^2$. Use the method shown previously for the Gaussian blob function to visualize the

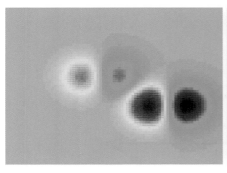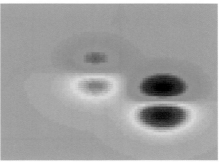

FIGURE 15.6
Visualizations of partial derivatives of the function z shown in Fig. 15.5. These were obtained using the `gradient` function. The left panel shows $\frac{\partial z}{\partial x}$, and the right panel shows $\frac{\partial z}{\partial y}$.

> partial derivatives of p, and visualize the function as a surface. Use a small range close to the origin, for example, $-1 \leq x, y \leq 1$. ∎

15.5.2 Integration

Integration of functions is used in a wide variety of contexts. In one dimension, the integral of a function can be used to provide the area underneath the graph of the function between a pair of specified limits. For a function of two variables, it can provide the volume beneath the surface representing the function, and so on. There are a number of methods for estimating the integral of a function numerically within MATLAB. This section will consider some of these methods with examples to illustrate them.

The trapezoidal rule

The trapezoidal rule is one of the simplest methods for finding the area under a curve. To obtain the estimate, the area to be integrated is divided into trapezia, and the areas of these are calculated and added together. An illustration of splitting the area to be integrated in this way is shown in Fig. 15.7. We have come across the trapezoidal rule before in Exercise 5.8, in which the task was to write code from scratch to implement the rule for a specific function. Here, we show how to use the built-in MATLAB function `trapz` to estimate an integral for an arbitrary function.

Assume that we have estimates of a function $f_1, f_2, \ldots, f_n$ at an evenly spaced set of locations $x_1, x_2, \ldots, x_n$, where $a = x_1$, and $b = x_n$, and Δ_x is the spacing between samples. Recall the formula for the trapezoidal rule for estimating the integral of f between a and b:

$$\int_a^b f(x)\, dx \approx \frac{\Delta x}{2} \left(f(x_1) + 2f(x_2) + 2f(x_3) + \ldots + 2f(x_{n-1}) + f(x_n) \right).$$

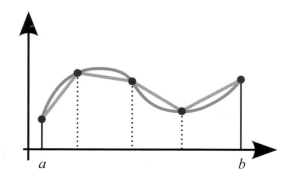

FIGURE 15.7
Estimating the area under a curve using the trapezoidal rule. By dividing the region beneath the curve (in blue) between a and b into trapezia (in yellow), we can sum their areas to approximate the integral of the function between a and b.

■ **Example 15.45**

015.D

Estimate the integral of the function $f(x) = \sin x$ between 0 and $\pi/2$.

First, we obtain a set of sample x locations. In this case, they divide the interval between 0 and $\pi/2$ into five equal sub-intervals:

```
>> x = linspace(0, pi/2, 6)
x =
        0    0.3142    0.6283    0.9425    1.2566    1.5708
```

Now we evaluate the function at the chosen locations:

```
>> f = sin(x)
f =
        0    0.3090    0.5878    0.8090    0.9511    1.0000
```

Finally, we pass the locations and the function values to the `trapz` function to obtain the estimate for $\int_0^{\pi/2} \sin x \, dx$:

```
>> est_area = trapz(x, f)
est_area =
    0.9918
```

■

This matches fairly closely with what we expect from the analytic expression for the integral:

$$\int_0^{\pi/2} \sin x \, dx = [-\cos x]_0^{\pi/2} = -\cos \frac{\pi}{2} - (-\cos 0) = 0 - (-1) = 1.$$

We can also confirm that the function `trapz` is carrying out its calculation according to the above rule by explicitly calculating it from the given values:

```
>> h = x(2)-x(1)
h =
    0.3142

>> h * ( f(1) + 2 * sum(f(2:end-1)) + f(end) ) / 2
ans =
    0.9918,
```

which matches the value obtained using `trapz`.

The `integral` function

A number of more accurate methods for numerically estimating integrals exist. Such methods can be described by the term *quadrature;* some quadrature approaches are implemented in MATLAB by the `integral` function.

To integrate a function `f`, of one variable, between limits `a` and `b`, we can use the expression `integral(f, a, b)`. This calculates a numerical approximation for the mathematical expression

$$\int_a^b f(x)\, dx.$$

Note that the first argument to the function `integral` is *another function* `f`. This raises the question: how can we pass a function as an argument to another function?

In MATLAB, this is achieved by using a *function handle* as a way to treat functions as if they were ordinary variables. We can illustrate this with an example.

Consider the function $f(x) = 1 + \sin 2x$. We can find the *exact* area under the graph analytically, between $x = 0$ and $x = \pi/3$, say:

$$\int_0^{\pi/3} 1 + \sin 2x \, dx = \left[x - \frac{1}{2}\cos 2x \right]_0^{\pi/3} = \frac{\pi}{3} + \frac{3}{4}.$$

■ Example 15.46

Now, we will use the `integral` function to estimate the integral above.   *015.D*

First, we can define the function by writing the following code and saving it in a file called *myFunc.m*:

```
function y = myFunc(x)
    y = 1 + sin(2 * x);
end
```

To pass the above function as an argument to the `integral` function, we need to obtain a *handle* that refers to `myFunc`. We can do this using the `@`

operator, placing it as a prefix before the function name, i.e., in the form @ myFunc. This allows us to estimate the required integral as follows:

```
>> est = integral(@ myFunc, 0, pi / 3)
est =
    1.7972
```

∎

We can compare this with the exact value of the integral, or to be precise, the best floating point representation of the exact value:

```
>> exact = pi/3 + 0.75
exact =
    1.7972
```

The difference between the two values is as follows:

```
>> abs(est − exact)
ans =
    0
```

which indicates that, to floating point accuracy, the estimate given by the integral function is indistinguishable from the exact value.

In general, we do not need to write the code for a function and save it in a .*m* file to use it as an argument in the integral function. The alternative is to define a so-called *anonymous* function in a line of code, and pass it directly to integral.

As a simple example, consider $g(x) = e^x$, and let us integrate it between $x = 1$ and $x = 5$. Analytically, we can obtain the exact value easily, so that we can compare against the estimate from MATLAB:

$$\int_1^5 e^x \, dx = \left[e^x \right]_1^5 = e^5 - e^1.$$

∎ Example 15.47

015.D

Use an *anonymous function* to estimate the integral above.

First, we create an anonymous function, once again using the @ operator, but this time, we use it when we *define* the function:

```
>> g = @(x) exp(x)
g =
  function_handle with value:
    @(x)exp(x)
```

Note that MATLAB describes the variable g as having a value of @(x)exp(x), which represents an anonymous function of a single variable x.

Now, we can estimate the integral required:

```
>> est = integral(g, 1, 5)
est =
   145.6949
```

∎

This time, we can pass the variable g directly as an argument to integral, because it is an anonymous function. Contrast this with Example 15.46, when we needed to prefix the name of the myFunc function with the @ operator when passing it to integral.

Once again, we can compare the estimated and (to floating point accuracy) "exact" values:

```
>> exact = exp(5) - exp(1);
>> abs(est - exact)
ans =
   2.8422e-14,
```

which shows the level of accuracy obtained by the integral function.

Note that the variable name does not matter when we are defining an anonymous function. In other words, we could have defined the previous function with any variable name we like,

```
>> g = @(myVar) exp(myVar)
g =
   function_handle with value:
     @(myVar)exp(myVar)

>> est = integral(g, 1, 5)
est =
   145.6949,
```

and the result would be the same.

∎ Activity 15.11

Define the function f by $f(x) = e^{3x} \cos 5x$. Let $g(x)$ be the derivative of f with respect to x. Find $g(x)$ on pen and paper using the product and chain rules. Estimate the integral of $g(x)$ in the interval $0 \le x \le 1$. Use $f(x)$ to obtain the exact integral, and find the error and relative error of the estimate. ∎

015.D

Integrating functions with more than one variable

The integral function has counterparts for integrating functions of two and three variables, which are named integral2 and integral3, respectively.

For example, consider the function of x and y, $z = 1 - x^2 - y^2$. This represents a paraboloid and is illustrated in Fig. 15.8. Analytically, we can evaluate the

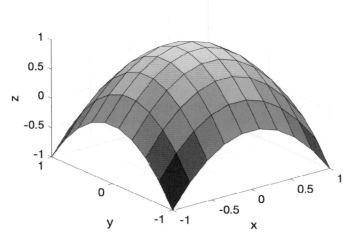

FIGURE 15.8
A function z of two variables (x and y). This function represents the paraboloid $z = x^2 + y^2$.

double integral of this function over arbitrary limits in x and y:

$$\int_c^d \int_a^b 1 - x^2 - y^2 \, dx \, dy \quad = \quad \left[\left[xy - \frac{x^3 y}{3} - \frac{xy^3}{3} \right]_a^b \right]_c^d. \tag{15.7}$$

Let us choose some specific limits, say, $-1 \leq x \leq 1$ and $-1 \leq y \leq 1$. Substituting these into the above expression, we can evaluate the exact integral:

$$\int_{-1}^1 \int_{-1}^1 1 - x^2 - y^2 \, dx \, dy \quad = \quad \frac{4}{3}$$

In this case, we can define a function of two variables anonymously in a very similar way to the single variable example. The most important difference is that we need to include the names of both variables in the list given to the @ operator.

■ Example 15.48

015.D

Define an anonymous function of two variables. We will use this later for a 2-D integral.

```
>> z = @(x,y) 1 - x.^2 - y.^2
z =
  function_handle with value:
    @(x,y)1-x.^2-y.^2
```

■

Note also how the dot operator was used above to ensure that the squaring operations are carried out *element-wise*. This is because the x and y values in the domain of the function are defined in two dimensions, i.e., we want the function to accept a 2-D array of values for each of x and y and act on each element in these arrays individually (as opposed to matrix-squaring the whole of each array, which is what would happen without the dot operator).

■ Example 15.49

Using the anonymous function z, obtain an estimate for the integral evalu- ated analytically in Eq. (15.7). *015.D*

```
>> est = integral2(z, −1, 1, −1, 1)
est =
    1.3333
```

This, once again, shows the level of agreement with the exact value of 4/3. ■

As we did for the first single variable function example, we can also use a function fully defined in code for the `integral` function.

■ Example 15.50

Use a non-anonymous function to evaluate the integral in Eq. (15.7). *015.D*

In this case, the following code would need to be saved inside a file:

```
function p = paraboloid(x, y)
        p = 1 − x.^2 − y.^2;
end
```

and the integral estimate could be obtained with the following call:

```
>> est = integral2(@paraboloid, −1, 1, −1, 1)
est =
    1.3333
```
■

For general interest, the code used to visualize the function in Fig. 15.8 is shown below:

```
>> x = linspace(−1,1,10);
>> y = linspace(−1,1,10);
>> [xx, yy] = meshgrid(x,y);
>> z = @(x,y) 1 − x.^2 − y.^2;
>> z_vals = z(xx,yy);
>> surf(xx, yy, z_vals)
```

15.5.3 Calculus summary

In this section, we have explored some of the numerical methods available in MATLAB for estimating the values of derivatives and integrals. For differentiation, we looked at difference methods with the `diff` function and the `gradient` function and compared the latter with convolution. We also looked at higher derivatives for single variable functions, and partial derivatives for functions of more than one variable.

For integration, we have looked at the simple `trapz` function that implements the trapezoidal rule and the more sophisticated `integral` function that makes use of the more powerful quadrature-based approximation. For functions of two and three variables, the functions `integral2` and `integral3` are available. Examples using `integral2` have been presented, and the use of `integral3` is very similar.

In the calculus estimation examples shown, we have mostly considered functions for which it is possible to analytically evaluate derivatives and integrals. This is only so that we can compare estimates with values we expect. In general, do not expect to be able to analytically obtain exact values, because functions are very likely to be much more complicated, and other methods should be used to assess the accuracy of the estimates obtained.

There are other ways to carry out calculus operations in MATLAB; notably it is possible to carry out symbolic differentiation and integration. These are considered beyond the scope of this book, because engineering approaches tend to focus on numerical approximations, and symbolic methods also require the symbolic math toolbox. Help on such symbolic methods can be found under *Symbolic Math Toolbox → Mathematics → Calculus*.

15.6 Differential equations

A differential equation expresses some property of the derivative(s) of an unknown function that we seek to find or approximate. A very simple differential equation of an unknown function y of an independent variable t is $\frac{dy}{dt} = 2t$, which can be written more compactly as $y' = 2t$. For this simple example, we can find the solution by inspection as $y(t) = t^2$.

It is also possible to have multiple unknown functions of the independent variable. For example, we might have two functions $y_1(t)$ and $y_2(t)$ related by the equations

$$y'_1 = y_2 \quad + 2y_1 - t$$
$$y'_2 = -y_2 + 3y_1 - 4$$

This form can be described as a *linear system* of differential equations.

Differential equations can be used to model a wide range of physical and biomedical systems and processes. Examples include the diffusion of heat through a body, the concentrations of chemicals or enzymes during a reaction or a metabolic process, the stresses of materials in a manufactured device, etc.

MATLAB has a range of tools for solving differential equations, which can be used to numerically estimate the values of one or more functions over a spatial or time domain. One notable example is the `ode45` function that can be used to solve linear systems of differential equations.

A good example of a linear system of differential equations is the model of disease spread in a population. This can be represented by a *compartment model*, in which the population is divided into different groups. One group represents the number of infected people, which can be written as a function of time $I(t)$. Another group represents the number of people susceptible to the disease, written as the function $S(t)$. The final group is written $R(t)$ and represents the number of individuals who have recovered from the disease.

The model of disease spread is known as the *SIR* model[1] and can be represented by a system of differential equations:

$$S' = -\beta S I$$
$$I' = \beta S I - \gamma I$$
$$R' = \gamma I$$

In the above equations, β and γ represent parameters of the model. The first equation of the three represents how the rate of change of S is $-\beta S I$, which depends on the product of the number of susceptible individuals S, and the number of infected individuals I. The parameter β encodes the rate of transmission between infected and susceptible individuals. The sign of the rate of change of S is negative, because the number of susceptible individuals decreases as they become infected.

The rate of change for infected individuals is represented by the second equation, which contains a term for the increase resulting from people becoming infected ($\beta S I$) and a term for the decrease resulting from people recovering from the infection, $-\gamma I$, which depends only on the current number infected, and a parameter (γ) to model how quickly people can recover. Finally, the rate of change for the recovered group is represented by an increase, due to infected people recovering (γI).

In MATLAB, we can represent the above system of differential equations for the SIR disease model with a single function.

```
function dydt = model_deriv_func(t, y)
% Model parameters
beta = 0.01;
gamma = 0.05;

% Components of disease model.
```

---

[1] See the following link for further details: https://www.maa.org/press/periodicals/loci/joma/the-sir-model-for-spread-of-disease-the-differential-equation-model.

```
S = y(1);
I = y(2);
R = y(3);

dSdt = -1 * beta * S * I;
dIdt = beta * S * I - gamma * I;
dRdt = gamma * I;

dydt = [dSdt; dIdt; dRdt];
end
```

This function has a specific signature that is required by the `ode45` function, with which we will later solve the system. The main requirement is that there are only two arguments to the function, `t` and `y`, which are provided in that order. The function must also have a single return argument to represent the derivative of `y` with respect to time.

For the purposes of the SIR model, the argument `y` in the above function contains all of the compartment variables as a vector, i.e., $\mathbf{y} = (S, I, R)^T$ at a specific time-step. The function extracts these variables from the vector and calculates the derivatives (using the hard-coded values of β and γ). The estimates of the derivatives use the SIR model equations, and they are returned as a single vector $\mathbf{y}'$, where $\mathbf{y}' = (S', I', R')^T$.

With the above derivative function for the SIR model in place, we can set up code to solve it using the `ode45` function. First, we can set up the initial conditions and the range of time values that we want to model. Assume there are 100 susceptible individuals and one infected individual at the start, with no persons in the recovered category:

```
% Start and end times in hours.
t0 = 0;
% Four days later.
tEnd = 4*24;

% Initial conditions
S0 = 100;
I0 = 1;
R0 = 0;
```

Now, we can assemble our variables into a 3×1 vector $\mathbf{y}_0$ that will be passed to the solver function:

```
y0 = [S0; I0; R0];
```

Next, we run the solver function with the following command:

```
[t,y] = ode45(@model_deriv_func, [t0 tEnd], y0);
```

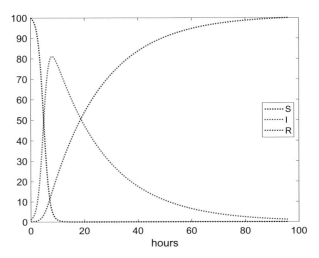

FIGURE 15.9
The progression of a disease using the *SIR* model, where S, I, and R represent susceptible, infected, and recovered individuals. The values of the variables were estimated by solving a linear system of differential equations with the `ode45` function (see text).

In the above call, we have used a function handle[2] to pass the derivative function for the system as the first argument. The second argument is an array containing the start and end times that we want to solve the system for, and the final argument is the vector containing the initial values of S, I, and R.

When the call to `ode45` is done, it returns an array of time values `t`, and an array of values for each of the compartment variables `y`. The size of the array `t` will be N, say, and because we have three variables in the SIR model, the size of `y` will be $N \times 3$ to represent the evolution of the values of S, I, and R over all time values. We can unpack the estimate values for the variables:

```
S = y(:,1);
I = y(:,2);
R = y(:,3);
```

Finally, we can plot the estimated solutions for the variables:

```
plot(t, S, ':k')
hold on
plot(t, I, ':r')
plot(t, R, ':b')
legend('S', 'I', 'R', 'location', 'east');
xlabel('hours');
```

Running the code above will result in the plot shown in Fig. 15.9, which illustrates how the number of susceptible individuals gradually decreases over time

---

[2]We previously used a function handle in a call to the `integral` function, see Example 15.46.

as they become infected. The number of infected individuals starts low, climbs to a peak, and subsequently decreases. The number of recovered individuals starts from zero and steadily rises as the disease passes through the whole population.

15.7 Summary and further resources

In this chapter, we have given an overview of some of the basic mathematical tools that are often needed in engineering applications. The treatment has necessarily been brief; dedicated textbooks on engineering mathematics will provide much greater levels of detail (e.g., [11], [12]).

There are many further areas of possible interest, and a small list of examples is listed below:

- Finding roots of equations, particularly for non-linear functions. See, for example, the `fzero` function in the MATLAB help.
- Interpolation and approximation of measured data. The help guide for the functions `griddedInterpolant` and `scatteredInterpolant` is a good place to start.
- Solving differential equations, which can be ordinary differential equations (ODEs) or partial differential equations (PDEs). MATLAB has a great variety of tools for solving ODEs and PDEs. This is covered in a large section in the MATLAB help under *Mathematics → Numerical Integration and Differential Equations*.
- Symbolic mathematical methods are described within the symbolic math toolbox.
- Transforms of functions, such as the Fourier transform and Laplace transform. These methods are used, for example, to analyze the spectra of data or to help make solving differential equations easier. A good place to start is *Mathematics → Fourier Analysis and Filtering* in the help guide.

15.8 Exercises

■ Exercise 15.1

015.A   Let two vectors be defined as

$$\mathbf{u} = \begin{pmatrix} 2 \\ 1 \\ 1 \end{pmatrix} \qquad \mathbf{v} = \begin{pmatrix} 1 \\ 5 \\ 2 \end{pmatrix}.$$

Carry out the following in MATLAB:

- Find a unit vector $\hat{\mathbf{u}}$ in the direction of $\mathbf{u}$.
- We can "project" $\mathbf{v}$ onto the line containing the vector $\mathbf{u}$. The distance of the "shadow" of $\mathbf{v}$ on this line is denoted by α (see the diagram below). We can obtain α by calculating $\hat{\mathbf{u}} \cdot \mathbf{v}$. Find the value of α.
- Using your value of α, find a vector in the direction of $\mathbf{u}$ with a length equal to α. Denote this vector by $\text{proj}_{\mathbf{u}}(\mathbf{v})$.
- Use your previous answer to find a vector $\mathbf{w}$, such that $\mathbf{v} = \mathbf{w} + \text{proj}_{\mathbf{u}}(\mathbf{v})$.
- Confirm that $\mathbf{w}$ is perpendicular to $\mathbf{u}$.
- Find the vector $\mathbf{z}$, where $\mathbf{z} = \mathbf{u} \times \mathbf{v}$.
- Confirm that $\mathbf{z}$ is perpendicular to each of $\mathbf{u}$ and $\mathbf{v}$.

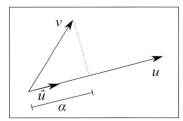

■

Exercise 15.2

For the pair of complex numbers $z_1 = 3 + 4i$ and $z_2 = -5 + 12i$, carry out the following steps in MATLAB:

015.B

- Calculate $z_1 + z_2$ and $z_1 - z_2$.
- Calculate $z_1 z_2$ and $\frac{z_2}{z_1}$.
- Confirm that the moduli of z_1 and z_2, and those of the product and quotient calculated above, obey the rules for multiplication and division (see Activity 15.4).
- Repeat the previous part to confirm that the arguments of z_1 and z_2, and those of the product and quotient, obey the corresponding rules for arguments under multiplication and division.
- Calculate the value of z_1^3 directly with the ^ operator and indirectly using its modulus and argument (see the polar form equation for powers at the end of Section 15.3). ■

Exercise 15.3

A system of linear equations is defined as

015.A, 015.C

$$x_1 + 3x_2 = -2$$
$$3x_1 - x_2 = 4$$
$$2x_1 + 2x_2 = 1$$

State whether the system is under-, over-, or exactly determined, and use MATLAB to find a solution.

If the solution is not exact, find the error of the solution. Select other vectors near the solution, and compare their error with the error of the solution. What do you notice? ∎

■ Exercise 15.4

015.D   In Example 11.10, we looked at a CT "scout" image (*CT_ScoutView.jpg*) and saw how we can extract a sub-region and apply a filter. In this exercise, we consider the image to be a function of two variables x and y. Load the image, and use MATLAB functions to estimate its partial derivatives in the x and y directions. You can use the same sub-region used in Example 11.10. You can visualize the image and its derivatives; they should look something like the illustration below:

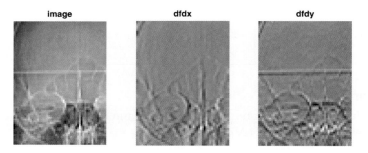

(Hint: In MATLAB, image reading functions tend to return arrays of integers, whereas processing functions require floating point values, so you will need to carry out a conversion step.) ∎

■ Exercise 15.5

015.D   For the function $h(x) = \ln(x + 3)$, estimate the area under its curve between $x = 1$ and $x = 4$ using quadrature.

Evaluate the exact integral.

(Hint: Use integration by parts to find the analytic integral.)

Find estimates for the integral using the trapezoidal rule for two different spacings along the x axis of 0.2 and 0.1.

For each of the three estimates, from quadrature and the two trapezoidal rule applications, find and compare the relative errors. ∎

FAMOUS COMPUTER PROGRAMMERS: TIMNIT GEBRU

Timnit Gebru was born and raised in Addis Ababa, Ethiopia. After receiving political asylum she attended high school in the USA and went on to study Electrical Engineering at Stanford University. In an early formative experience in the USA, a black female friend of Timnit's was assaulted in a bar and Timnit phoned the police to report the incident. Instead of investigating the alleged assault, her friend was arrested in what Timnit called "a blatant example of systemic racism."

Subsequently, Timnit Gebru has gone on to become one of the most influential figures in the field of the ethics of artificial intelligence (AI). AI refers to computer programs, often based on machine learning (see Chapter 14), which aim to exhibit what humans would refer to as 'intelligent' behavior. Timnit has highlighted many cases of potential bias and misuse of AI systems as well as the lack of diversity of researchers working in the field, and has become a passionate advocate of the fair and ethical use of AI. She is one of the founders of 'Black in AI' (https://blackinai.github.io/), a community of black researchers working in AI. She co-authored one of the most influential papers in the field of fair AI: in the 'Gender Shades' paper [13], Timnit and Joy Buolamwini highlighted racial and gender bias in commercial video-based gender classification AI tools, and showed that the bias was due to lack of representation of women and non-white ethnicities in the training data used to develop the tools.

Over the years Timnit has worked for Apple, Microsoft, and Google. She was famously sacked by Google in 2020 for sending an internal email that accused Google of "silencing marginalized voices." She has since gone on to found DAIR (Distributed AI Research Institute, https://www.dair-institute.org/), an institute aiming to promote the responsible and ethical use of AI.

"You want people in AI who have compassion, who are thinking about social issues, who are thinking about accessibility"

Timnit Gebru

References

[1] O. Bernard, A. Lalande, C. Zotti, F. Cervenansky, X. Yang, P. Heng, I. Cetin, K. Lekadir, O. Camara, M.A. Gonzalez Ballester, G. Sanroma, S. Napel, S. Petersen, G. Tziritas, E. Grinias, M. Khened, V.A. Kollerathu, G. Krishnamurthi, M. Rohe, X. Pennec, M. Sermesant, F. Isensee, P. Jager, K.H. Maier-Hein, P.M. Full, I. Wolf, S. Engelhardt, C.F. Baumgartner, L.M. Koch, J.M. Wolterink, I. Igum, Y. Jang, Y. Hong, J. Patravali, S. Jain, O. Humbert, P. Jodoin, Deep learning techniques for automatic MRI cardiac multi-structures segmentation and diagnosis: is the problem solved? IEEE Transactions on Medical Imaging 37 (11) (2018) 2514–2525.

[2] M. Studenski, Effective dose to patients and staff when using a mobile PET/SPECT system, Journal of Applied Clinical Medical Physics 14 (3) (2013).

[3] C. Le Tourneau, J.J. Lee, L.L. Siu, Dose escalation methods in phase I cancer clinical trials, Journal of the National Cancer Institute 101 (10) (2009) 708–720.

[4] Y. Nakamoto, M. Osman, C. Cohade, L. Marshall, J. Links, S. Kohlmyer, R. Wahl, PET/CT: comparison of quantitative tracer uptake between germanium and CT transmission attenuation-corrected images, Journal of Nuclear Medicine 43 (9) (2002) 1137–1143.

[5] D. Persaud, J.P. O'Leary, Fibonacci series, golden proportions, and the human biology, HWCOM Faculty Paper 27 (2015).

[6] C.J. Mode (Ed.), Applications of Monte Carlo Methods in Biology, Medicine and Other Fields of Science, InTech, 2011.

[7] A. King, R. Eckersley, Statistics for Biomedical Engineers and Scientists: How to Visualize and Analyze Data, 1st edition, Academic Press, 2019.

[8] S. McKenzie, Vital Statistics: An Introduction to Health Science Statistics, 1st edition, Churchill, Livingstone, 2013.

[9] L. Faes, X. Liu, S.K. Wagner, D.J. Fu, K. Balaskas, D.A. Sim, L.M. Bachmann, P.A. Keane, A.K. Denniston, A clinician's guide to artificial intelligence: how to critically appraise machine learning studies, Translational Vision Science and Technology 9 (2) (2020) 7.

[10] K. Simonyan, A. Zisserman, Very deep convolutional networks for large-scale image recognition, arXiv preprint, arXiv:1409.1556, 2014.

[11] K.F. Riley, M.P. Hobson, S.J. Bence, Mathematical Methods for Physics and Engineering, Cambridge University Press, 2006.

[12] A. Jeffrey, Mathematics for Engineers and Scientists, Chapman and Hall/CRC, 2004.

[13] J. Buolamwini, T. Gebru, Gender shades: intersectional accuracy disparities in commercial gender classification, in: Sorelle A. Friedler, Christo Wilson (Eds.), Proceedings of the 1st Conference on Fairness, Accountability and Transparency, in: Proceedings of Machine Learning Research, vol. 81, PMLR, 2018, pp. 77–91.

Index